EXERCISES IN PHYSICAL GEOGRAPHY

THE UNIVERSITY OF GEORGIA

GEORGE BROOK, AUTHOR
P. W. SUCKLING, AUTHOR
R. JAMES HEYL, AUTHOR
R. R. DOYON, CONTRIBUTING AUTHOR AND CARTOGRAPHER

CPC CONTEMPORARY PUBLISHING COMPANY
508 St. Mary's Street Raleigh, North Carolina 27605

CREDITS

Drawings from

Marland P. Billings, *Structural Geology*, 2nd ed., † 1954, p. 48. Reprinted by permission of Prentice-Hall, Inc., Englewood Cliffs, N.J.

C. F. Stewart Sharpe, *Landslides and Related Phenomena*, † 1938, pp. 37, 54, 62. Columbia University Press. Reprinted by permission of the author.

A. N. Strahler, *Physical Geography*, 4th ed., † 1975, p. 438, p. 553. Reprinted by permission of the author and John Wiley and Sons.

Fluvial Processes in Geomorphology by Luna B. Leopold, M. Gordon Wolman and John P. Miller. W. H. Freeman and Company. Copyright † 1964.

Cox, C. B. and Moore, P. D. (1973) *Biogeography* (3rd edition) John Wiley and Sons, New York.

Map and Article from

Earthquake Information Bulletin, July-August 1976, V. 8, No. 4, pp. 14-15, U.S. Geological Survey.

Peter A. Rona, *Plate Tectonics and Mineral Resources* † July 1973, p. 89. Scientific American. Reprinted by permission of the Publisher.

Robinson, H. (1972) *Biogeography*, Macdonald and Evans Ltd., London.

Gerasimov, I. B. (1969) *Degradation of the Last European Ice Sheet*, in Wright, H. E. (ed.) *Quaternary Geology and Climate* (National Academy of Sciences, Washington).

Brown, J. H. and Gibson, A. C. (1983) *Biogeography*, C. V. Mosby Co., St. Louis.

Illies, J. (1974) *Introduction to Zoogeography*. MacMillan, London.

Final Piece in the Gondwana Game, Vol. 106 No. 4, July 27, 1974. Science News. Reprinted by permission of the Publisher.

Drawings and Table from

K. W. Butzer *Geomorphology from the Earth* 1976 p. 24, p. 84, p. 87, p. 342, p. 388. Harper & Row.

Printed in the United States of America.

ISBN: 0-89892-054-X

PREFACE

The exercises in this manual have been selected from INTRODUCTION TO LANDFORMS by George A. Brook and R. James Heyl, first published in 1978, and STUDIES IN WEATHER AND CLIMATE by Philip W. Suckling and Roy R. Doyon, published in 1981. George Brook has added four new exercises on soils and biogeography, resulting in an extensive selection of material suitable for use in an introductory physical geography course. The large number of exercises (31) allows an instructor to choose those most suitable for his/her own course.

All exercises are designed to be undertaken in laboratory or tutorial classroom sessions. Each has an introduction followed by comprehensive questions. It should be noted that although this manual is independent of any particular textbook, it does not replace a course textbook. Instead, the exercises are intended to supplement the lecture and textbook material. Many of the exercises require the use of a calculator, lens stereoscope, colored pencils, ruler, protractor, or world atlas.

The authors wish to thank Chuck Grantham of Contemporary Publishing Company for the opportunity to produce EXERCISES IN PHYSICAL GEOGRAPHY.

Athens, Georgia
January 1984

George A. Brook
Philip W. Suckling
R. James Heyl
Roy R. Doyon

TABLE OF CONTENTS

EXERCISE 1
Geographic Features of the Earth's Surface 1
EXERCISE 2
Topographic Maps 19
EXERCISE 3
Geological Maps and Sections 27
EXERCISE 4
Aerial Photographs 37
EXERCISE 5
Composition of the Atmosphere and Earth-Sun Relations 45
EXERCISE 6
Energy in the Earth Atmosphere System 57
EXERCISE 7
Temperature and Humidity 67
EXERCISE 8
Vertical Air Motion, Clouds and Precipitation 81
EXERCISE 9
Horizontal Air Motion and the General Circulation of the Atmosphere 95
EXERCISE 10
The Mid-Latitude Cyclone 109
EXERCISE 11
The Tropical Cyclone 123
EXERCISE 12
Weather Map Analysis 135
EXERCISE 13
Climate Data and Climatic Regimes 143
EXERCISE 14
Climate Classification 155
EXERCISE 15
Climates of the World 165
EXERCISE 16
Global Tectonics 183
EXERCISE 17
Landforms of Intrusive and Extrusive Vulcanism 195
EXERCISE 18
Structural Landforms 203
EXERCISE 19
Hillslope Form and Mass Wasting 215
EXERCISE 20
Fluvial Landscapes 229
EXERCISE 21
The Drainage Basin 241
EXERCISE 22
Stream Discharge and Flood Frequency 249

EXERCISE 23
Karst Landscapes 255
EXERCISE 24
Glaciers and Glacial Landforms 267
EXERCISE 25
Periglacial Landforms 283
EXERCISE 26
Coastal Landforms 291
EXERCISE 27
Arid and Semiarid Landscapes 303
EXERCISE 28
The Physical Properties and Erodibility of Soils 313
EXERCISE 29
Soil Horizonation, Mapping and Classification 327
EXERCISE 30
Vegetation Structure and Plant and Animal Distributions 343
EXERCISE 31
Quaternary Paleoenvironments 361

APPENDIX A
Selected Bar Scales
APPENDIX B
Tangent Tables
APPENDIX C
Geologic Time Scale
APPENDIX D
Conversion Factors for Units of Measure
APPENDIX E
Surface Weather Map
APPENDIX F
Global Climates According to the Koppen Classification System

EXERCISE **1**

GEOGRAPHIC FEATURES OF THE EARTH'S SURFACE

In this exercise, some basic geographical aspects of the earth's surface are studied. The chapter begins by examining aspects of the earth's grid of latitude and longitude and time. The remaining questions refer to physical features of the earth's surface such as the map locations of the principal land masses and water bodies, the location and extent of principal mountain ranges and the position and features of the earth's ocean currents. Later, it will be seen that these geographic features of the earth's surface have special significance with respect to the weather and climate in different parts of the world.

Specific locations on the globe are designated and effectively communicated using the earth grid of latitude and longitude. Latitude is expressed in terms of degrees, minutes and seconds of arc distance, measured north or south of the equator. The equator is at a latitude of 0°. Latitude increases in value towards the poles, reaching a maximum value of 90° N at the north pole and 90°S at the south pole. Thus, polar regions are known as the "high latitudes" and equatorial areas are known as the "low latitudes." Longitude is expressed in terms of degrees, minutes and seconds measured westward or eastward from the "prime meridian" which runs through Greenwich, England. The prime meridian is at a longitude of 0°. Longitude increases in value westward (° W) and eastward (° E) from the prime meridian, reaching a maximum of 180° at the "international date line." "Time" is related to longitudinal location with solar noon (literally midday) occurring at the longitude where the sun is highest in the sky.

Name______________________________ Date________________

Instructor______________________________ Section________________

QUESTIONS

1.1 The circumference of our planet is about 40,075 km (24,900 miles). What distance across the earth's surface is spanned by 1° of latitude?

____________________________km

1.2 What distance is the equator from either the north or south pole?

____________________________km

1.3 Nome, Alaska is located at 64° 30′N, 165° 20′W. Cairo (Al Qahirah), Egypt is located at 30° 30′N, 31° 17′E. How much closer to the equator is Cairo compared to Nome?

____________________________km

1.4 (a) Approximately what distance is spanned by 1° of longitude at the equator?

____________________________km

(b) What happens to the distance spanned by 1° of longitude as one moves poleward from the equator?

1.5 Using an atlas, find the latitude and longitude of the following locations:

(a) Athens, Georgia ______

(b) Athens, Greece ______

(c) Vancouver, British Columbia ______

(d) London, England ______

(e) Buenos Aires, Argentina ______

(f) Bombay, India ______

(g) South Pole ______

(h) Denver, Colorado ______

(i) Nairobi, Kenya ______

(j) Honolulu, Hawaii ______

(k) Washington, D. C. ______

(l) Tokyo, Japan ______

(m) Salisbury, Zimbabwe ______

(n) Mexico City, Mexico ______

Name____ Date____________________

Instructor____________________________________ Section__________________

1.6 The distribution of land and water bodies and the proximity to large water bodies have a profound effect on climate. The location of major topographic features, especially mountain ranges, is of great importance to weather and climate. A general knowledge of the geographic features of the world will be beneficial when considering the general climatic regions of our planet. With the aid of an atlas or world wall map, complete the questions on the following pages using the three world maps provided.

(a) On the first world map (Fig. 1-1) mark the following principal land and water features:

CONTINENTS AND OCEANS

Africa	Europe	Atlantic O.
Antarctica	North America	Indian O.
Asia	South America	Pacific O.
Australia	Arctic O.	

OCEAN CURRENTS

(Indicate direction of movement and whether cold or warm)

Agulhas (Mozambique)	East Australian Drift	North Equatorial
Alaska	Falkland	Oyashio (Kamchatka)
Benguela	Gulf Stream	Peru (Humboldt)
Brazil	Japan (Kuroshio)	South Equatorial
California	Labrador	West Australian Drift
Canaries	North Atlantic Drift	West Wind Drift

MOUNTAIN RANGES

Alaska Range	Ethiopian Highlands	Rocky Mountains
Alps	Great Dividing Range	Sierra Nevada
Andes	Great Rift Valley Highlands	Southern Alps of New Zealand
Appalachians	Himalayas	Urals
Caucasus Mtns.	Madagascar Highlands	Verkhoyansk Range
Coast Range	Mexican Highlands	Yemen Highlands
Drakensberg	Ranges of Central America	

DESERTS

Atacama	Ogaden	Arabian (Syrian, An Nafud, Ar Rub al Khali)
Gobi	Patagonian	Australian (Great Sandy, Gibson, Great Victoria, Simpson)
Great Indian	Sahara	
Kalahari	Takla Makan	deserts of Turkestan region
Mojave		

Name: ______________________ **Date:** ______________

Instructor: ____________________ **Section:** ____________

PRINCIPAL LAND and WATER FEATURES

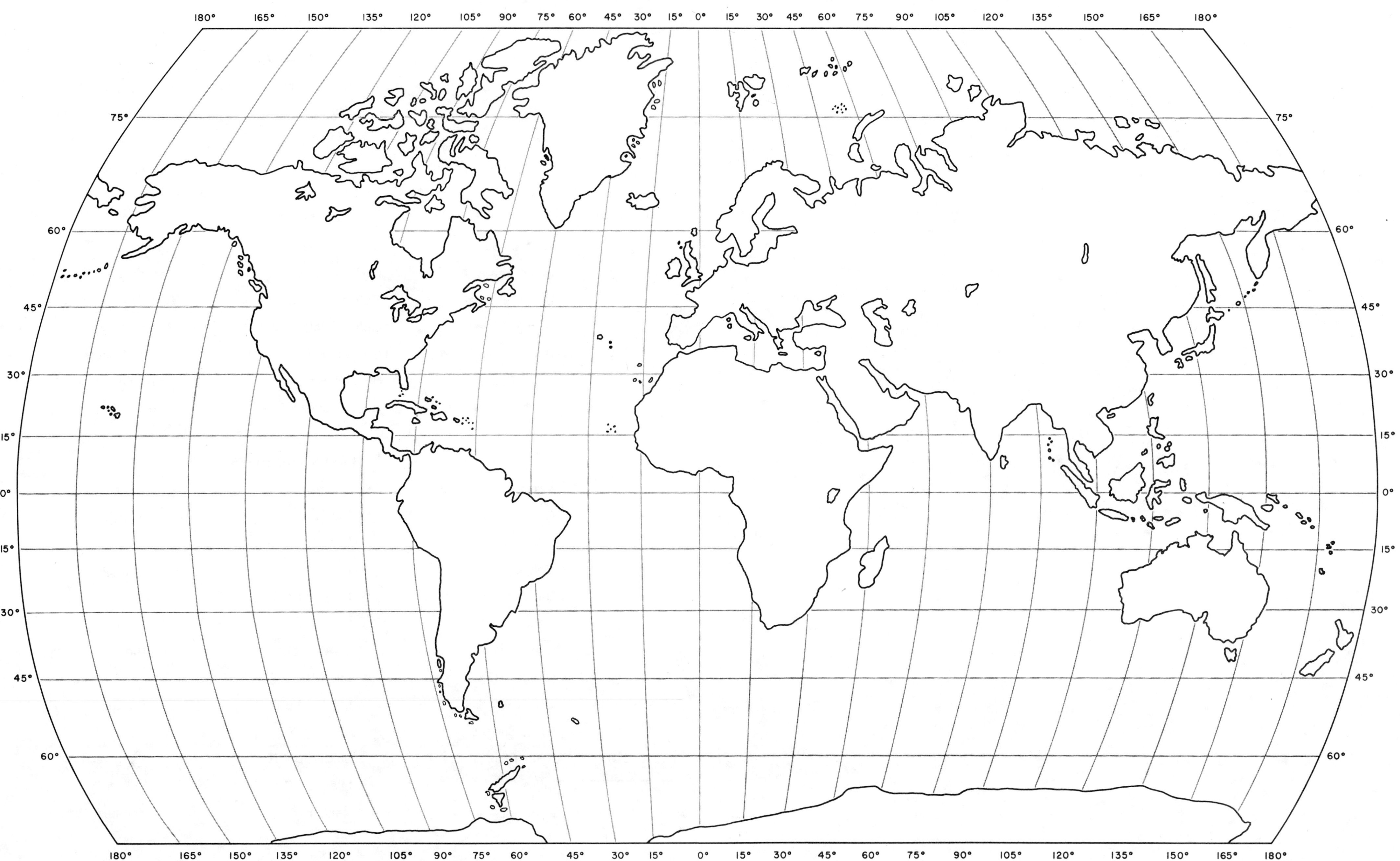

Figure 1-1

(b) On the second world map (Fig. 1-2) label the following islands, water bodies, lakes and rivers:

ISLANDS

Aleutians	Greenland	Madagascar
Baffin	Hawaii	Newfoundland
Borneo	Iceland	New Guinea
Ellesmere	Java	Tasmania

WATER BODIES

Arabian Sea	Caspian Sea	North Sea
Baltic Sea	East China Sea	Persian Gulf
Bay of Bengal	Gulf of Alaska	Red Sea
Beaufort Sea	Gulf of California	Sea of Japan
Bering Sea	Gulf of Mexico	Sea of Okhotsk
Black Sea	Hudson Bay	South China Sea
Caribbean Sea	Mediterranean Sea	Weddell Sea

LAKES AND RIVERS

Amazon R.	L. Erie	Niger R.
Colorado R.	L. Huron	Nile R.
Columbia R.	L. Michigan	Ob R.
Congo R.	L. Ontario	Orange R.
Danube R.	L. Superior	Orinoco R.
Darling R.	L. Victoria	Rhine R.
Euphrates R.	L. Winnipeg	Rio de la Plata
Ganges R.	Lena R.	Rio Grande
Great Bear L.	Mackenzie R.	St. Lawrence R.
Great Slave L.	Mekong R.	Volga R.
Hwang Ho (Yellow R.)	Mississippi R.	Yangtze R.
Indus R.	Missouri R.	Yukon R.
L. Baikal	Nelson R.	Zambezi R.

Name: ______________________ Date: ______________

Instructor: ____________________ Section: ____________

ISLANDS, WATER BODIES, LAKES and RIVERS

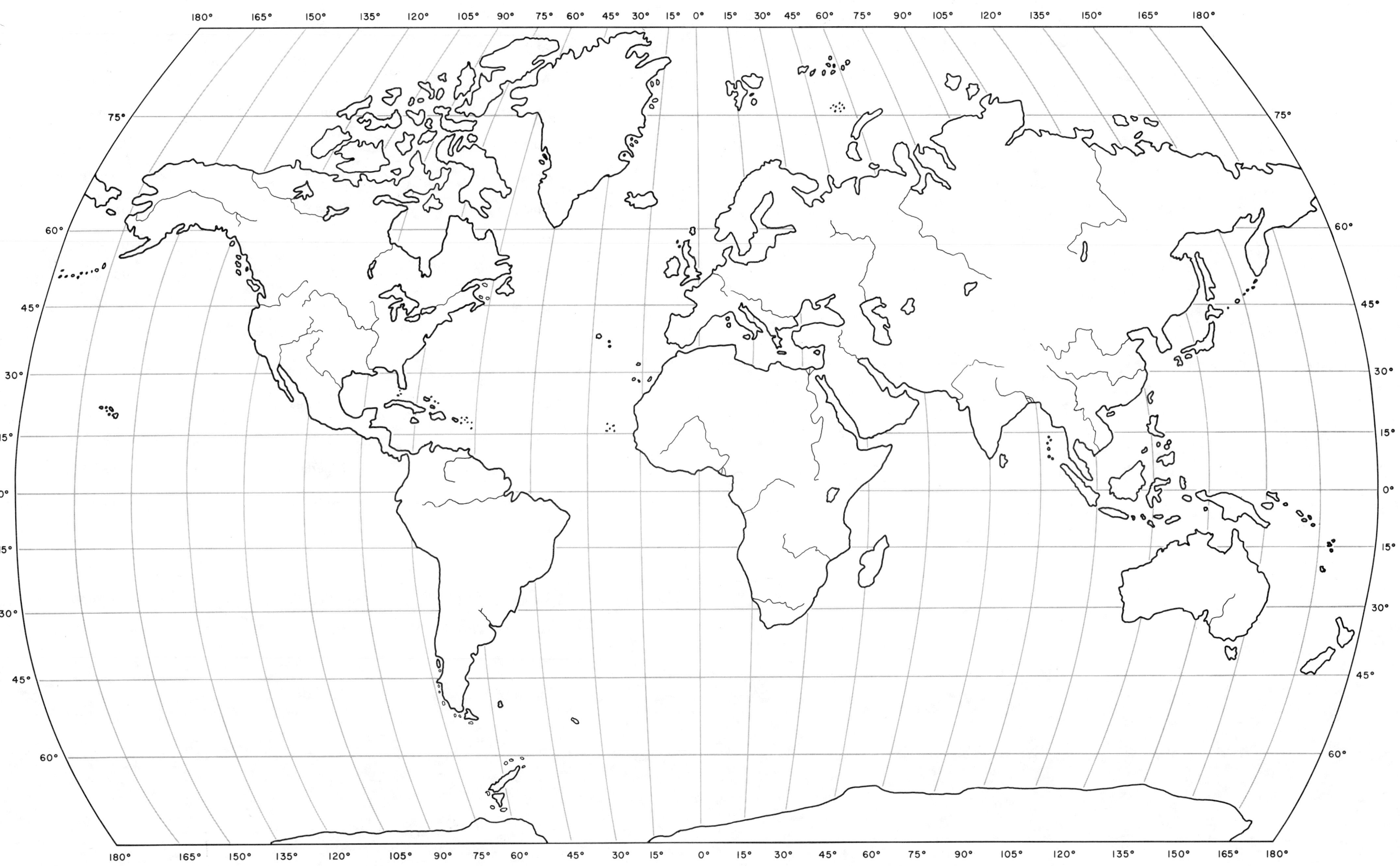

Figure 1-2

(c) On the third world map (Fig. 1-3) label the following countries: (Note that this is a partial list. Study the world map in great depth on your own. Several countries have been omitted because of space limitations.)

Afghanistan	India	Romania
Algeria	Indonesia	Saudi Arabia
Argentina	Iran	Somalia
Australia	Iraq	South Africa
Bolivia	Italy	Soviet Union
Brazil	Japan	Spain
Burma	Kenya	Sri Lanka
Canada	Libya	Sudan
Chad	Malaysia	Sweden
Chile	Mexico	Tanzania
China	Mongolia	Thailand
Colombia	New Zealand	Turkey
Cuba	Niger	United Kingdom
Egypt	Nigeria	United States
Ethiopia	Norway	Uruguay
Finland	Oman	Venezuela
France	Pakistan	Vietnam
Germany (East-GDR)	Paraguay	Yugoslavia
Germany (West-FRG)	Peru	Zaire
Hungary	Philippines	Zimbabwe
Iceland	Poland	

Name: ______________________ Date: __________

Instructor: ______________________ Section: __________

SELECTED COUNTRIES

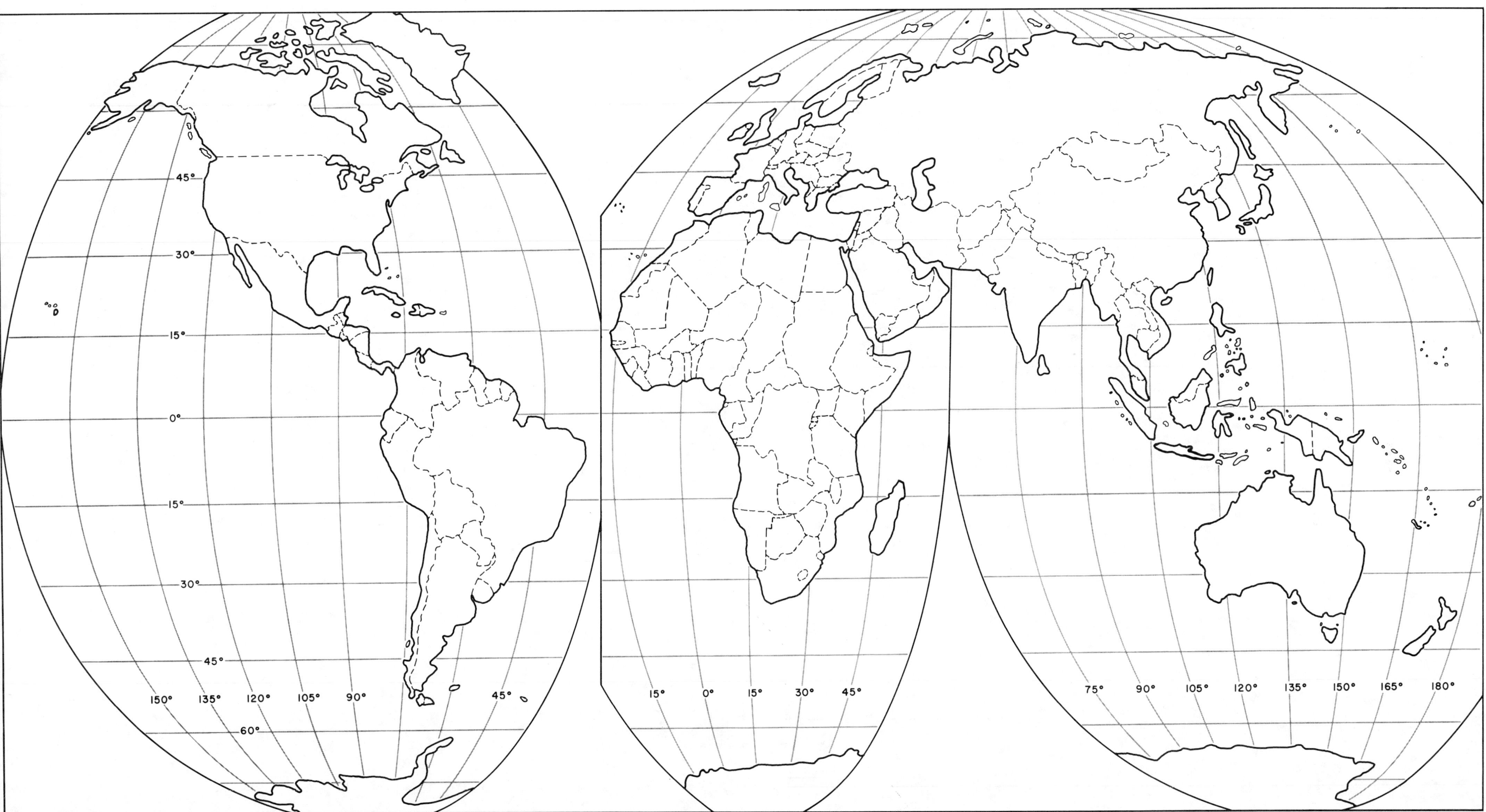

Figure 1-3

EXERCISE **2**

TOPOGRAPHIC MAPS

Planimetric maps depict spatial information such as house locations, political boundaries and forests. *Hypsometric maps* also include information about landscape form. Height is displayed by use of *contours* which are lines joining points of equal height above sea level (a.s.l.). Hypsometric maps are more commonly referred to as *topographic maps*. Colors and symbols are used to depict the natural and man-made characteristics of the landscape (Figure 2.1 (in pocket)).

Most maps are smaller than the ground area they represent. The relationship between map distance and actual ground distance defines the *map scale* which may be expressed as a ratio such as 1:500 or 1:2,000,000. Such ratios are known as *representative fractions* (R.F). A representative fraction of 1:500 simply means that one inch on the map represents a distance of 500 inches on the earth's surface, or alternatively that one centimeter on the map represents 500 centimeters or 5 meters on the ground. *Verbal scales* such as one inch to one mile or one centimeter to five kilometers are often used for convenience when measuring distances on maps. A verbal scale of one inch to one mile is equivalent to a ratio scale of 1:63,360. Most maps also include *bar scales* which are visual representations of the verbal scale. A series of bar scales are included in Appendix A. If a 1:10,000 scale map has the same dimensions as a 1:20,000 scale sheet it will cover only one quarter of the area. Similarly a 1:50,000 scale map will cover 100 times the area covered by a 1:5,000 scale map.

The scale of a map determines how much detail can reasonably be included on it. A scale of 1:125,000 is twice as large as a scale of 1:250,000. Generally maps are classified into large, medium and small scale and although there is no standard practice the U.S. Army Map Service defines maps with scales ranging from 1:600,000 down to 1:100,000,000 or smaller as *small-scale maps*; those of scale 1:600,000 to 1:75,000 as *medium-scale maps* and those of scale greater than 1:75,000 as *large-scale maps*. Small-scale maps are much less detailed than are larger scale maps.

Because parallels of latitude and meridians of longitude frequently form the boundaries of individual U.S. map sheets these are often referred to as *quadrangles*. Seven standard scales make up the *National Topographic Map Series:*

Series	Representative Fraction (RF)
7.5 minute	1:24,000
7.5 minute	1:31,680
15 minute	1:62.500
Alaska	1:63,360
30 minute	1:125,000
1:250,000	1:250,000
1:1,000,000	1:1,000,000

Individual maps may be referred to by the *sheet name* (eg. Pittsburgh Quadrangle), or the *sheet number* (eg. 101-NW). They may be identified by the geographic coordinates of the lower left hand corner of the map and the series name. For example N4015-W7645/15 indicates that the map is one of the 15 minute series and that the latitudinal and longitudinal coordinates of the southeast corner are N.40° 15′ and W. 76° 45′ respectively.

Information contained in the margins of a topographic map constitutes the *map legend.* Frequently included is a location diagram showing the map area in relation to a county, state or country; or an index diagram giving the names and numbers of adjoining sheets of the same series (on some maps the same information is displayed around the edges of the mapped area). Most maps give details of when and how the map was initially constructed, subsequent revisions, and the map projection used. This information is extremely important to the user for there is no point in using a map last revised in 1952 to plot the distribution of forested land in 1976.

The *geographic grid* consisting of intersecting lines of latitude and longitude is used on almost all topographic maps with *true north,* fixed by the North Pole, to the top of the map. The angular difference between true and *magnetic north*—the *magnetic declination,* is also given so that maps can be correctly oriented in the field using a compass. Points on a map can be located if the latitudinal and longitudinal coordinates are given. It is generally sufficient to quote only in degrees and minutes so that a full locational statement might read 45°46′N., and 145°30′W. In this manual a square grid has been superimposed on many maps and photographs so that points can be rapidly located or specified in terms of simple coordinates. For example, points X and Y in Figure 2.2 are located at B.5-2.4 and D.5-1.5 respectively. In addition many figure headings include a statement which gives directional information. The figure heading of Figure 2.2 for example, states that A.0-1.0 is in the southeast corner of the figure indicating that north is to the right (as indicated by the north arrow).

A variety of methods are used to depict vertical relief on maps including hachuring, contouring, layer shading and tinting, and hill shading. *Spot heights* (eg. x1692) and *bench marks* (eg. xB.M. 1622) are often used in conjunction with other methods to provide precise height information on accurately surveyed points. Brass plates pinpoint bench mark locations in the field. The use of contours is the most common method of depicting relief. Modern photogrammetric methods and machines enable extremely accurate and rapid contouring to be carried out from air photographs with a minimum of ground control. *Index or key contours* are shown on maps in a heavy weight brown line, *intermediate contours* are depicted in a lighter weight brown line, and *supplementary contours* are shown in a light weight dashed brown line. Where natural hollows exist in a landscape contours are ticked in the direction of the depression (eg.).

Figure 2.2

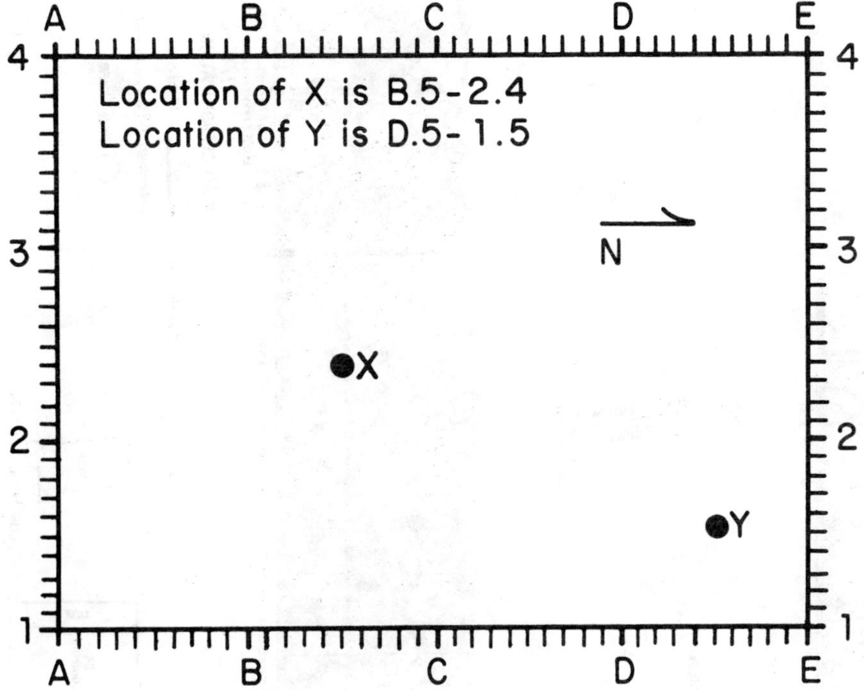

Locating and referencing points with the grid used in this manual (A.0-1.0 is in the southeast corner of the figure).

Hachures are lines drawn down the slope in the direction of the steepest gradient and are conventionally drawn closer together where the slope is steeper. Hachure maps lack absolute elevation information but often do depict minor and often important details of the landscape lost in contour maps because of the contour interval. *Layer shading or tinting* is the shading or coloring of areas of a map lying within a certain altitudinal range. A most effective pictorial method of depicting relief is to shadow the hills on a contour map creating the illusion of a three-dimensional surface.

One of the main advantages of contour maps is that topographic profiles can be drawn from them. In the construction of profiles the vertical scale is generally made larger than the horizontal scale to enhance the relief. The relationship between the vertical and horizontal scales is the *vertical exaggeration.* If the horizontal scale is 1:63,360 and the vertical scale is 1:12,000 the exaggeration is 5.28 times. Ground slope can also be obtained by measuring the horizontal distance or *horizontal equivalent* (H.E.) between two points and noting the difference in elevation or vertical interval (V.I.). Gradient may be expressed verbally as the ratio V.I./H.E. with V.I. reduced to unity so that if H.E. is 500 yards and V.I. is 150 feet the gradient will be 150/1,500 or 1 in 10. Gradient may also be measured as an angle such as 1° or 10°. The tangent of the slope angle is given by decimal V.I./H.E. Alternatively a slope can be expressed as a percentage; for example a slope of 1 in 10 is also a slope of 10 in 100 or 10%. (Appendix B).

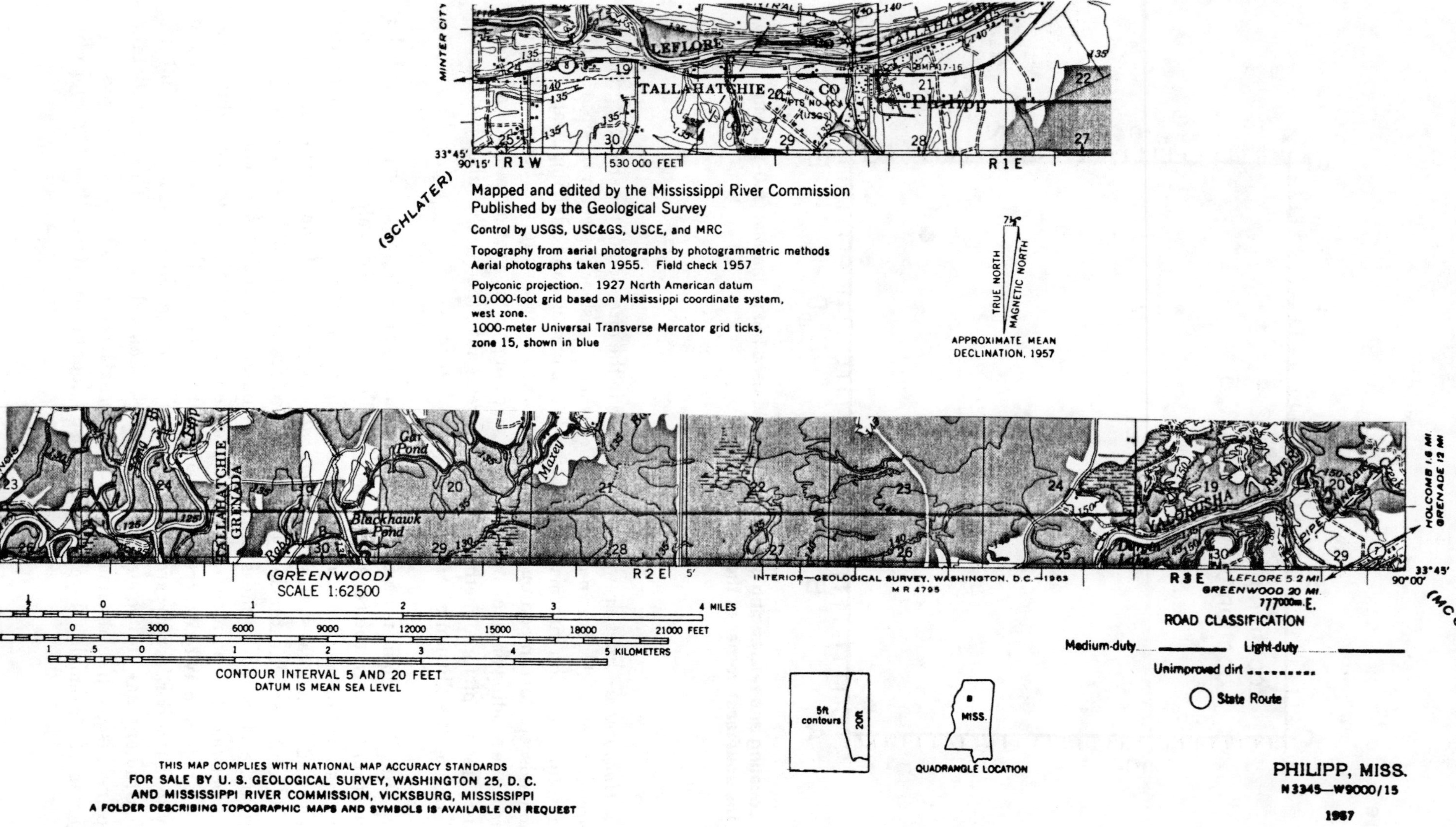
MINTER CITY
LEFLORE
TALLAHATCHIE
CO
Philipp
33°45'
90°15'
R 1 W
530 000 FEET
R 1 E
(SCHLATER)
Mapped and edited by the Mississippi River Commission
Published by the Geological Survey
Control by USGS, USC&GS, USCE, and MRC
Topography from aerial photographs by photogrammetric methods
Aerial photographs taken 1955. Field check 1957
Polyconic projection. 1927 North American datum
10,000-foot grid based on Mississippi coordinate system, west zone.
1000-meter Universal Transverse Mercator grid ticks, zone 15, shown in blue
TRUE NORTH
MAGNETIC NORTH
APPROXIMATE MEAN DECLINATION, 1957
TALLAHATCHIE
GRENADA
Gar Pond
Blackhawk Pond
Maxey
YALOBUSHA RIVER
(GREENWOOD)
SCALE 1:62500
R 2 E
INTERIOR—GEOLOGICAL SURVEY, WASHINGTON, D.C.—1963
M R 4795
R 3 E
LEFLORE 5.2 MI.
GREENWOOD 20 MI.
777000m E.
HOLCOMB 1.6 MI.
GRENADE 12 MI.
33°45'
90°00'
(MC CARLEY)
4 MILES
21000 FEET
5 KILOMETERS
CONTOUR INTERVAL 5 AND 20 FEET
DATUM IS MEAN SEA LEVEL
ROAD CLASSIFICATION
Medium-duty
Light-duty
Unimproved dirt
State Route
5ft contours
20ft
MISS.
QUADRANGLE LOCATION
THIS MAP COMPLIES WITH NATIONAL MAP ACCURACY STANDARDS
FOR SALE BY U. S. GEOLOGICAL SURVEY, WASHINGTON 25, D. C.
AND MISSISSIPPI RIVER COMMISSION, VICKSBURG, MISSISSIPPI
A FOLDER DESCRIBING TOPOGRAPHIC MAPS AND SYMBOLS IS AVAILABLE ON REQUEST
PHILIPP, MISS.
N3345—W9000/15
1957

Figure 2.3 The legend of the Philipp, Mississippi topographic map.

Name______________________ Date______________

Instructor______________________ Section______________

QUESTIONS

2.1 Examine Figure 2.3, the Philipp map legend and complete the following table.

	Philipp Quadrangle		Philipp Quadrangle
Map Series		Revisions (if any)	
Mapping Agency		Representative Fraction	
Publishing Agency		Verbal Scale in inches per mile	
Geographical Grid/ Series Identity		Verbal Scale in centi-meters per kilometer	
Source of Topographic Information		Name of Map to South	
Map Projection		Name of Map to Southwest	
Date Issued		Contour interval	
Magnetic Declination and Date		Area Covered (sq. miles)	

2.2 Examine Figures 2.4, 23.4, and 24.2 and complete the following table.

Map	**Location**	**Geographic or Grid Coordinates**
Lake Wales	Lake Serena	
Lake Wales		B.5-1.5
Seward	Willard Island in Blackstone Bay	
Philipp	Bear Lake	
Philipp		D.5-3.0
Philipp	Macel	

2.3 Choose examples of large, medium, and small scale maps from the figures in this manual.

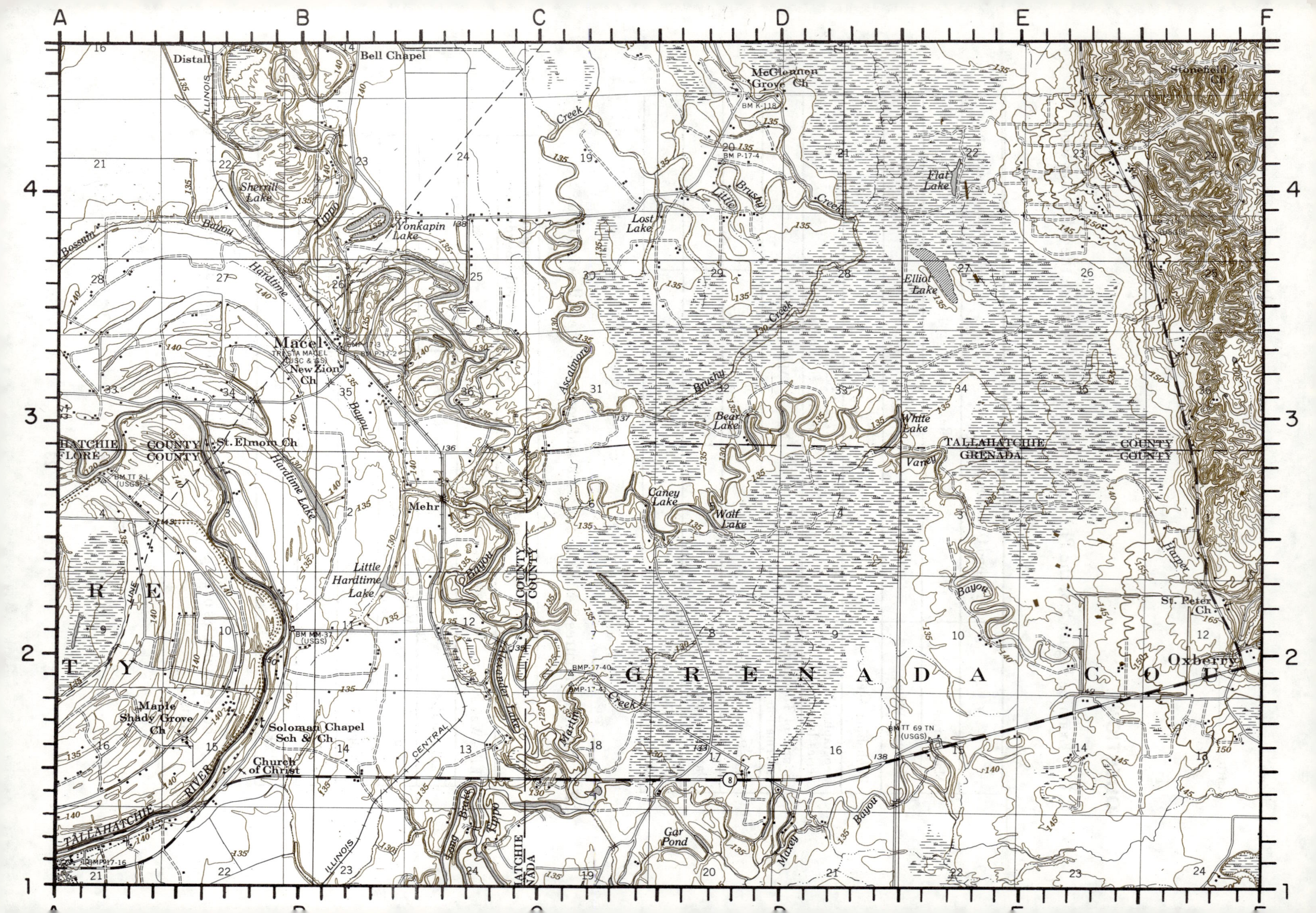

Bell Chapel
Distall
McClennen Grove Ch
Stonefield Ch
Creek
Little Brushy Creek
Lost Lake
Flat Lake
Elliot Lake
Sherrill Lake
Tippo
Yonkapin Lake
Bossum
Bayou
Hardtime
Macel
New Zion Ch
Ascalmore
Brushy Creek
Bear Lake
White Lake
TALLAHATCHIE COUNTY
GRENADA COUNTY
Varney
St. Elmom Ch
Hardtime Lake
Mehr
Caney Lake
Wolf Lake
Little Hardtime Lake
Harper
St. Peter Ch
Oxberry
G R E N A D A C O U
R E
T Y
Maple Shady Grove Ch
Soloman Chapel Sch & Ch
Church of Christ
Alexander Lake
Martin Creek
CENTRAL
ILLINOIS
TALLAHATCHIE RIVER
Gar Pond
Maxey Bayou
Long Brake

Name______________________________ Date______________

Instructor______________________________ Section______________

2.4 Two maps of the same dimensions have scales of 1:5,000 and 1:20,000 respectively. How much bigger is the area covered by the smaller scale map compared to that covered by the larger scale sheet?

2.5 A topographic profile has a horizontal scale of 1:25,000 and a vertical scale of 1 inch to 100 feet. What is the vertical exaggeration?

2.6 The horizontal distance between two points on a gravel road is 520 yards and they differ in elevation by 500 feet. The road slopes evenly from one location to the other. What is the average gradient (verbal) along it?

2.7 The map of a small area surveyed by plane table methods includes lakes, rivers, roads, forested areas, buildings and a number of accurately determined spot heights (Figure 2.5). Complete this map by drawing contours at intervals of 50 feet (i.e. 600 ft., 650 ft. etc.).

2.8 Construct a topographic profile from B.1-1.0 to C.6-3.5 on the Menan Buttes Map (Figure 17.2). Use a horizontal scale of 1:24,000 and a vertical scale of 1 inch to 400 feet. Calculate the vertical exaggeration.

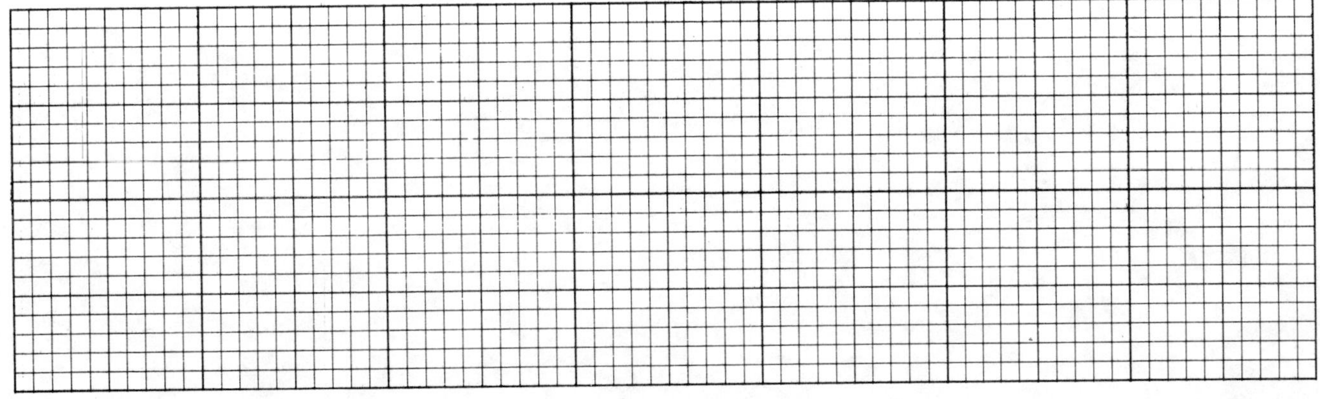

Figure 2.4 **Philipp, Mississippi (1:62,500, A.0-1.0 in southwest corner).**

Figure 2.5

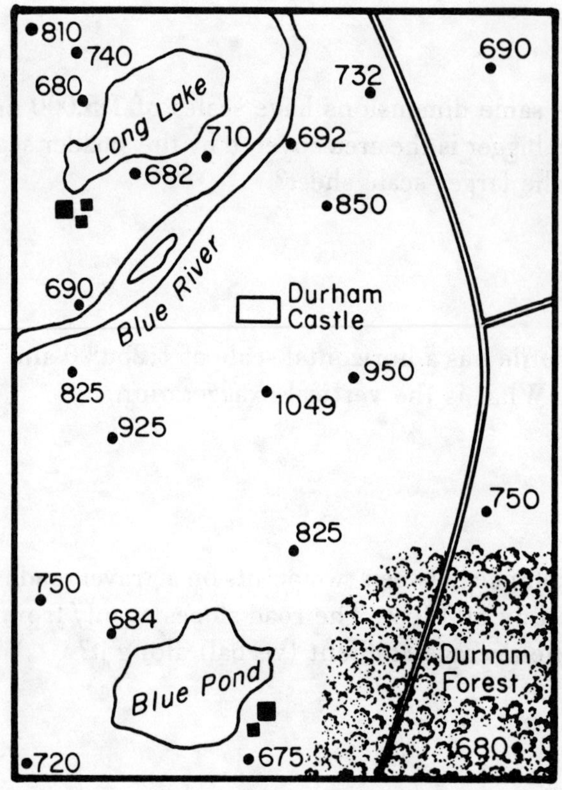

Small area surveyed by plane table methods.

2.9 Using the topographic profile you constructed in answering question 2.8, calculate the southwest slope of the volcano. Express your answer verbally and in angular and percentage terms.

EXERCISE **3**

GEOLOGICAL MAPS AND SECTIONS

Whereas the topographic map shows both planimetric and hypsometric data of the earth's surface the prime purpose of the geologic map is planimetric information about the outcrop of different rocks at the surface. In most cases further structural information is given in the form of symbols. Geologic or structure sections show the subsurface configuration of rock bodies and layers as they would appear in the sides of a deep, vertical-walled canyon. The *dip* of a layer of rock or strata is directed towards the maximum acute angle between the bedding and a horizontal plane; the acute angle itself is the magnitude of the dip. The *strike* of a bed is the direction of a line formed by the intersection of the bedding and a horizontal plane. The directions of dip and strike are always at right angles to one another and on the map are indicated by the flattened T symbol ⊤10°. The long line indicates the direction of strike, the short line and number the direction and magnitude of the dip in degrees. Vertical and horizontal beds are indicated by the symbols +90° and ⊕ where the position of the 90° may be used to indicate the upper surface, or horizon, of the rock layer.

Sedimentary and metamorphic rock layers are commonly *folded.* An *anticline* is a fold that is convex upward, a *syncline* one that is concave upward. The *axial plane* of a fold divides it as symmetrically as possible so that the *limbs* dip evenly away from it. A symmetrical fold is one with the axial plane almost vertical so that the limbs dip in opposite directions but by the same magnitude. In an asymmetrical fold the axial plane is inclined and the limbs dip by different amounts. The *axis* of a fold is the intersection of the axial plane with any bed. In some folds the axis is horizontal, in others it is inclined and the folds are then said to *plunge* (Figure 3.1). When horizontal folds are eroded, the beds on opposite limbs parallel one another and do not converge. When plunging folds are eroded, beds which outcrop at the surface converge in the direction of the *nose* which in an anticline is in the direction of the plunge and in the syncline away from it. A *doubly plunging fold* is one that reverses its direction of plunge; *basins* and *domes* are sub-circular double plunging folds. When *antiform* structures are eroded the oldest rocks are at the center of the outcrop pattern, in a *synform* structure the youngest rocks occupy the center of the outcrop (Figure 3.2).

Figure 3.1

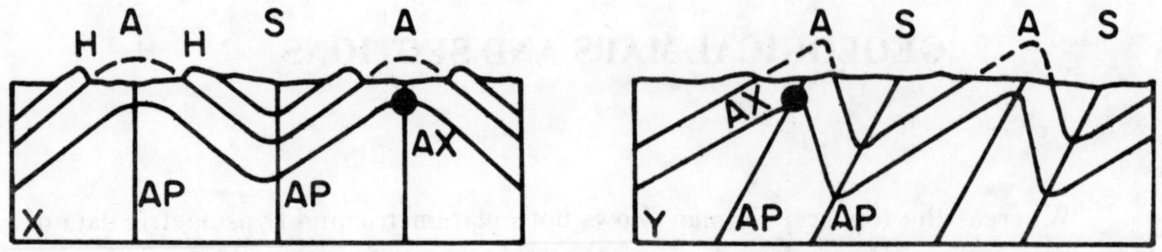

Eroded symmetrical (X) and asymmetrical (Y) anticlines and synclines (AP = axial plane, AX = axis, A = anticline, S = syncline, H = hogback ridge of hard rock).

Figure 3.2

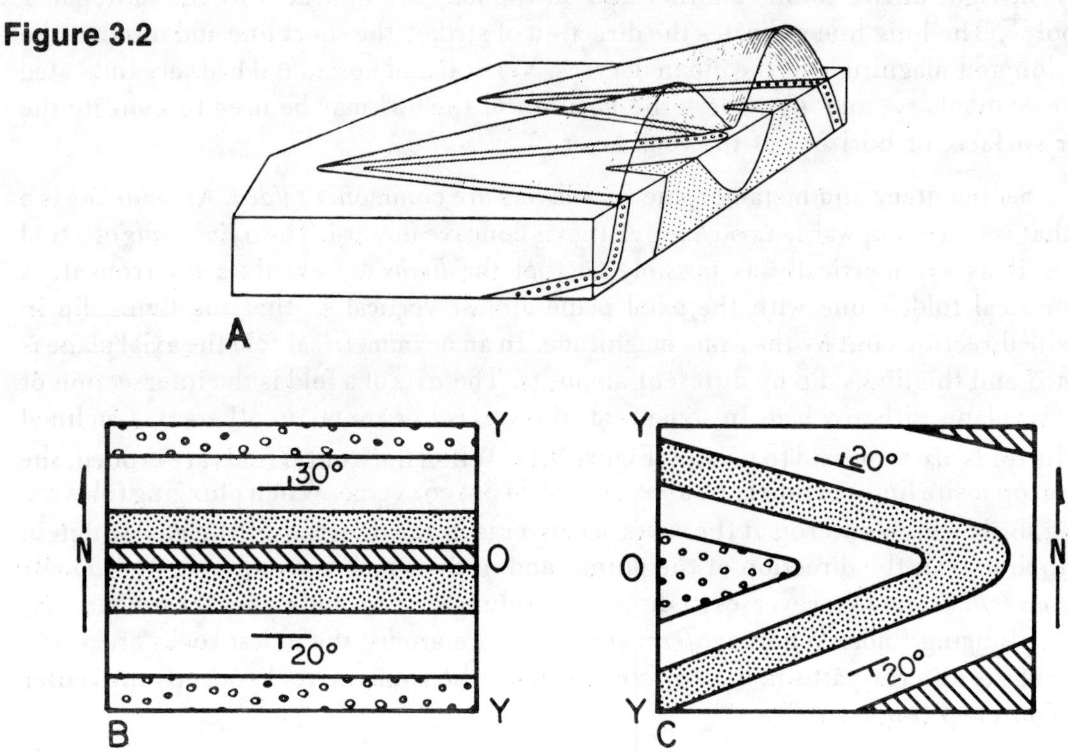

Surface outcrop patterns of horizontal folds (B) and plunging folds (A and C). The outcrop patterns of B and C are of anticlinal structures. Y = youngest rock, O = oldest rock (diagram A from Billings 1954).

Faulting is the rupturing of rocks and is always accompanied by a displacement along the plane of breakage; where no such movement occurs *joints* are produced. The *fault plane* has a dip and a strike and where it intersects the ground surface is the *fault line* or *fault trace*. According to the nature and relative direction of displacement, several types of fault are recognized. In vertical faulting one block, the *upthrown block*, is raised relative to the other or *downthrown block*. The displacement produces a *fault scarp* at the surface. When a fault scarp is eroded so that the topographic scarp no longer parallels the fault trace, the scarp is called a *fault line scarp*. When the fault plane dips in the direction of the downthrown block the fault is *normal*; when it dips towards the upthrown block it is *reverse*. In *transcurrent* or *strike-slip* faulting movement is predominantly horizontal so that fault scarps are rarely present. Although strike-slip faults are usually linear in plan, vertical faults may be sinuous. When the dip of a reverse fault is less than about 40° both vertical and horizontal motion of blocks takes place and the fault is generally known as a *thrust fault*.

Figure 3.3

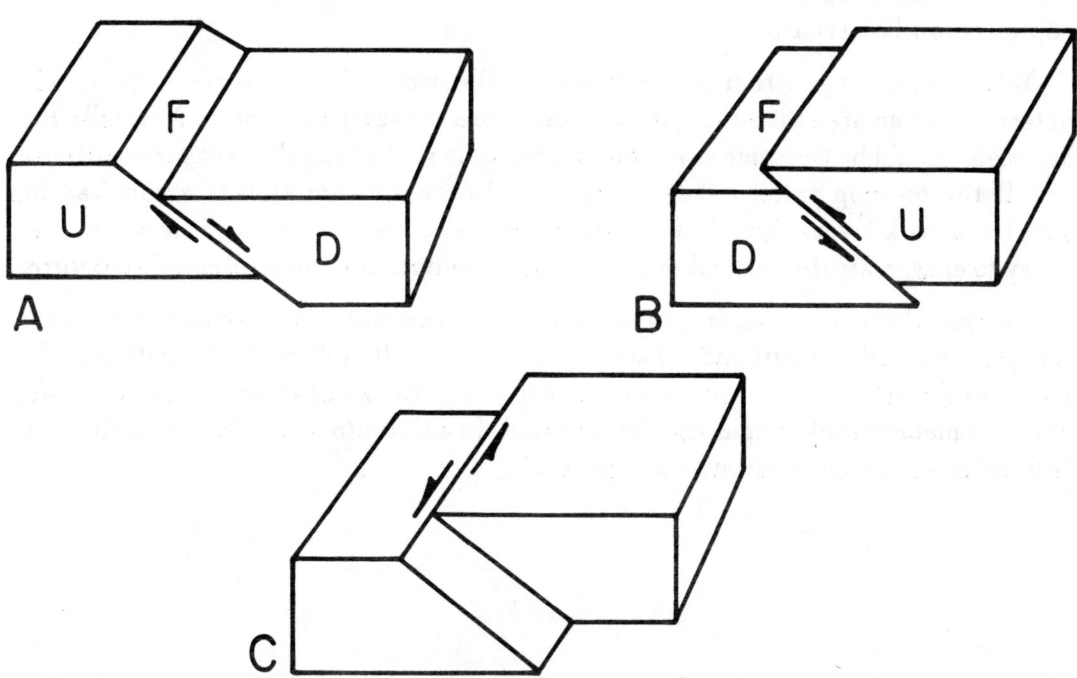

Normal (A), reverse (B) and strike slip (C) faults. U = upthrown block, D = downthrown block, F = fault scarp

If rock strata are horizontal, each stratigraphic horizon is everywhere at the same elevation and will thus follow the topographic contours. The width of the outcrop depends upon the thickness of the bed, and where this is constant will entirely depend upon the topographic slope, being greatest where slopes are gentle. The outcrop of a horizontal bed forms a V as it crosses a valley and the apex of the V points upstream. The top and bottom of a vertical bed will appear on a geologic map as straight lines parallel to the strike of the layer. Topography has no control upon the outcrop pattern of vertically dipping structures such as dikes.

In areas of dipping beds where fold structures have been eroded, the outcrop pattern of beds that dip upstream forms a V that points upstream and the limits of the outcrop do not parallel the topographic contours. The outcrop pattern of beds that dip downstream at an angle greater than the stream gradient form a V with the apex pointing downstream. The outcrops of beds that dip downstream at an angle less than the stream gradient form a V that points upstream.

The outcrop pattern of any planar bed can be predicted if a topographic map is available, if the dip and strike are known, and if the location of one exposure is determined. For example, if a bed outcrops at X and is known to strike east to west and to dip north at 45°, it is possible to determine its position at any place in the area. The upper or lower surfaces of the bed can be represented by *structure contours* which like topographic contours indicate the configuration of a surface. At each point where a structure contour intersects a topographic contour of the same elevation the bed must outcrop. By connecting known outcrop points for the upper and lower horizons an outcrop map can be prepared.

When preparing structure sections to illustrate the subsurface geological characteristics of an area the first stage is to prepare a topographic profile. Normally the vertical scale should be the same as the horizontal scale so that angular measurements are correct. If the outcrop pattern at the surface and rock dips are known, angles can be measured and rock layers directly plotted. If the strata are horizontal, it is sometimes necessary to exaggerate the vertical scale for best presentation of the geological structure.

An *unconformity* is a surface of erosion or non-deposition that separates younger strata from older rocks. Uplift and subaerial erosion generally follow the formation of the older rock and finally the younger strata are deposited. Rocks of sedimentary, volcanic, plutonic and metamorphic origin may be involved. In a *disconformity,* the formations on opposite sides of the unconformity are parallel.

Name ______________________ Date ______________

Instructor ______________________ Section ______________

QUESTIONS

3.1 Examine Figure 3.4:

(a) What is the dip of the sedimentary strata?

(b) Briefly describe the relationship between the outcrop pattern and the topography.

(c) Draw a topographic profile and geologic section along line A-B. Use the same horizontal scale as in the figure with no vertical exaggeration.

Figure 3.4

Topographic map and geologic outcrop pattern

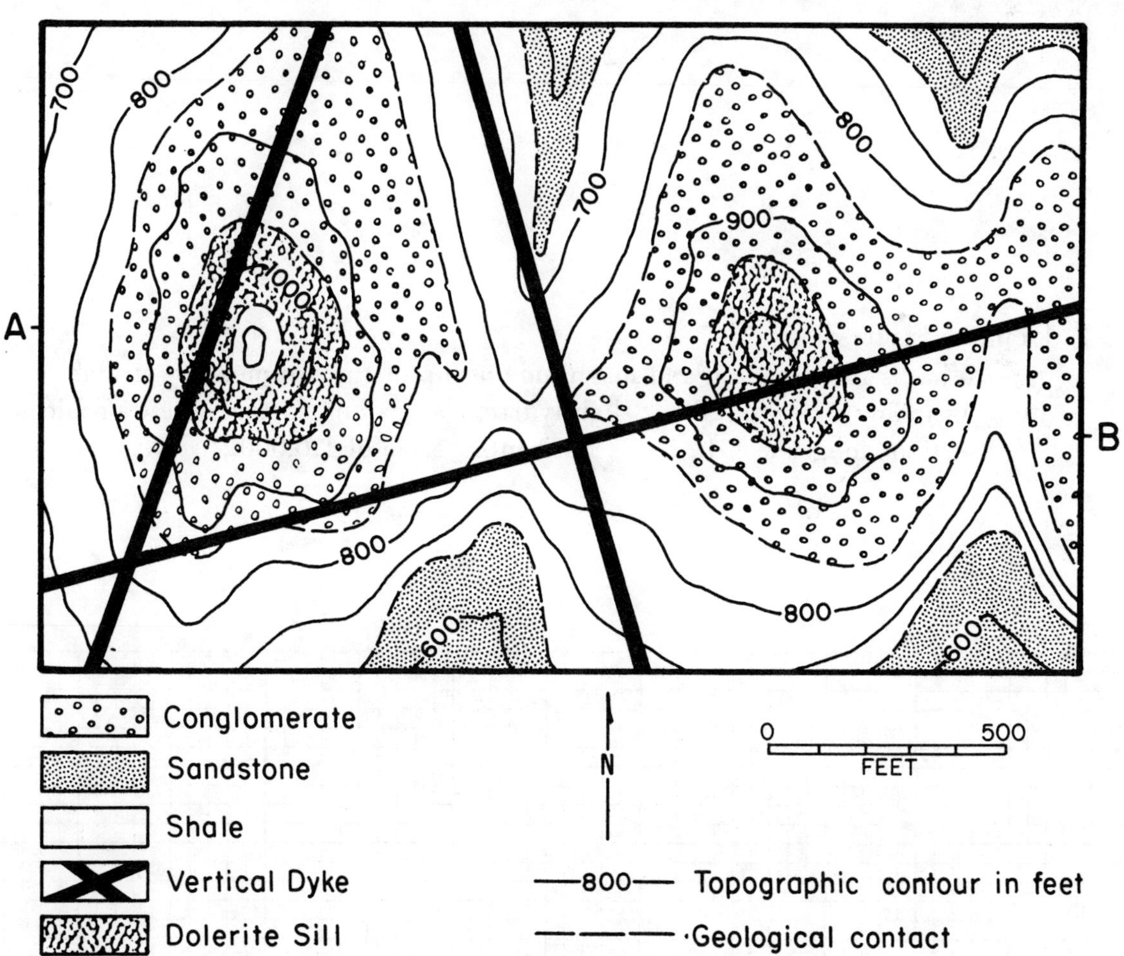

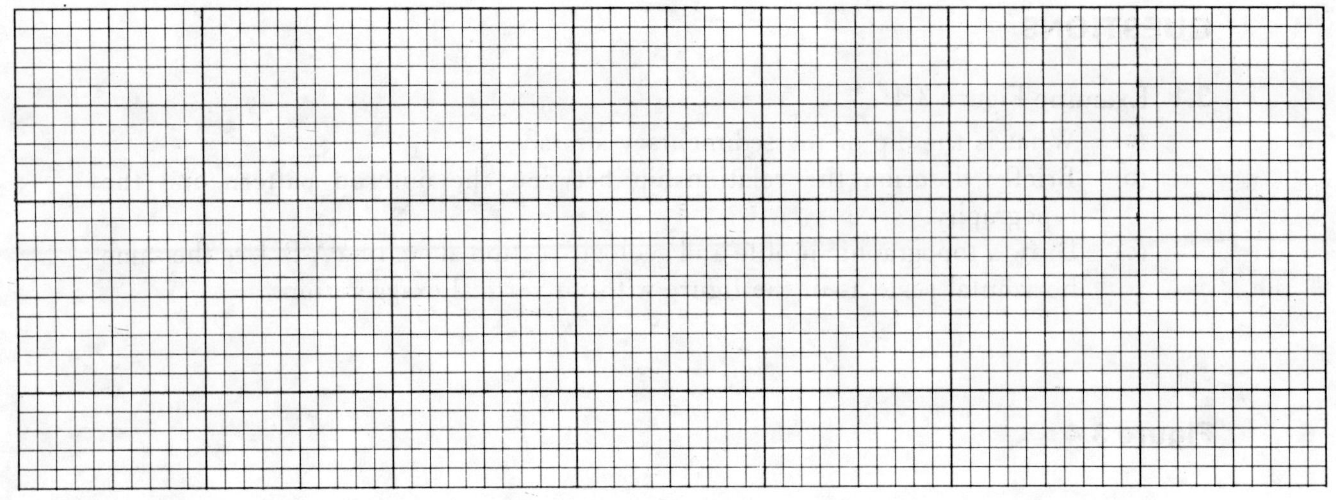

3.2 Examine Figure 3.5:

(a) What is the direction and magnitude of dip of the sedimentary strata?

(b) Assuming a horizontal but slightly irregular ground surface, draw a geologic section along line A-B. Use a horizontal scale of 1:6,000.

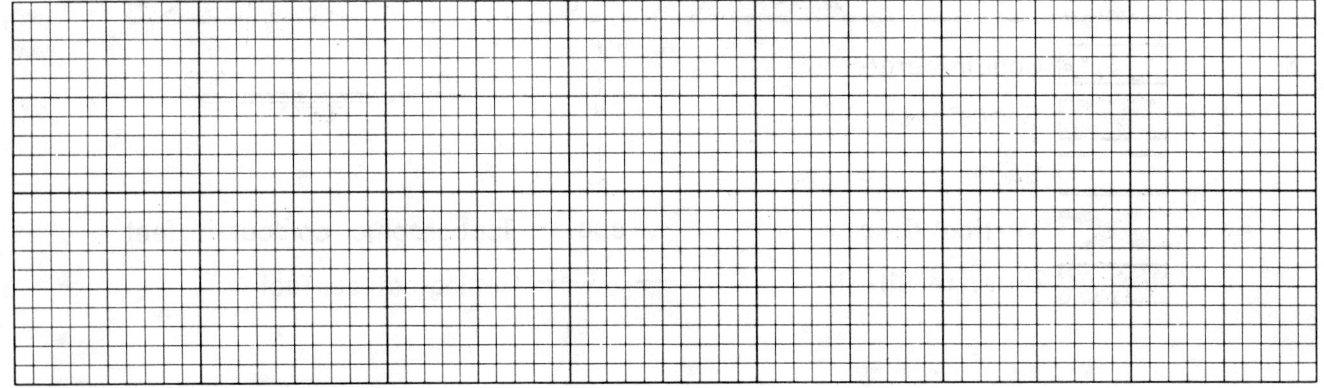

Name________________________________ Date________________

Instructor ____________________________ Section______________

Figure 3.5

Outcrop pattern of dipping rocks

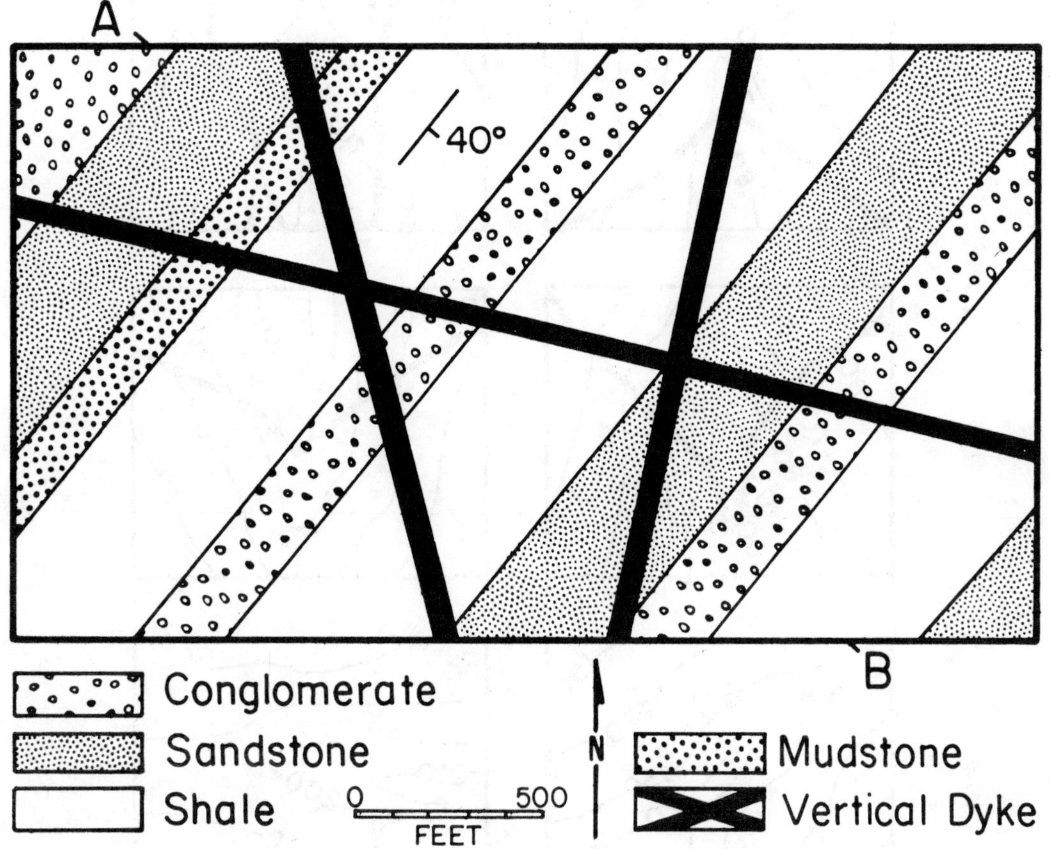

3.3

(a) In A, B, C and D of Figure 3.6 the heavy lines represent the outcrop of a thin layer of limestone. Determine the angle and direction of dip in each case.

(b) In E and F of Figure 3.6, X and Y mark two points where a thin coal layer outcrops at the surface. In E the layer dips south at an inclination of 1 in 10; in F it dips east with an inclination of 1 in 5. Mark in the surface outcrop of the coal layer in each case.

Figure 3.6

The outcrop pattern of a thin rock layer in different topographic situations

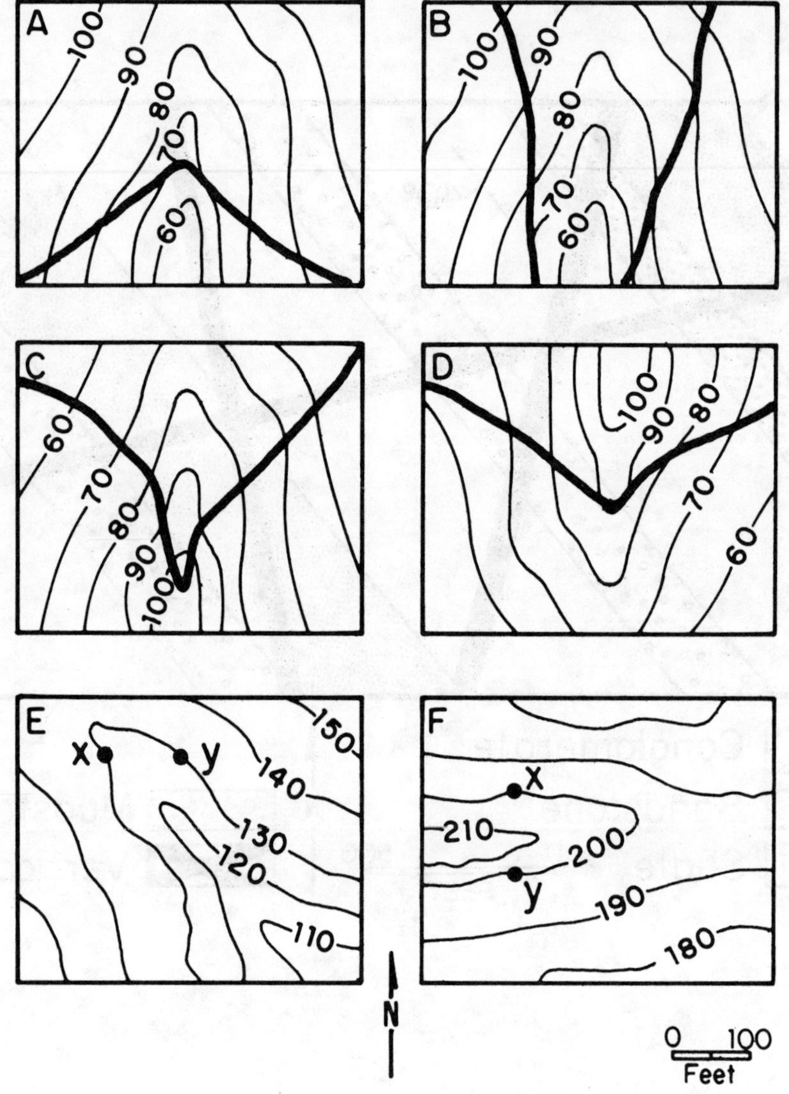

3.4 Examine Figure 3.7:

(a) In the figure the base of a horizontal bed of sandstone 150 feet thick is exposed at an elevation of 1,950 feet on the northeast slope of Long Ridge. Above the sandstone is 300 feet of shale and below it 100 feet of conglomerate, 300 feet of limestone and 300 feet of dolomite. Produce a geologic map of the area coloring or shading the various strata.

(b) At point X northwest of White Mountain the upper surface of a narrow dyke is exposed at an elevation of 1,700 feet. The dyke dips at an angle of 60° to the north. Plot the outcrop of this dyke on your geologic map.

Name____________________ Date__________

Instructor ____________________ Section__________

Figure 3.7

Plotting the outcrop pattern of dipping rocks on a topographic map

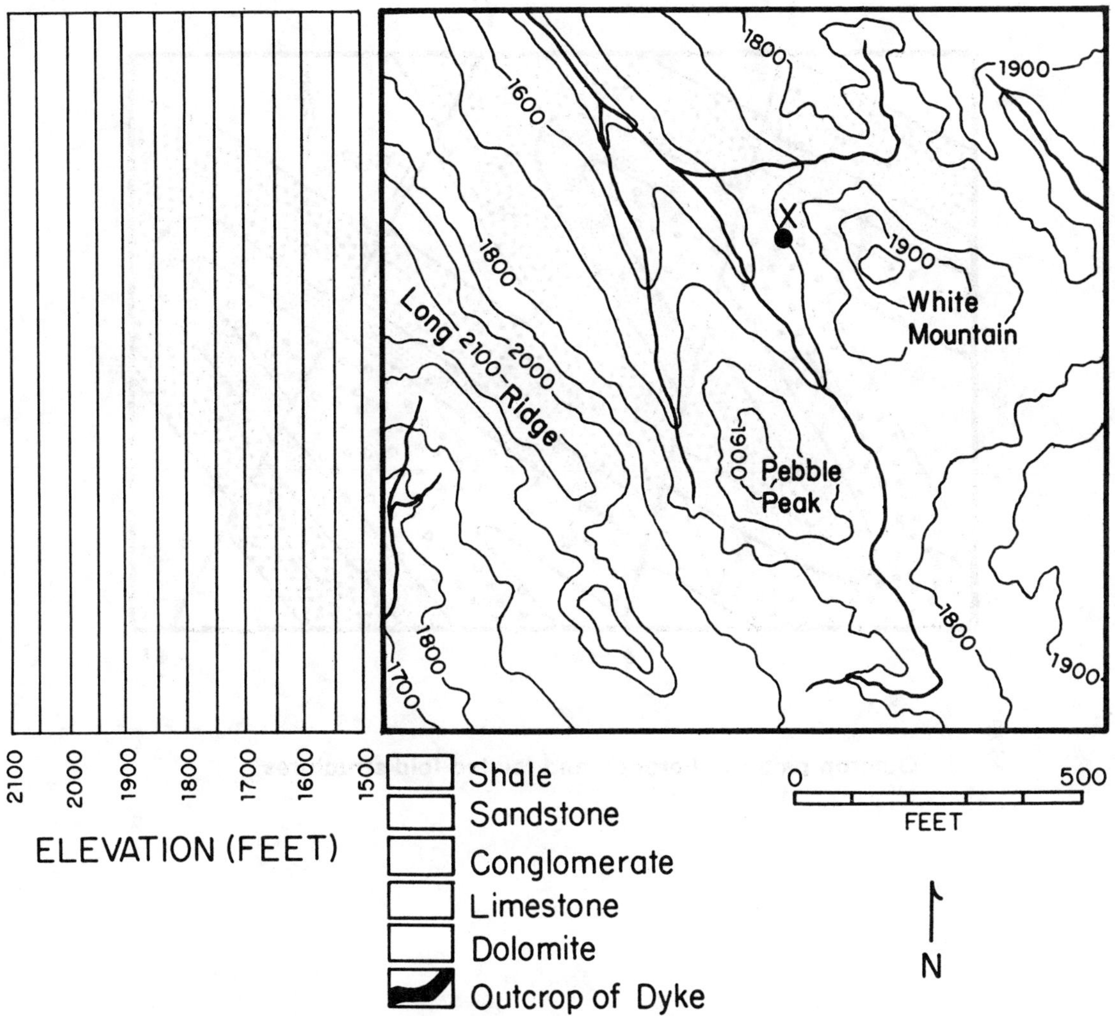

3.5 Figure 3.8 shows the outcrop pattern of a series of eroded plunging anticlines and synclines which have been faulted. Determine which features are anticlines and which synclines and mark their direction of plunge. If one of the faults dips at 60° and the other is vertical what kinds of faults are they and how do you know?

Figure 3.8

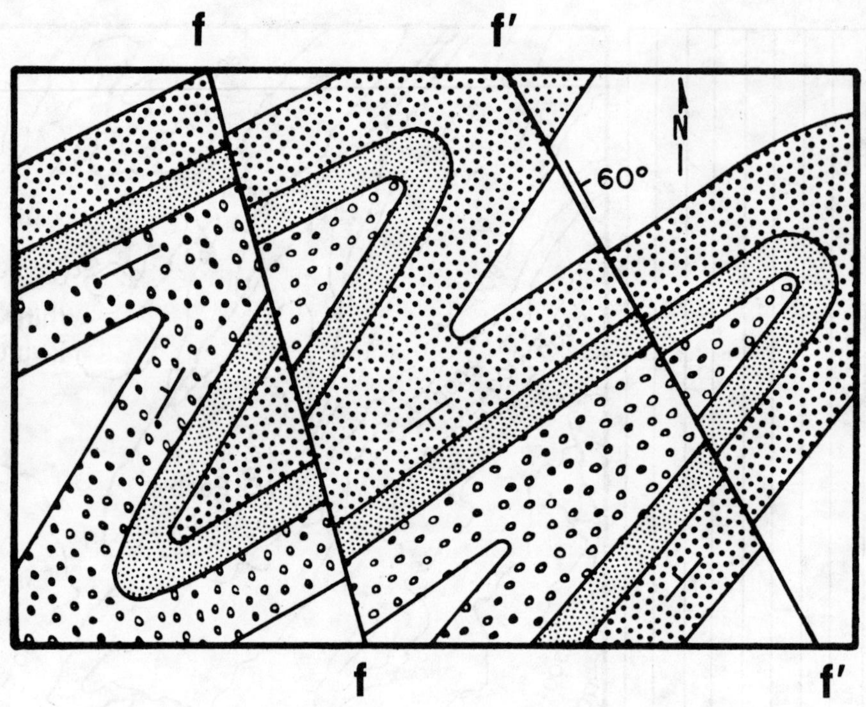

Outcrop pattern of eroded and faulted fold structures

EXERCISE **4**

AERIAL PHOTOGRAPHS

Most areas of the earth's surface have been photographed from either aircraft or satellites. These photographs have been widely used in topographic and geological mapping, land use survey, mineral exploration and in military reconnaissance. Many areas have been flown several times so that some sets of photographs show changes in the landscape that have occurred within the last 50 years. Many such changes are related to earthquake, volcanic, flood and hurricane action.

Two types of aerial photograph are generally used in landscape interpretation. *Vertical photographs* are taken with the axis of the camera as nearly vertical as possible and are by far the most widely used forms for mapping purposes. *Oblique photographs* are taken with the axis of the camera lens intentionally directed between the horizontal and the vertical. A *high oblique* photograph is one whose field of view includes a portion of the horizon, a *low oblique* does not.

One of the great advantages of using low and high altitude aerial photographs is that the earth's surface can be viewed in three dimensions, that is, stereoscopically. Aerial photographs are generally taken so that they *overlap* by 60% in the direction the aircraft is moving, that is along the *flight line*. Complete coverage of an area is achieved by overlapping at right angles to the flight lines by 20%, this is called the *sidelap* (Figure 4.1). For *stereoscopic vision* two overlapping photographs are adjusted so that the same feature on each photograph is separated by a distance equal to that between the observer's eyes (approximately 2.5 inches). If this arrangement of the photographs is viewed through a *lens stereoscope* the terrain will be seen in three dimensions but with the relief and slope of the landscape exaggerated. This is because the observer is effectively looking at the ground as if one eye were more than a mile from the other. In this manual pairs and triplets of photographs have already been arranged for stereoscopic study. If a lens stereoscope is correctly placed over adjacent photographs so that the left eye sees the same feature on the left photo as the right eye sees on the right photo, a three-dimensional image will come into view.

The *scale* of aerial photography, as taken, depends upon the elevation of the photographic platform, be it an aircraft or a satellite, above the ground surface (H), and the focal length of the camera lens (f) (Figure 4.2). If these two values are known the scale of the photography can be calculated using the formula:

$$\frac{f}{H} = \frac{1}{\text{scale factor (Sf)}}$$

Figure 4.1 Overlap and sidelap in aerial photographs

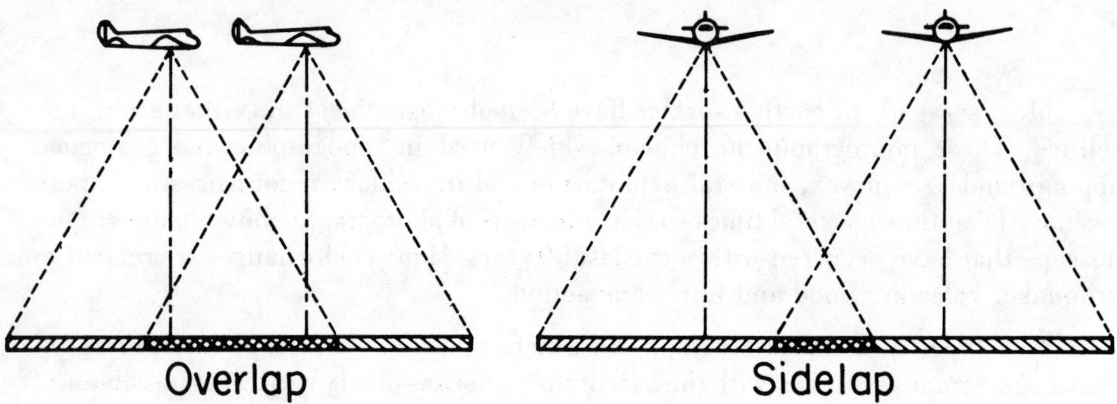

Figure 4.2 Scale and ground coverage of aerial photography

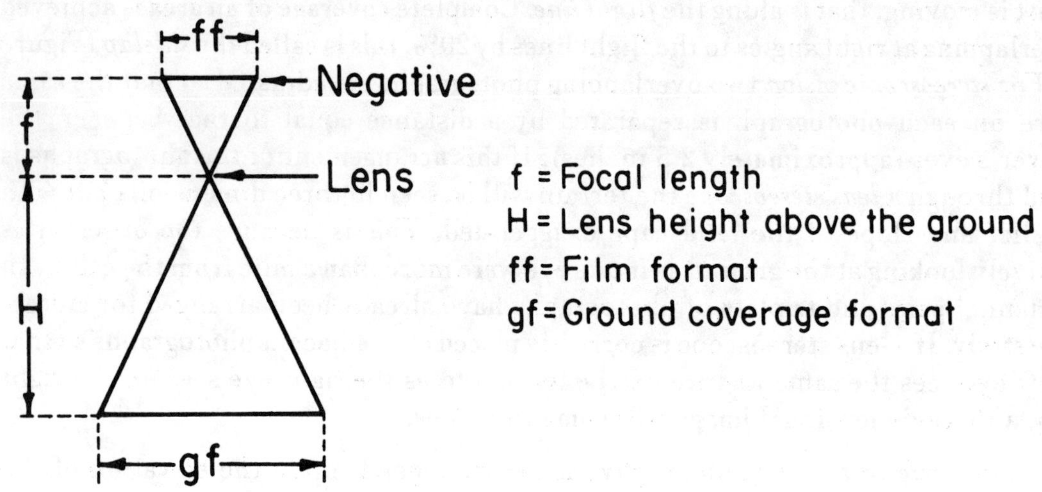

where f and H are in the same units. When they are not known the photographic scale can be determined by comparison with a map of the same area and of known scale. If the length of an object (such as a road) is measured on both map and photograph the photographic scale factor is given by

$$\text{Sf of photograph} = \frac{\text{dimensions of object on map x Sf of map}}{\text{dimensions of object on photograph}}$$

where dimensions are in the same units. Scale can also be calculated if the actual size of an object visible in the image is known. For example, if a road is known to be 83 feet 4 inches wide on the ground and 0.05 inches on the photograph then the photographic scale factor 1:20,000 is given by

$$\text{Sf of photograph} = \frac{\text{dimensions of object on ground}}{\text{dimensions of object on photograph}}$$

where dimensions are in the same units. If the scale of the photography is known as well as the initial film negative format size (Figure 4.2) the ground coverage can be calculated from the relationship

Ground format (gf) = film format (ff) x photographic Sf

Figure 4.3 Details of relief displacement in a single aerial photograph

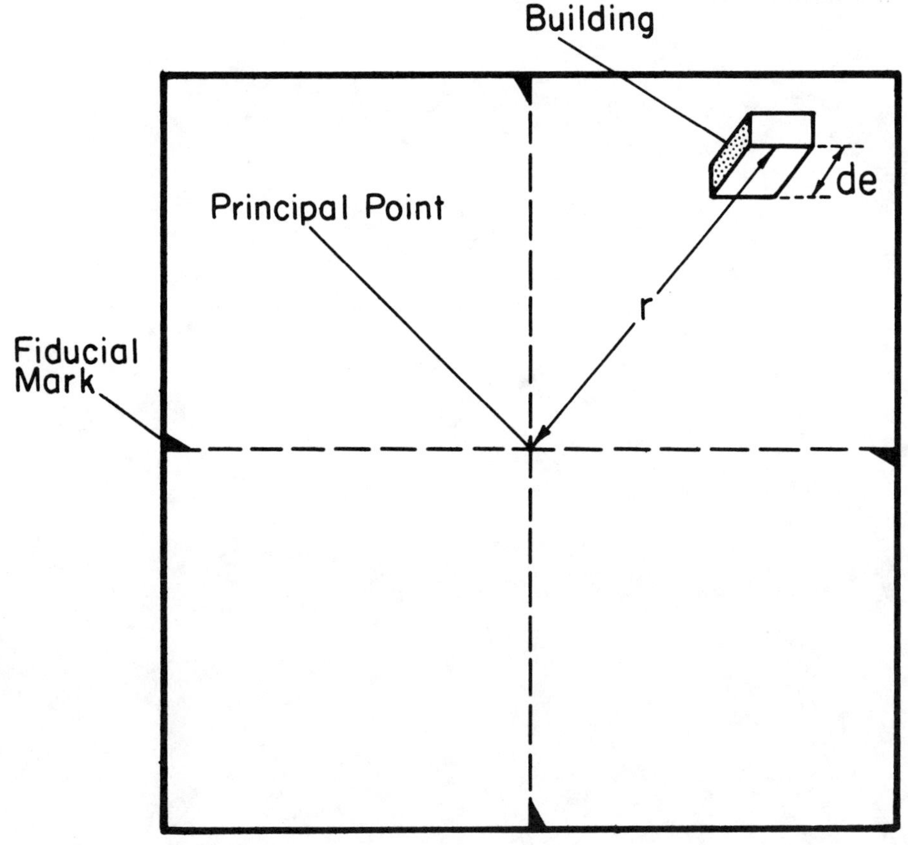

Photographs differ from maps in that the scale is not constant across the photograph and there are relief displacements which make the photo planimetrically incorrect. Scale variations are caused by the ground being hilly or mountainous instead of flat. Features on the top of a mountain are imaged at a larger scale than features in a valley nearby because the mountain is closer to the camera. All features on a map are shown at the same scale regardless of their relative elevations. Away from the center of a photograph (which is called the *principal point*) the top of an object such as a tall building will not occupy exactly the same position as its base. This *relief displacement* is more apparent at increasing distance from the principal point and is due to the camera viewing the terrain at an increasingly oblique angle. The amount of displacement d_e is given by

$$d_e = \frac{h}{H} \cdot r$$

where h is the height of the object, H is the flying height of the aircraft and r the radial distance from the principal point to the top of the object (Figure 4.3). H and h must be in the same units (e.g. feet); d_e will then be in the units of r (e.g. inches). If vertical photographs of downtown Los Angeles were taken from 5,000 feet elevation, and the Crocker-Citizen Plaza, which is 42 stories and 620 feet high, was located 6 inches from the center of one photograph, the relief displacement would be 0.74 inches. The top of the building would be 0.74 inches from its base so that its sides would be visible. If the flying height of photography is known and d_e and r can be measured on the photographs, the heights of objects can be approximately calculated. *Photomosaics* when made from several aerial photographs should be constructed from the central portions where relief displacements are at a minimum.

Name ______________________ Date ______________

Instructor ______________________ Section ______________

QUESTIONS

4.1 Some photographs in this manual are not vertical. What kinds of photographs does the manual contain? Support your answer by giving examples.

4.2 The U-2 reconnaissance aircraft flies at 70,000 feet altitude above sea level. If a U-2 is used to photograph the ground surface and the film format is kept constant at 9 inches square, the scale of the photography obtained and the area of the ground covered will depend upon the focal length of the camera lens. The greater the focal length the larger the photographic scale but the smaller the area covered. Demonstrate this by completing Table 4.1.

Table 4.1 Effect of camera focal length on area covered and photographic scale

Aircraft Altitude Above Ground (feet)	Focal Length of Camera (inches)	Photographic Scale Factor	Ground Coverage Format with 9″x9″ Film.
70,000	6		
70,000	8.25		
70,000	18		
70,000	36		

4.3 Cameras used in aerial photography commonly have focal lengths of 6.0 or 8.25 inches. If a camera of 6 inch focal length and a constant film negative format of 9 inches square is used to photograph an area, the photographic scale and the area covered will depend upon the altitude of the aircraft above the ground. Demonstrate this by completing Table 4.2.

Table 4.2 Effect of flying height above ground on photographic coverage and scale

Aircraft Altitude Above Ground (feet)	Focal Length of Camera (inches)	Photographic Scale Factor	Ground Coverage Format with 9″x9″ Film.
60,000	6		
30,000	6		
10,000	6		
5,000	6		

4.4 An area of varied terrain is photographed from an aircraft flying at 20,000 feet above sea level. The focal length of the camera lens is 6 inches and the film negative format 9 inches square. The area covered is broadly divisible into three regions (i) a broad lowland of average elevation 1,000 feet (ii) a high plateau with an elevation of 4,000 feet and (iii) a mountainous area at 10,000 feet. How will the scale and coverage of the photography vary from one physiographic region to the next and what problems does this pose if a map is to be prepared from the photographs?

Name ____________________ Date ____________________

Instructor ____________________ Section ____________________

4.5 Use the topographic maps in Figures 17.2 and 23.2 to calculate the scale of the photographs in Figures 17.3 and 23.3.

4.6 An automobile is clearly visible on the aerial photograph of an urban area. On the photo the automobile is 0.05 inches long but its true length is known to be 14 feet. What is the scale of the photograph, and assuming it was taken with a 6 inch focal length lens, what was the aircraft flying height?

4.7 With reference to Figures 17.2, 17.3, 23.2, and 23.3, what are the advantages and disadvantages of aerial photographs and topographic maps in the study of the earth's surface?

4.8 (a) Measure d_e and r in Figure 4.3 and given that the photograph was taken from an altitude of 5,000 feet, calculate the height of the building.

(b) Vertical aerial photography of a steep hill 800 feet high, which is thought to be a volcanic plug, is taken from an altitude of 10,000 feet above the ground. The plug is 5 inches from the principal point of the best photograph which was taken with a 6 inch focal length lens. The top of the plug is mapped directly from the photograph at the same scale. What is the error in ground distance of the position of the plug on the map to its actual position on the ground?

4.9 Skylab, the first American space station, was launched in 1972 and returned to Earth in 1979. While in orbit Skylab astronauts took thousands of photographs of the Earth using the Earth Resources Experimental Package (EREP). Skylab orbited the Earth every 90 minutes at an altitude of 235 miles. The EREP included a camera assembly with six high precision multispectral cameras of 6 inch focal length which utilized a film format 2.25 inches square. What is the size of the area covered by each photograph? The EREP also included an Earth terrain camera with a film format of 5 inches square. If the area covered by each frame is 59 miles on a side, determine the scale of the photography and the focal length of the Earth terrain camera.

4.10 Several 70 mm. color photographs of southern Arizona and northwestern Sonora were taken by Virgil Grissom and John Young during the first manned Gemini flight. It was found that most of the geological detail shown on a 1:375,000 map could be seen on a photo whose scale, as taken, was approximately 1:2,250,000. If the focal length of the camera used was 80 mm. what was the distance in kilometers between the astronauts and the earth when the photographs were taken?

EXERCISE 5

COMPOSITION OF THE ATMOSPHERE AND EARTH-SUN RELATIONS

The primary ingredient of the earth's atmosphere is dry air which is composed of the following gases:

gas	percent by volume
Nitrogen (N_2)	78.084
Oxygen (O_2)	20.946
Argon (Ar)	0.934
Carbon dioxide (CO_2)	0.0325
Neon (Ne)	0.00182
Helium (He)	0.000524
Methane (CH_4)	0.00015
Krypton (Kr)	0.000114
Hydrogen (H_2)	0.00005

These gases can be referred to as the "permanent gases" or "non-variable gases" since they are found in the same relative proportions throughout the atmosphere. It should be noted, however, that the carbon dioxide content has been rising during the last century or so due to man's burning of fossil fuels. In that respect, CO_2 is a variable gas through time.

Mixed in with the above gases are the "variable gases." These are gases which vary in amount from place to place or time to time. These variable gases may include local pollutants such as nitrous or sulphur oxides. However, the two most significant variable gases in the atmosphere as a whole are water vapor (H_2O) and ozone (O_3). Water vapor is concentrated in the lower portion of the atmosphere near the earth's surface. It varies spatially and temporally (in time) from approximately 0% to 4% of the atmosphere by volume. Ozone is limited to trace amounts found primarily as a layer of the atmosphere around 25 km altitude. This ozone layer is of vital importance to life on earth since ozone absorbs the harmful ultraviolet radiation from the sun.

The atmosphere also contains non-gaseous materials such as liquid cloud water droplets, dust and other suspended particulate matter.

The earth revolves in an orbit around the sun at a mean distance of 150 million kilometers (93 million miles) being closest to the sun (perihelion) around January 3 at a distance of 147.5×10^6 km and furthest from the sun (aphelion) around July 4 at a distance of 152.5×10^6 km. However, these small fluctuations in earth to sun distance do not cause the seasons. The seasons result from a variation in sun angles caused by the 23½° tilt of the earth's polar axis from the perpendicular of the plane of the ecliptic (the plane upon which the earth revolves around the sun). In fact, the north pole is always pointing towards the North Star or Polaris regardless of the earth's position in its orbit around the sun. This results in the sun's direct rays shining over different latitudes at different times of the year. Also, daylengths will vary seasonally. Fig. 5-1 illustrates this concept showing that the north pole is tilted "away" from the sun on approximately December 22 creating the beginning of winter in the northern hemisphere (summer in the southern hemisphere since the south pole is tilted "towards" the sun). The situation is reversed on approximately June 21. These dates are known as the winter and summer solstices respectively (opposite in the southern hemisphere). Around March 21 and September 23, the polar axis still points towards Polaris (the North Star) but neither "towards" or "away" from the sun. During these two days, all locations receive 12 hours daylight and 12 hours darkness (except the poles which experience 24 hours twilight) and hence these days are known as the equinoxes. Also shown in Fig. 5-1 is a line called the "circle of illumination" which is the boundary between daylight and darkness. Latitudes that are entirely on the daylight side of the circle of illumination are experiencing 24 hours continuous daylight (e.g. the north pole region on June 21). On the other hand, latitudes that are entirely on the darkness or shaded side of the circle of illumination are experiencing 24 hours continuous darkness (e.g. the north pole region on December 22).

As a result of this 23½° tilt of the earth's axis from the perpendicular to the plane of the ecliptic, the latitude at which the noonday sun is directly overhead (called the solar declination) varies between 23½° N (the Tropic of Cancer) on approximately June 21 to 23½° S (the Tropic of Capricorn) on approximately December 22. Values of the solar declination for specific dates can be estimated from Fig. 5-2. Thus, the angle of the noonday sun above the horizon (called the noonday solar altitude or elevation) for any location varies with the seasons according to the following relationship:

$$\theta = 90^\circ - \phi + \delta$$

where θ is the solar elevation (in degrees), ϕ is the latitude of the location under consideration (in degrees) and δ is the solar declination (in degrees). Solar declination is considered positive (+) if the noonday overhead sun is in the same hemisphere (northern or southern) as the location under consideration and negative (—) if in the opposite hemisphere. Note also that the time when the sun is at its highest point in the sky is called solar noon. When using the above equation in tropical latitudes, values of Θ greater than 90° may occur for some dates. When this occurs, the noonday sun is actually "behind" you when looking towards the equator. Under these circumstances, Θ should be recalculated as:

$$\Theta = 90^\circ - (\text{originally calculated } \Theta - 90^\circ).$$

This results in a value of solar elevation with reference to the nearest horizon and will never exceed 90°.

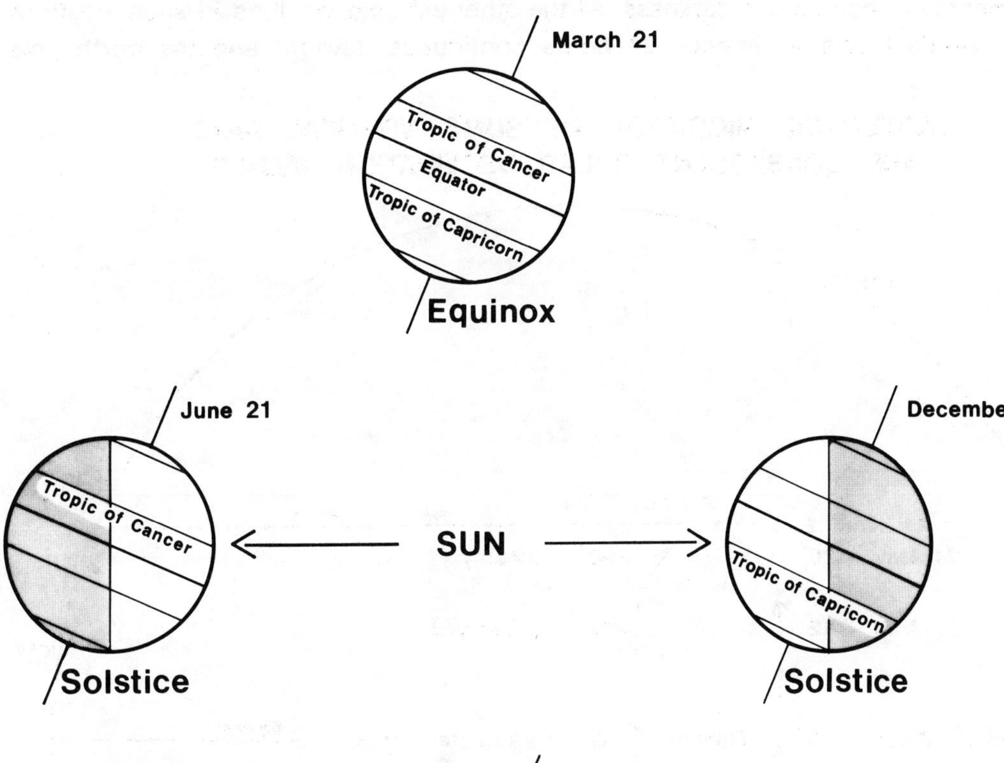
EARTH - SUN RELATIONSHIPS
March 21
Tropic of Cancer
Equator
Tropic of Capricorn
Equinox
June 21
Tropic of Cancer
Solstice
SUN
December 22
Tropic of Capricorn
Solstice

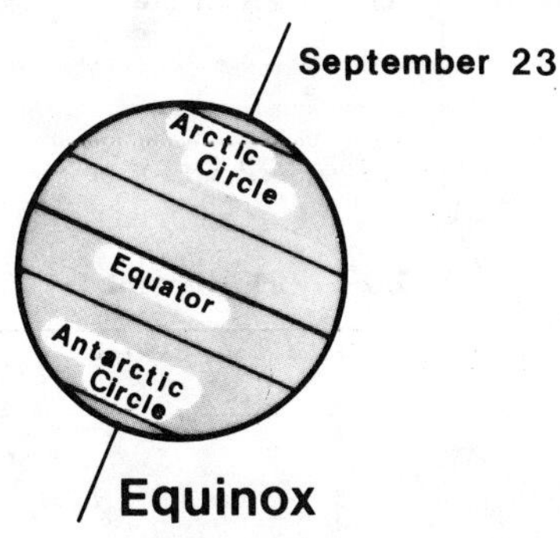
September 23
Arctic Circle
Equator
Antarctic Circle
Equinox

Figure 3-1

The length of daylight is influenced by the sun angle and consequently also has seasonal variation. For example, on December 22 when $\delta = 23\frac{1}{2}°S$, a location in the northern polar areas beyond 66½° N (the Arctic Circle) would experience a noonday solar elevation of zero (or a negative value) implying that the sun never rises on that day. Indeed, as mentioned earlier when discussing the circle of illumination in Fig. 5-1, such Arctic locations would be experiencing 24 hours darkness on December 22 and, in fact, the north pole itself experiences 6 months of continuous darkness. At the other extreme, on June 21 such locations above the Arctic Circle experience 24 hours continuous daylight and the north pole

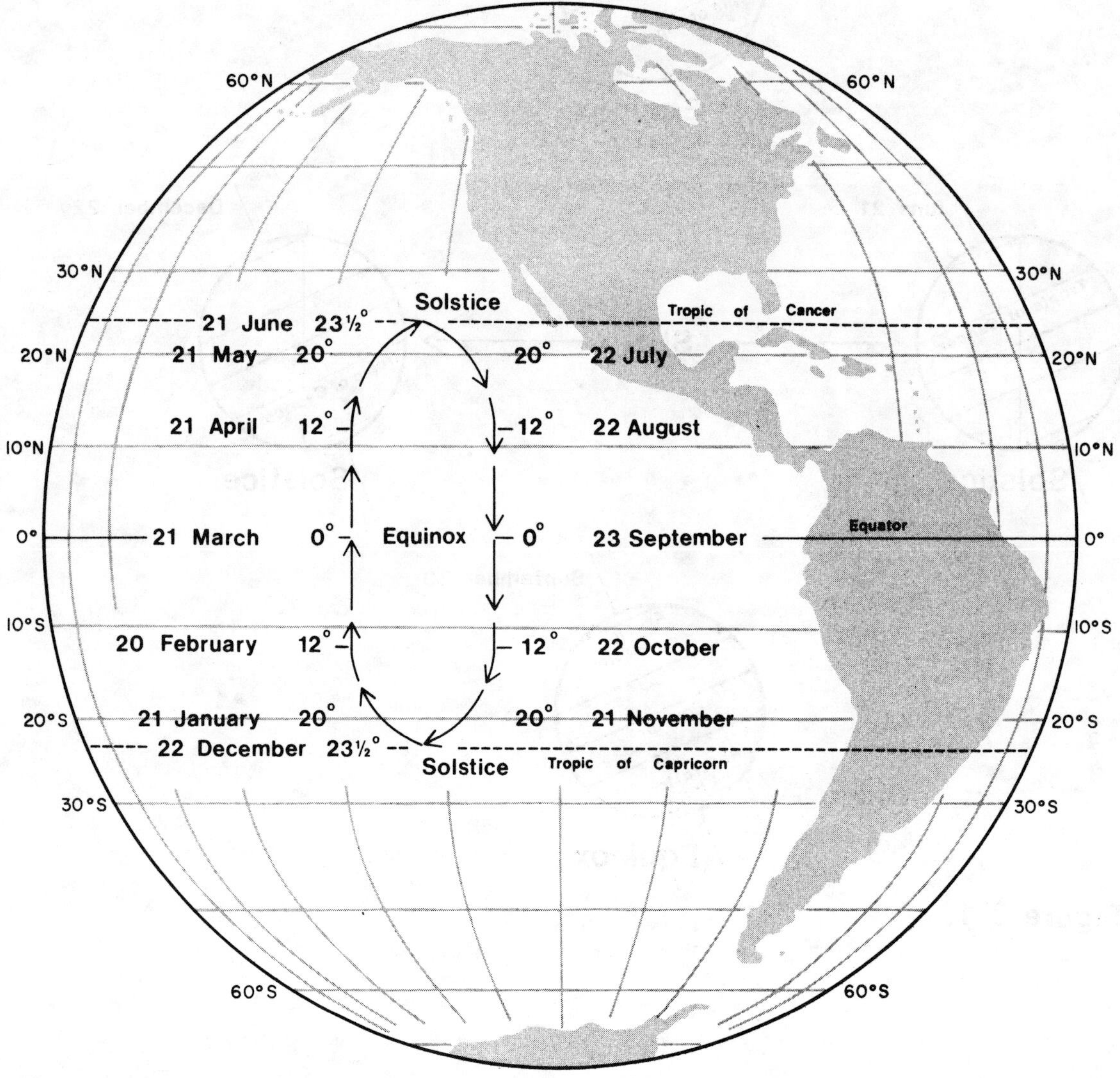

Figure 5-2

experiences 6 months continuous daylight between the equinoxes. Similar events occur beyond 66½° S (the Antarctic Circle) during their winter and summer. In fact, at the equator, lengths of daylight are constant at 12 hours year round and the further poleward the location, the greater the annual variation in length of daylight.

Name________________________________ Date____________________

Instructor______________________________ Section____________________

QUESTIONS

5.1 Complete the following:
Two gases which comprise approximately 99% of the dry air in the atmosphere are ____________________ and ____________________. The percentage content of the atmosphere represented by ____________________ has increased with mankind's increased use of fossil fuels. A variable gas found at about 25 km altitude which absorbs harmful ultraviolet radiation is ____________________ .

5.2 On average, a person breathes 10 m^3 (283 ft^3) of air each day. What volume of this is oxygen?

5.3 Cite a reason why the capacity of ozone to absorb ultraviolet radiation is so important?

5.4 The earth's atmosphere has been divided into a number of named layers, identified primarily by the nature of the vertical temperature distribution of the atmosphere. The diagram in Fig. 5-3 shows the general temperature structure of the atmosphere. Label the various layers (i.e. mesosphere, troposphere, stratosphere and thermosphere) as well as the boundaries between layers (i.e. stratopause, mesopause and tropopause).

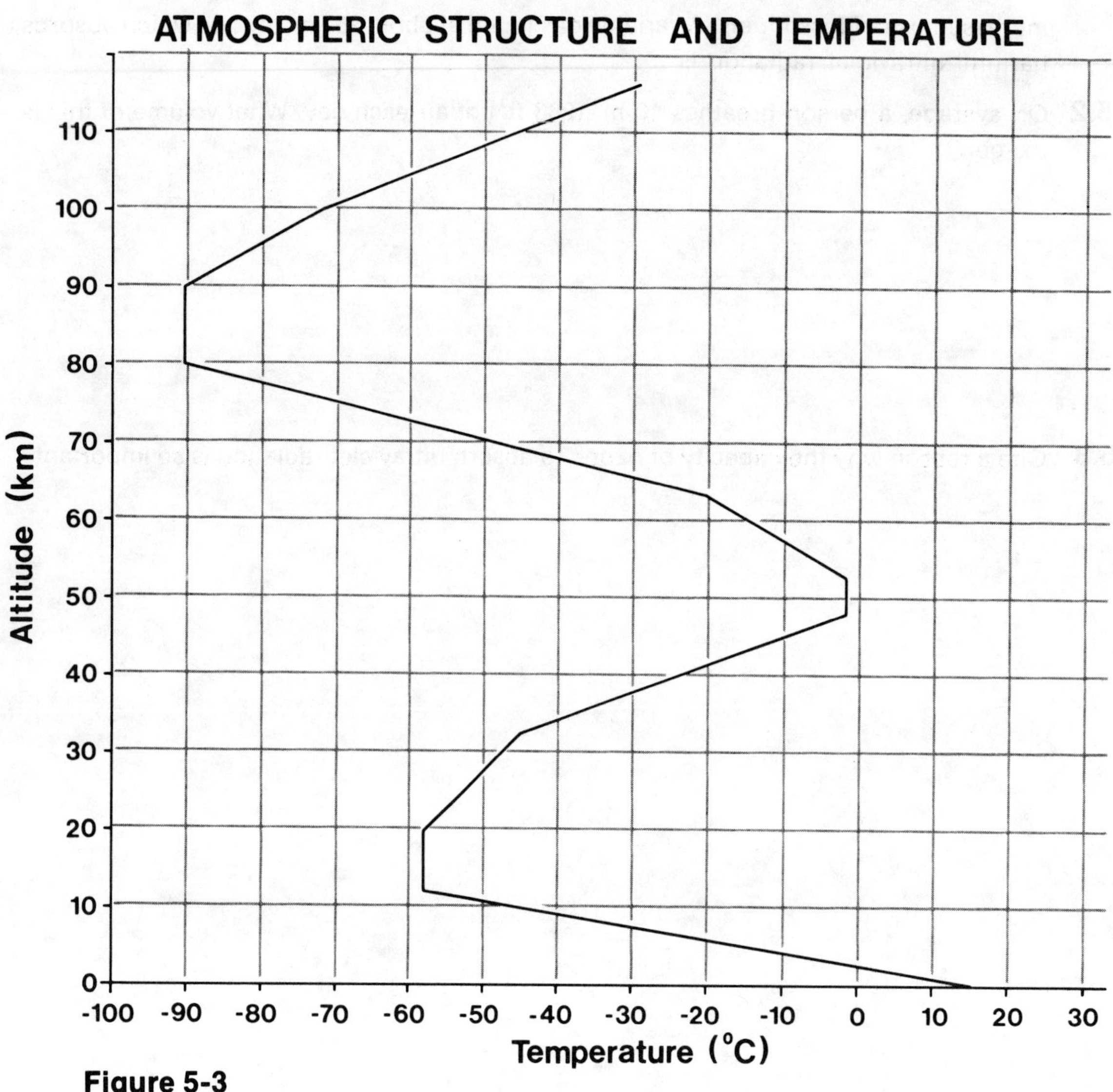

Figure 5-3

5.5 The vertical atmospheric temperature structure varies from location to location. For example, the tropopause is located at an altitude of about 8 km in polar areas and 17 km in tropical areas with an average height of 10 km. The vertical temperature structure also can vary seasonally.

The diagram in Fig. 5-4 shows the average temperature structure of the lowest 60 km of the atmosphere, for 40° N latitude, during winter and summer. Complete the following statements. (Note that Celsius degrees are used. Further discussion of temperature units can be found in Exercise 7).

Name________________________ Date____________

Instructor________________________ Section____________

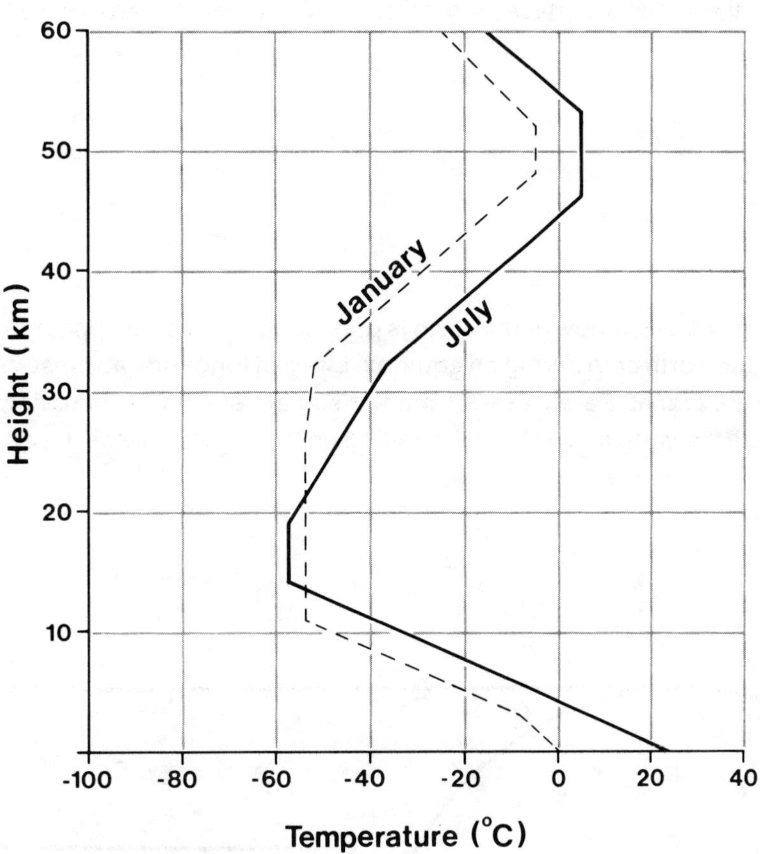

Figure 5-4

(a) The difference in the height of the tropopause between summer and winter is ____________ km.

(b) The depth of the troposphere is ____________ km in winter and ____________ km in summer.

(c) The stratopause is located at ____________ km altitude in summer.

(d) The average rate of temperature change in the troposphere is ____________ °C km^{-1} in winter and ____________ °C km^{-1} in summer. (Note that the rate of temperature change with height is known as a "temperature lapse rate.")

(e) The difference in winter and summer temperature is ______________ °C at the surface, ______________ °C at the tropopause and ______________ °C at the stratopause.

5.6 The earth continually encounters small meteors as it moves through space. However, few of these penetrate the entire atmosphere to reach the surface. What happens to meteors as they travel through the earth's atmosphere that prevents most of them from colliding with the earth's surface? What indicates that the frequency of meteor collisions with the moon's surface is greater than that for the earth's surface?

5.7 In the diagram in Fig. 5-5, a view of the earth is presented as it would appear from a point in space above the north pole during an equinox. Lines of longitude at constant intervals are marked on the planet. Parallel rays from the sun are shown, illuminating the earth from the side with the vertical ray of the sun falling on the equator (since it is an equinox).

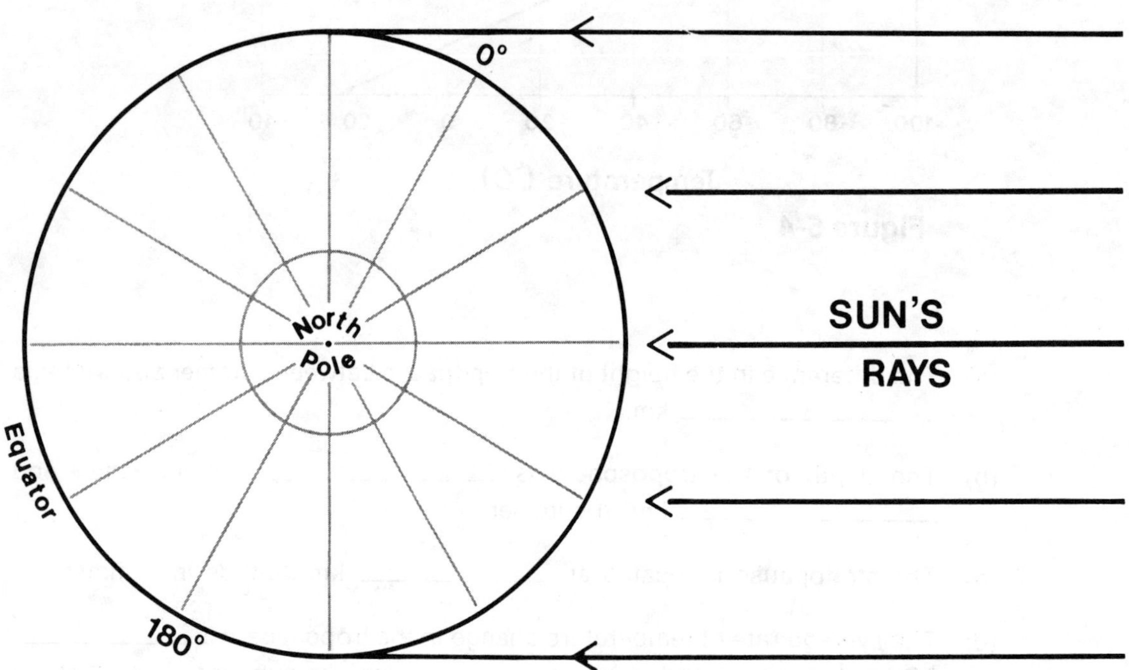

Figure 5-5

Name____________________ Date__________

Instructor____________________ Section__________

(a) On Fig. 5-5 carefully draw the "circle of illumination" and shade in the part of the earth that is in darkness. Add arrows around the earth indicating the direction of rotation as seen from above the north pole.

(b) During one day the earth rotates through an angle of ________ degrees. It rotates through an angle of ________ degrees in one hour. With the earth and sun related as shown in Fig. 5-5, solar noon is located at ________ degrees longitude. Along the prime meridian the time is ________. Along ________ degrees longitude it is midnight. In ________ hours it will be solar noon at the international date line.

5.8 There are 24 standard time zones each centered on a standard meridian and extending 7½° either side of this meridian. The time zone for the prime meridian (0° longitude), is called Greenwich Mean Time (GMT) and the standard meridians occur in 15° increments east and west of the prime meridian. Determine the date and the time for each of the following locations if it is 6:30 a.m. on October 27 in London, England. (The latitude and longitude of these locations were determined in Exercise 1.)

(a) Washington, D. C.

(b) Vancouver, British Columbia

(c) Buenos Aires, Argentina

(d) Salisbury, Zimbabwe

(e) Denver, Colorado

(f) Honolulu, Hawaii

(g) Tokyo, Japan

5.9 The length of time that the sun is above the horizon has an important influence upon the amount of solar radiation which will be received at a location during the day. The following table can be used to determine the length of time (in hours and minutes) of daylight for a given latitude.

LATITUDE AT WHICH YOU ARE	DURATION OF SUNLIGHT December 22	March 21 or September 23	June 21
90°	0:00	Sun at horizon	6 months
80°	0:00	12:00	4 months
70°	0:00	12:00	2 months
66½°	0:00	12:00	24:00
60°	5:33	12:00	18:27
50°	7:42	12:00	16:18
40°	9:08	12:00	14:52
30°	10:04	12:00	13:56
23½°	10:35	12:00	13:25
20°	10:48	12:00	13:12
10°	11:25	12:00	12:35
Equator	12:00	12:00	12:00

Using this table, determine the approximate daylengths for the following locations on the dates indicated (use interpolation when necessary):

	December 22	June 21
Quito, Ecuador (0° latitude)	________	________
Havana, Cuba (23½° N)	________	________
Miami, Florida (25° N)	________	________
Athens, Georgia (34° N)	________	________
Pittsburgh, Pennsylvania (40° N)	________	________
Minneapolis, Minnesota (45° N)	________	________
Winnipeg, Manitoba (50° N)	________	________
Anchorage, Alaska (61° N)	________	________
Fort Yukon, Alaska (66½° N)	________	________
Thule, Greenland (77° N)	________	________

Name____________________________ Date______________

Instructor__________________________ Section______________

5.10 Refer to Fig. 5-2.

(a) During which months of the year does the solar declination change:

1. at the most rapid rate?

2. at the slowest rate?

(b) Between May 21 and July 22, the noonday sun migrates across how many degrees of latitude? ____________

(c) The noonday solar elevation is 80° or greater at the equator for approximately how many continuous days? ____________.

5.11 (a) How does the varying rate of solar declination affect the length of daylight at any given location?

(b) During what seasons in the mid-latitudes does the length of daylight change:

1. the slowest?

2. the fastest?

5.12 Using the information in Fig. 5-2, calculate the solar elevation at noon for each of the following sites on the dates indicated. If the sun is below the horizon, indicate so. The latitude of many of these were determined in Exercise 1 and others are given in question 9 above.

Location	Dates					
	Dec. 21	Feb. 20	Apr. 21	June 21	Aug. 22	Oct. 22
Mexico City	_____	_____	_____	_____	_____	_____
Miami, Florida	_____	_____	_____	_____	_____	_____
Denver, Colorado	_____	_____	_____	_____	_____	_____
Vancouver, British Columbia	_____	_____	_____	_____	_____	_____
Anchorage, Alaska	_____	_____	_____	_____	_____	_____
Thule, Greenland	_____	_____	_____	_____	_____	_____
North Pole	_____	_____	_____	_____	_____	_____
Nairobi, Kenya	_____	_____	_____	_____	_____	_____
Salisbury, Zimbabwe	_____	_____	_____	_____	_____	_____
Buenos Aires, Argentina	_____	_____	_____	_____	_____	_____

EXERCISE **6**

ENERGY IN THE EARTH-ATMOSPHERE SYSTEM

Radiant energy from the sun provides virtually all of the heat energy for our planet. This energy or "radiation" takes the form of electromagnetic waves which travel in straight lines. Because a medium is not necessary for their propagation, waves of radiant energy are able to pass through space and impact upon the earth and its atmosphere. This radiant energy is the motive power and the ultimate source of the earth's weather.

Radiation can be characterized by its "wavelength." Fig. 6-1 shows part of the "electromagnetic spectrum" and indicates the wavelengths of some commonly known forms of energy from Gamma rays to radio waves.

All bodies radiate energy over a range of wavelengths. The amount of energy and wavelengths of this energy depend on the temperature (in degrees Kelvin — see Exercise 7) of the body. Higher temperatures result in more energy emitted and this radiation is of shorter wavelengths compared to the case for lower temperatures. For example, the sun has a temperature of about 6000°K whereas the earth-atmosphere system has temperatures around 300°K. The energy radiated for two bodies with such temperatures is illustrated in Fig. 6-2. Note that there is a convenient break between the radiation emitted by these two bodies. Hence we refer to the sun's radiation as "shortwave" or "solar" radiation (wavelengths of 0.1 to 4 μm) and that being radiated by the earth and its atmosphere as "longwave" radiation (4 to 100 μm).

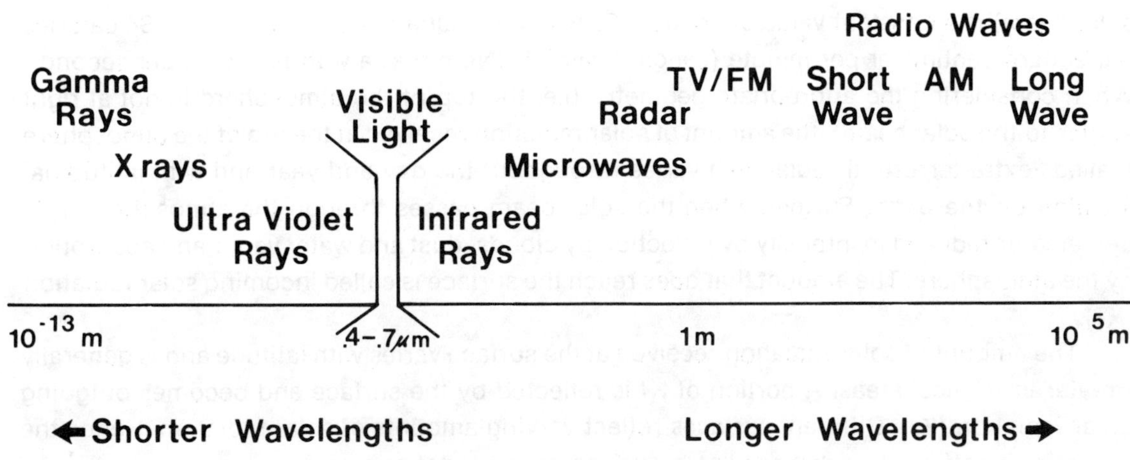

Figure 6-1

ELECTROMAGNETIC SPECTRA: SUN AND EARTH ATMOSPHERE SYSTEM

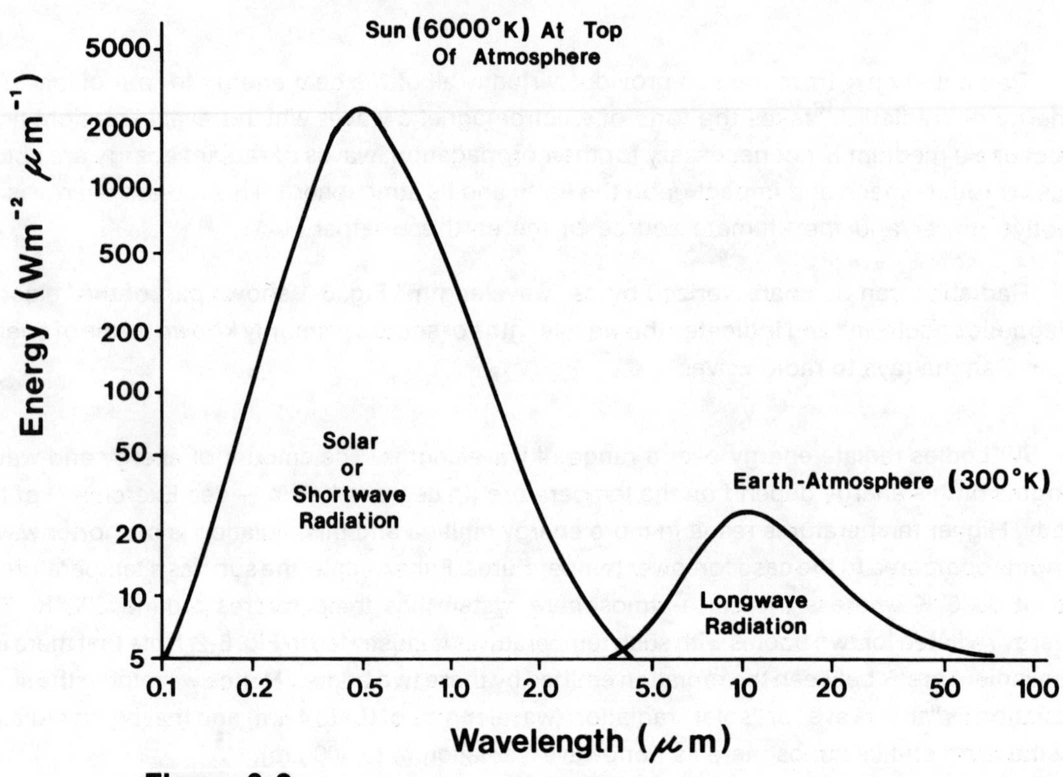

Figure 6-2

The intensity of solar radiation reaching the top of the atmosphere (at right angles to the solar beam) is a constant value of about 1353 watts per square meter (Wm^{-2}) or 1.94 calories per square centimeter per minute (cal cm^{-2} min^{-1}). (Note that a watt is a joule per second.) When considering the appropriate geometry (i.e. the top of the atmosphere is not at right angles to the solar beam), the amount of solar radiation received at the top of the atmosphere (called "extra-terrestrial radiation") varies throughout the day and year and with latitudinal location on the earth. Further, when the solar beam passes through the atmosphere, it is depleted or reduced in intensity by reflection by clouds, dust and water vapor and absorption by the atmosphere. The amount that does reach the surface is called incoming solar radiation ($K\downarrow$).

The amount of solar radiation received at the surface varies with latitude and is generally greater in tropical areas. A portion of $K\downarrow$ is reflected by the surface and becomes outgoing solar radiation ($K\uparrow$). Different surfaces reflect varying amounts of solar radiation. Hence the reflection coefficient or "albedo" of a surface may be defined as:

$$\text{albedo} = \frac{K\uparrow}{K\downarrow} \times 100\%.$$

The earth's surface and atmosphere also emit radiation but in the longwave portion of the electromagnetic spectrum. The net amount of radiative energy available at the surface (called net radiation Q*) can be derived as:

$$Q^* = K\downarrow - K\uparrow + L\downarrow - L\uparrow$$

where $L\downarrow$ is the incoming longwave radiation from the atmosphere and $L\uparrow$ is the outgoing longwave radiation emitted by the surface.

A similar "radiation balance" can be derived for the atmosphere alone as well as for the whole earth-atmosphere system combined. Fig. 6-3 illustrates the values of net radiation by latitude for (1) the surface (Q^*_{sfc}), (2) the atmosphere (Q^*_{atm}), and (3) the combined earth-atmosphere system (Q^*_{e-a}). By analysing this diagram, note two important imbalances:

(1) Q^*_{sfc} is generally positive everywhere while Q^*_{atm} is negative. This implies that there must be a **vertical** transport of energy from the surface to the atmosphere.

(2) Q^*_{e-a} is generally positive in tropical latitudes and negative in polar latitudes. This implies that there must be a **horizontal** transport of energy from tropical regions to polar regions.

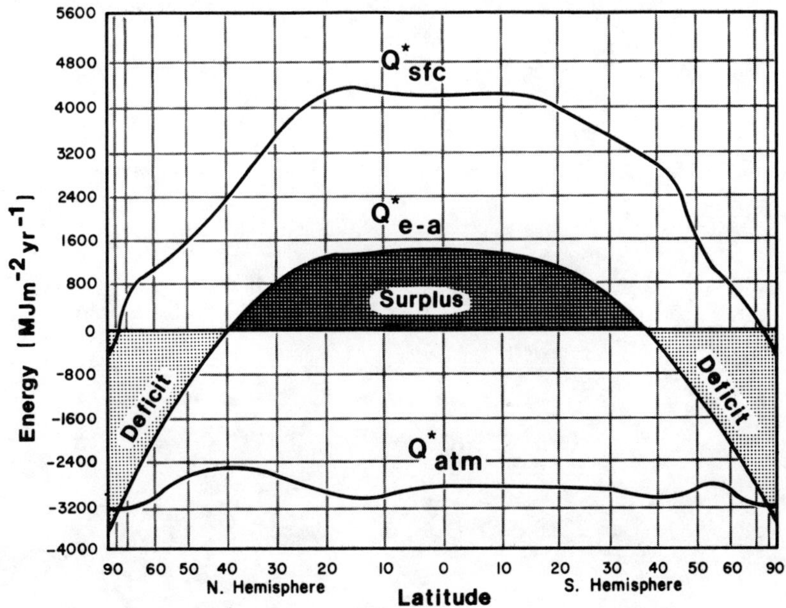

Figure 6-3

In order to comprehend the **vertical** transport of energy, the "energy balance" of the surface must be considered. The net radiation available at the surface can be used primarily to heat the air (called the sensible heat flux — Q_H), evaporate water (called the latent heat flux — Q_E) or heat the ground (called the subsurface heat flux — Q_G). A small amount may also be used for photosynthesis. The first two accomplish the necessary vertical transport of energy from the surface to the atmosphere by physically moving warm air upwards (Q_H) or evaporating water at the surface using energy and subequently condensing it in the atmosphere releasing the energy (Q_E). Energy may also be transported horizontally towards the poles as sensible or latent heat in air currents or as sensible heat in ocean currents. These vertical and horizontal transports of energy are the basis of weather and climate.

Name____ Date____________________

Instructor____________________________________ Section__________________

QUESTIONS

6.1 The table below lists average values of albedo for different surfaces:

Surface	**Albedo (%)**
Forest	5-10
Grass	20-25
Dry Sand	20-30
Old Snow	50-60
Fresh Snow	80-90
Water	3-5
[Water (sun very near horizon)	50-80]

Note the peculiar behavior of water. Normally, water has a very low albedo. However, when the sun is very low in the sky, the values of albedo jump to very high magnitudes. This represents only a small portion of time, so water generally can be taken to have a low albedo.

(a) What type of surface is most reflective?________________________________

Which surface has the lowest albedo? ________________________________

(b) Near the poles, there is always a low percentage of solar radiation **absorption.** What kinds of surfaces account for this low absorption?

(c) Briefly describe the seasonal variation in albedo to be expected for a broadleaf deciduous forest. These trees drop their leaves in the fall, and the forest experiences a lasting snow cover during the winter months.

6.2 Recall that the surface radiation balance can be expressed as:

$$Q^* = K\downarrow - K\uparrow + L\downarrow - L\uparrow$$

The following table consists of measured data obtained for a dwarf apple orchard near Simcoe, Ontario, Canada on June 25, 1973 and August 14, 1973. Terms are expressed in the units of watts per square meter (Wm^{-2}).

June 25 — **Solar Time**

	06	**08**	**10**	**12**	**14**	**16**	**18**
$K\downarrow$	210	560	830	940	770	510	200
$K\uparrow$	50	110	160	170	140	100	50
$L\downarrow$	300	300	300	300	300	310	320
$L\uparrow$	390	410	470	490	490	460	450

August 14 — **Solar Time**

	06	**08**	**10**	**12**	**14**	**16**	**18**
$K\downarrow$	60	290	330	340	330	280	30
$K\uparrow$	20	60	60	60	60	55	10
$L\downarrow$	330	330	330	330	340	340	350
$L\uparrow$	400	420	430	440	440	440	420

(a) Calculate Q^* for the times and dates indicated.

	06	**08**	**10**	**12**	**14**	**16**	**18**
June 25:							
August 14:							

On the graph in Fig. 6-4, plot Q* versus time for each of these days. (Join the points representing June 25 with a continuous line and the points representing August 14 with a dashed line.)

(b) After consideration of the given data, compare the general weather conditions on these two days. (note especially cloud cover and surface temperature.)

Name____ Date____________________

Instructor____________________________________ Section__________________

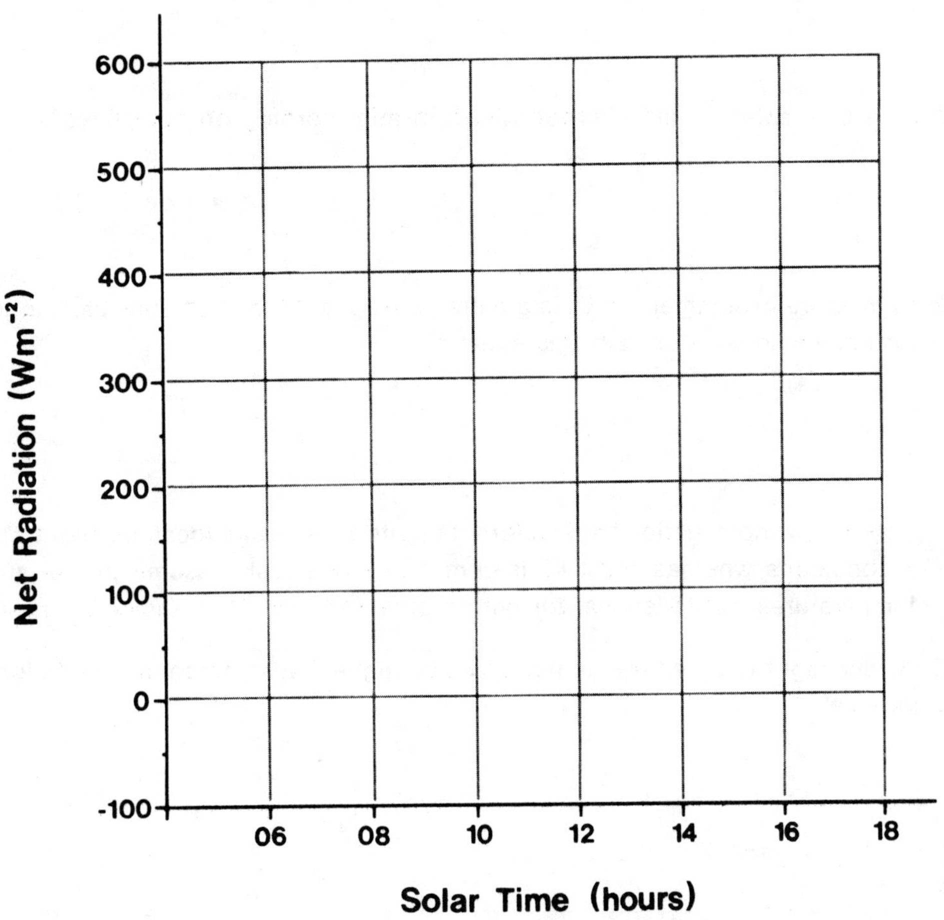

Figure 6-4

(c) Calculate the albedo for the times and days indicated.

	06	08	10	12	14	16	18
June 25:							
August 14:							

Approximately what was the average albedo for the orchard (this should be weighted to the times of day when K↓ was greatest)?

____

Was there much of a difference between the two days?

(d) What causes L↑ generally to be larger than L↓ on both days?

(e) Why is L↑ greater in mid-afternoon than in mid-morning on both days?

(f) Both days are in summer, yet L↓ is greater on August 14 than on June 25. Suggest a reason or reasons to explain this situation.

6.3 Consider the following situation for 2 different nights at the same location. Night #1 has clear sky conditions whereas night #2 is completely overcast. Assume that at sunset surface temperatures were identical for both nights and thus L↑ was initially the same.

(a) On which night will the "sky" temperature be higher? What factor accounts for this difference?

(b) Will the value of Q* be positive or negative at the surface on each night?

Night #1______________________________; Night #2________________________________

In terms of actual quantity, which night would have the most positive or negative Q*?

(c) Later during the night (i.e. shortly before sunrise), which night will have the coolest surface temperatures? (Note that, in spring and fall, such a night would represent the night with the greater risk of frost.)

Name______________________________ Date______________

Instructor____________________________ Section__________

(d) What name is given to the effect illustrated above?

6.4 The global energy budget may be considered to be in a steady state situation where Q^* for the whole earth-atmosphere system averaged for the **whole** planet for an entire year is zero (i.e. energy coming in = energy going out). If this wasn't the case, then our planet as a whole would be heating or cooling. By considering the global radiation balance

$$Q^*_{e-a} = K\downarrow - K\uparrow - L\uparrow$$

where the terms on the right are fluxes through the top of the atmosphere, answer the following:

(a) Suggest means by which the **whole** earth-atmosphere system might become cooler.

(b) Suggest two means by which the **whole** earth-atmosphere system might become warmer.

6.5 Although Q^*_{e-a} must be zero for an entire year, net radiation for the earth's **surface** alone is generally positive (gaining energy) on an annual basis while for the **atmosphere** alone, net radiation is negative (losing energy) (see Fig. 6-3). Thus, there must be a vertical transport of energy from the surface to the atmosphere to overcome this apparent imbalance.

(a) If this did not occur, what would happen to the earth's surface temperature over time?

(b) By what 2 processes is this imbalance overcome?

6.6 Consider the surface radiation balance:

$$Q^*_{sfc} = K\downarrow - K\uparrow + L\downarrow - L\uparrow .$$

(a) With the solar input at the top of the atmosphere unchanged, suggest a change in the atmospheric composition that might warm the earth's **surface** and explain how this compositional change will warm the surface.

(b) With the solar input at the top of the atmosphere unchanged, suggest a change in the atmospheric composition that might cool the earth's **surface** and explain how this compositional change will cause the surface to become cooler.

6.7 (a) Questions in this chapter have illustrated that solar radiation receipt varies considerably being generally greater in tropical areas. Since solar radiation is a major term in the radiation balance, how does the average annual net radiation of the whole earth-atmosphere system (Q^*_{e-a}) vary from equator to poles? (Note: Averaged for the whole globe it again must be zero. But, there is indeed a latitudinal variation — see Fig. 6-3.)

(b) To overcome this imbalance, there must be a horizontal transport of energy from tropical towards polar latitudes. By what mechanisms is this accomplished?

EXERCISE 7

TEMPERATURE AND HUMIDITY

This exercise defines and considers two important elements of weather: temperature and humidity.

Atmospheric temperature is one of the most frequently considered of the many meteorological elements. The influence of air temperature on the growth and well-being of the earth's life forms is paramount. Temperature is basically a measure of the molecular activity (how fast the molecules are moving) of a substance. Therefore reference points upon which to compare molecular activities or temperatures are needed. The freezing and boiling points of water are used where 32° (Fahrenheit) or 0° (Celsius) is the freezing point and 212° (Fahrenheit) or 100° (Celsius) is the boiling point. Another scale sometimes used in science is the Kelvin or absolute scale where 0° (Kelvin) is "absolute zero" or the temperature for no molecular movement. This temperature is equal to −273° on the Celsius scale.

Water — the chemical compound H_2O — is a basic necessity for the life forms of the earth. At a given moment, only about 1/100,000 part of the earth's supply of water is found in the atmosphere, yet it is this atmospheric water that brings moisture to the surface of the earth and makes it habitable. In the hydrologic cycle, water is constantly being moved from the earth's surface into the troposphere where it gives rise to clouds and precipitation, thereby bringing the water back to the surface again. Not only does the hydrologic cycle provide a means by which water is physically transferred, it also provides the means by which energy can be exchanged in important amounts between the earth's surface and the atmosphere. The amount of water held in the atmosphere in a vapor state is called atmospheric humidity. (Water in a cloud is liquid or solid and not part of humidity.)

Humidity can be expressed in several ways. One measure of humidity is "vapor pressure" (e). Every gas existing in air is contributing to part of the pressure being exerted by the air. The vapor pressure is the part of the total atmospheric pressure being exerted by the water vapor in the air. Therefore, it is a measure of the **actual** vapor content or absolute humidity. If the air is holding the maximum amount of vapor it is capable of holding, it is said to be saturated and the vapor content can be expressed as the saturated vapor pressure (e_s). There is a significant relationship between the temperature of the air and its capacity to hold moisture (in the vapor state). The higher the temperature, the greater the capacity of the air to hold moisture (or higher e_s). This relationship between temperature (T) and e_s is shown in Fig. 7-1.

RELATIONSHIP BETWEEN AIR TEMPERATURE
AND SATURATION VAPOR PRESSURE

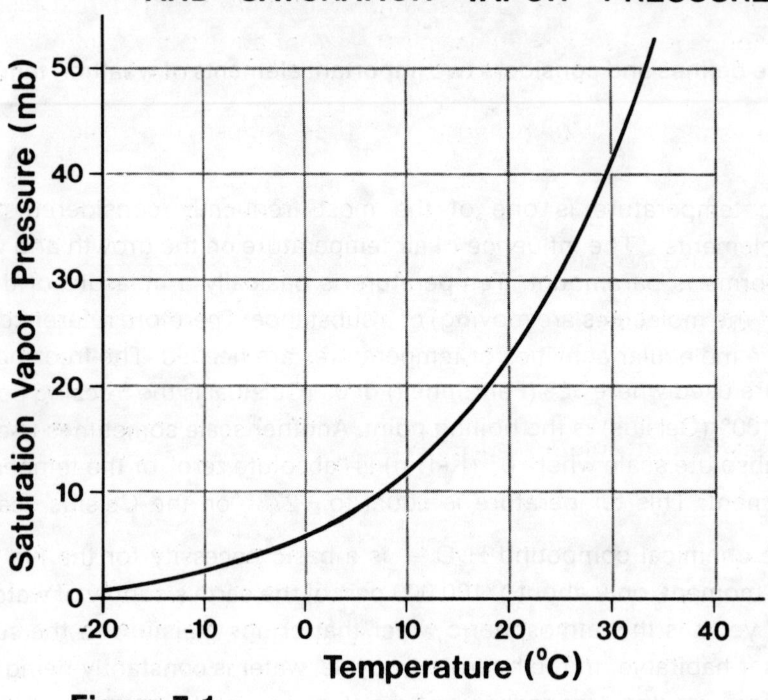

Figure 7-1

A common measure of humidity is "relative humidity"(RH). This is the amount of water vapor actually in the air relative to the maximum amount the air is capable of holding at that temperature and is usually expressed as a percentage. Thus,

$$RH = \frac{e}{e_s} \times 100\%.$$

RH is therefore temperature dependent and is a measure of the degree of saturation (100% RH equals saturation). It is not, however, a measure of the actual water vapor content.

Name____________________ Date____________

Instructor____________________ Section____________

QUESTIONS

7.1 Three temperature scales are in common use: Fahrenheit (°F), Celsius or centigrade (°C), and Kelvin or absolute (°K). The scientific world has adopted the Celsius scale for most uses although Kelvin is sometimes most appropriate in some equations and laws. In this manual, the Celsius scale generally will be used.

(a) Give the values of the following in the three temperature scales:

	°F	°C	°K
Freezing point of water	______	______	______
Boiling point of water	______	______	______

(b) What does 0°K (absolute zero) refer to?

(c) Using the conversion equations listed below, complete the following table.

$$°C = \frac{5}{9}(°F - 32)$$

$$°F = \frac{9}{5}(°C) + 32$$

$$°K = °C + 273$$

$$°C = °K - 273$$

	°F	°C	°K
Sun's temperature	______	______	6000
Earth's surface	______	______	288
Human body temperature	98.6	______	______
Hot desert temperature	120	______	______
Cold arctic temperature	______	—40	______
Swimming weather	80	______	______
Sweater weather	______	10	______
Skiing weather	______	—5	______

7.2 Tabulated below are vertical temperature data obtained from a radiosonde (Balloon) ascent over a location. Using the graph provided in Fig. 7-2, draw the "temperature profile" for this data.

Altitude (m)	0	250	500	750	1000
Temperature (° C)	20	22.5	17.5	15	15

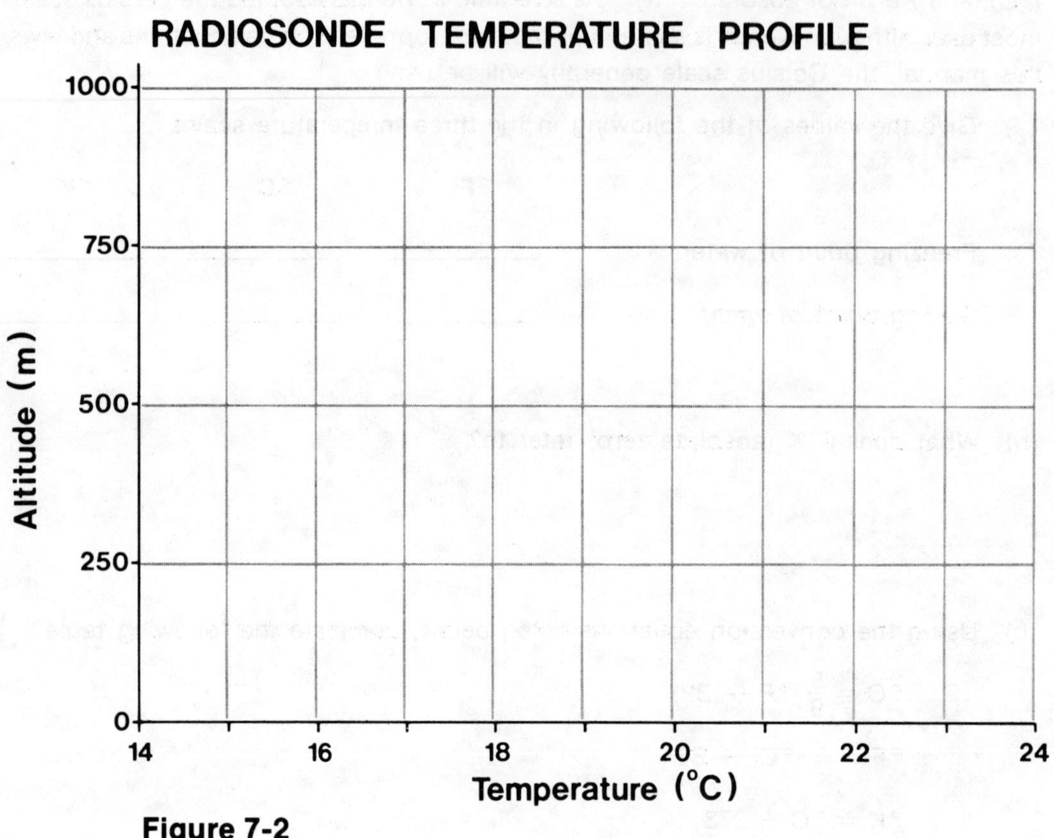

Figure 7-2

(a) Calculate the "temperature lapse rate" (i.e. temperature change with height) for each of the 4 layers illustrated in Fig. 7-2. Express your answers in units of °C per 100 m.

(b) What is the special name given to the type of temperature profile depicted in the lowest of these 4 layers?

Name____________________ Date____________

Instructor____________________ Section____________

7.3 When land and water bodies have equal values of net radiation, they are absorbing equal amounts of energy. However, their surface temperature responses will be quite different. The temperature of the land surface will show a distinctly greater increase than will the temperature of the water surface. Conversely, during periods when net radiation is small or negative, cooling will occur but at different rates. The land will cool more rapidly than the water. This phenomena is referred to as the "differential heating and cooling" of land and water.

(a) With respect to each of the physical properties listed below, explain how land and water surfaces are different, and indicate how this difference affects the rate of warming or cooling of the two types of surfaces when net radiation loads are equal.

1 — Specific heat of surface materials

2 — Surface transparency

3 — Surface mobility

4 — Potential for evaporation at the surface

(b) Considering the above, should the northern or southern hemisphere be expected to have the greater seasonal temperature change? What main factor causes this hemispheric difference?

7.4 Below are the normal average monthly temperatures for two locations at approximately the same latitude: Vancouver, British Columbia, Canada (49° N, 123° W) and Winnipeg, Manitoba, Canada (50° N, 97° W). Note that a normal average monthly temperature is the average of daytime maximums and nighttime minimums throughout the month and over several years while the normal annual temperature is the average of the 12 monthly values. Also, annual temperature range refers to the difference between the highest and lowest **average monthly** values.

Month	**Normal Average Temperature (° C)**	
	Vancouver	**Winnipeg**
January	2	—18
February	4	—16
March	6	—8
April	9	3
May	12	11
June	15	17
July	17	20
August	17	19
September	14	13
October	10	7
November	6	—4
December	4	—14

Name________________________ Date____________

Instructor______________________ Section____________

(a) On the diagrams provided (Fig. 7-3), plot the monthly temperature data for the locations connecting data points with smooth lines.

Figure 7-3

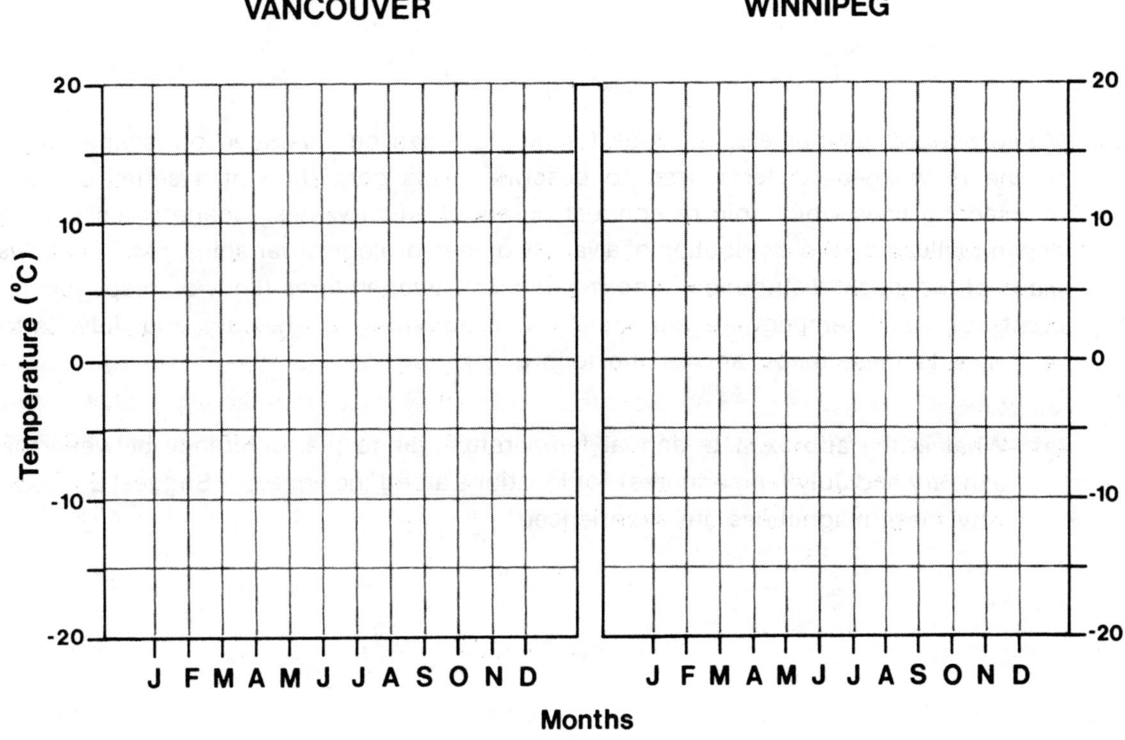

(b) From the data determine the following:

	Vancouver	Winnipeg
Normal annual temperature	______	______
Annual temperature range	______	______

(c) Explain the difference in magnitude of the annual ranges of temperature for these two locations. Which city has the more "continental" climate?

(d) The peak inputs of solar radiation (K↓) and net radiation (Q*) occur in June for northern hemisphere locations while the minimum energy inputs are in December. With this in mind, why should the hottest temperatures for Vancouver and Winnipeg occur in July and August and the coldest temperatures in January and February?

7.5 Meteorological phenomena, such as temperature can be illustrated on isoline maps. Isoline is the generic term used to describe maps consisting of a series of non-intersecting lines which join or connect points of equal values. Isoline maps can be drawn to illustrate the distribution of a variety of meteorological variables. Fig. 7-4 shows maps of the global distribution of normal monthly temperatures (lines on maps joining points of equal temperature are known as isotherms) for January and July. With reference to these maps, answer the following:

(a) What is the approximate annual temperature range (i.e. difference between the January and July temperatures) for locations along the equator? Suggest a reason why these magnitudes are experienced.

(b) Name two general regions where annual temperature ranges are the largest. Suggest 2 reasons why these areas have such large annual ranges.

(c) What is the annual temperature range for locations around 60° S latitude? How do these compare with values found at 60° N latitude? What factor accounts for this difference?

Name____________________ Date____________

Instructor____________________ Section____________

(d) What is the temperature difference between the U.S. Gulf of Mexico coast region and the Lake Superior area in January?__________ What is the difference in July?__________ What causes this seasonal change in temperature differences?

(e) At what latitude does the 20° C isotherm cross the west coast of South America in January?__________ At what latitude does it cross the east coast?__________ What factor causes this latitude difference?

(f) What is the annual temperature range for the following locations?

Athens, Georgia	______	Athens, Greece	______
Vancouver, British Columbia	______	London, England	______
Buenos Aires, Argentina	______	Bombay, India	______
South Pole	______	Denver, Colorado	______
Nairobi, Kenya	______	Honolulu, Hawaii	______
Washington, DC	______	Tokyo, Japan	______
Salisbury, Zimbabwe	______	Mexico City, Mexico	______

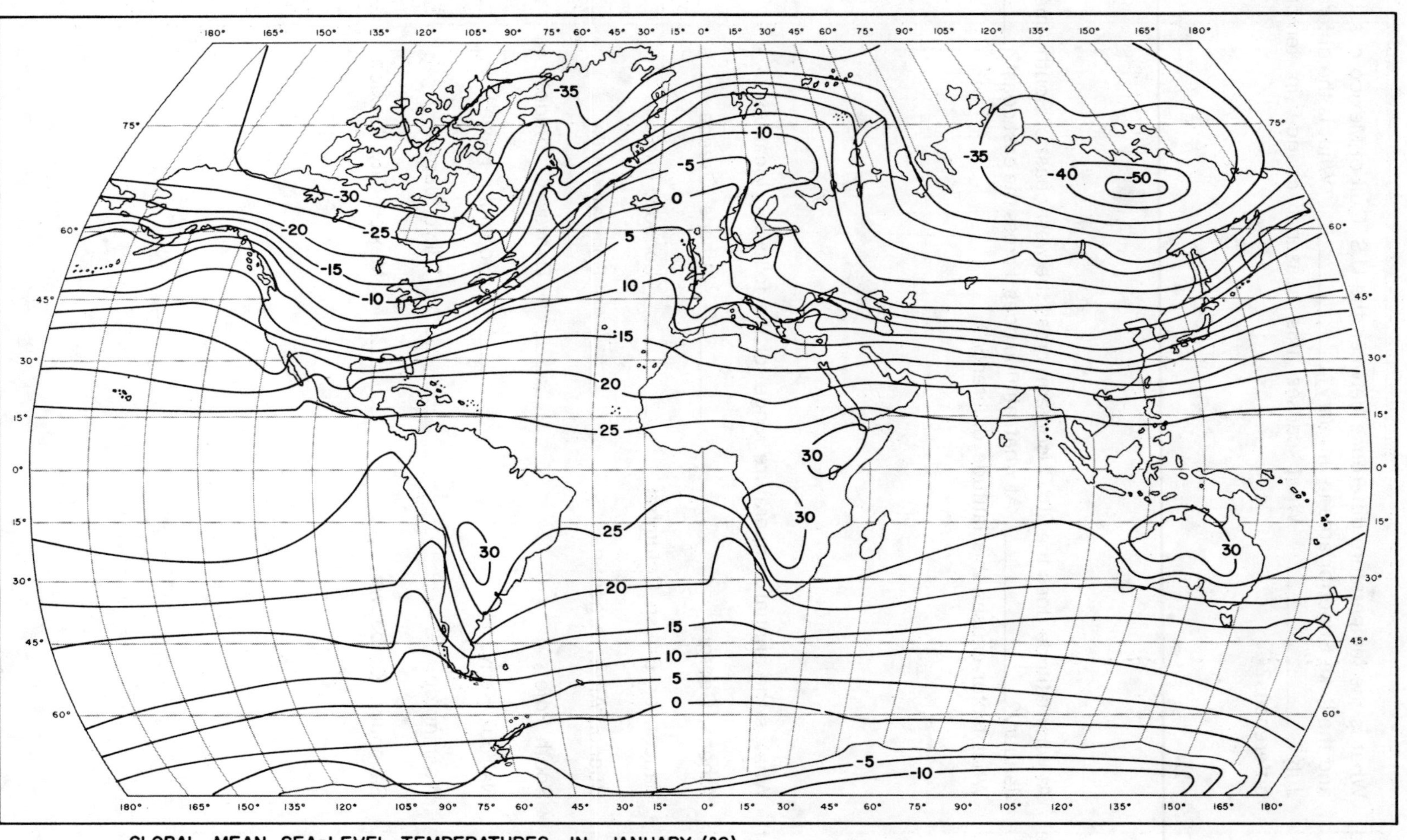

GLOBAL MEAN SEA-LEVEL TEMPERATURES IN JANUARY (°C)

Figure 7-4A

Name________________________ Date____________

Instructor____________________ Section__________

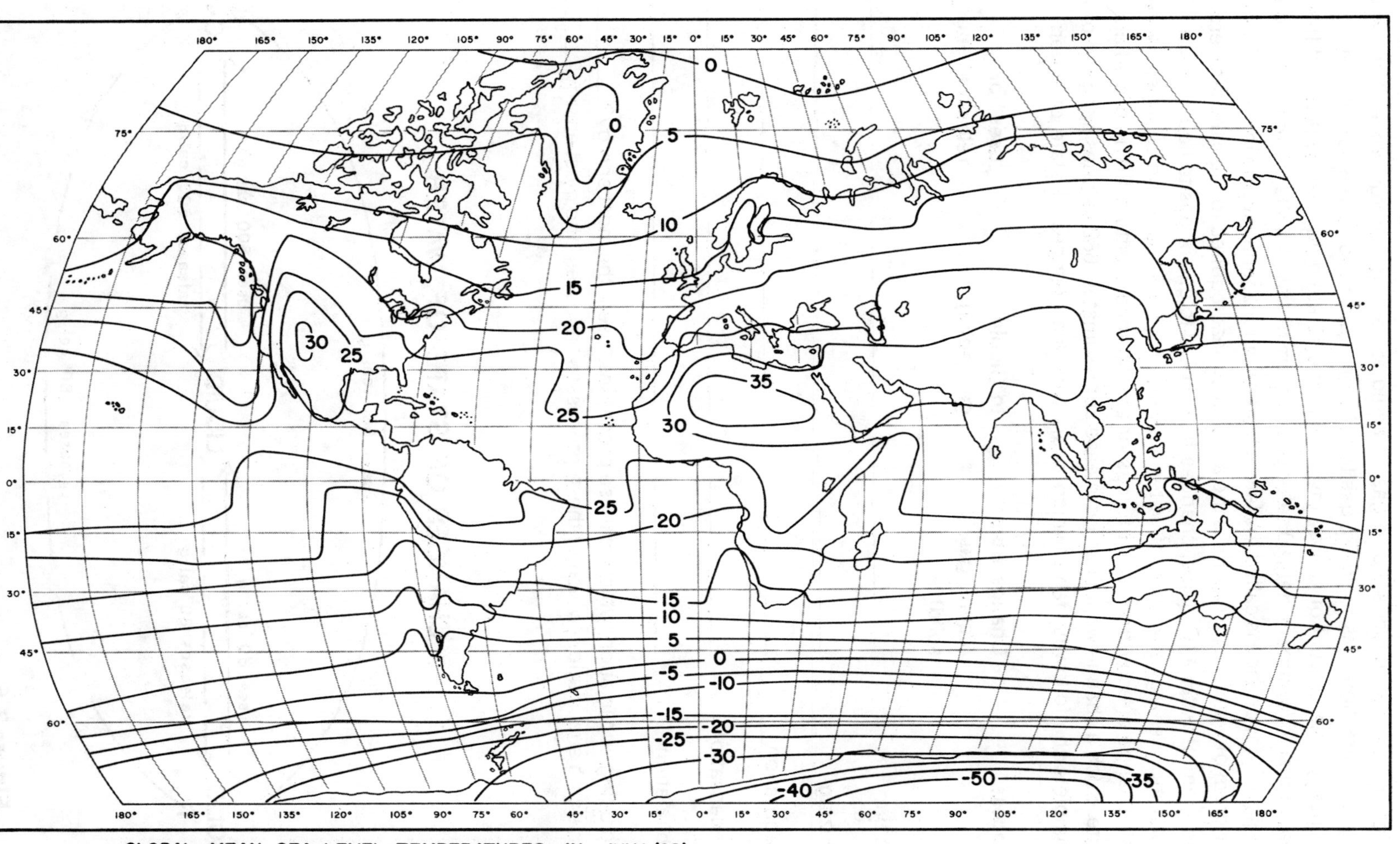

GLOBAL MEAN SEA-LEVEL TEMPERATURES IN JULY (°C)

Figure 7-4B

7.6 When water changes from one state (liquid, gas or solid) to another, energy is either used or released. This is shown in Fig. 7-5 where changes going towards the right in the diagram (solid to liquid, solid to gas or liquid to gas) use energy while changes in the opposite direction (gas to liquid, gas to solid or liquid to solid) release an equal amount of energy. The amount of energy involved in the solid/liquid change is called the latent heat of fusion (approximately 80 calories per gram of water) while the amount involved in the liquid/gas change is called the latent heat of vaporization (approximately 600 calories per gram of water). One calorie is the amount of heat required to raise the temperature of one gram of water 1°C. Thus, changes from liquid to gas or gas to liquid use or release much more energy per gram of water than that for the solid/liquid changes. Water can also change directly from solid to gas or gas to solid omitting the intermediate liquid state. Such a change would use or release approximately 680 calories per gram of water. Complete the following table which deals with the changes of state of water:

Process	**Change from (liquid, gas, solid)**	**to (liquid, gas, solid)**	**"uses" or "releases" energy**
melting	________	________	________
freezing	________	________	________
evaporation	________	________	________
condensation	________	________	________
sublimation	________	________	________

Excluding sublimation, which of these processes "uses" the most energy?____________
Excluding sublimation, which of these processes "releases" the most energy?__________

CHANGES OF STATE OF WATER

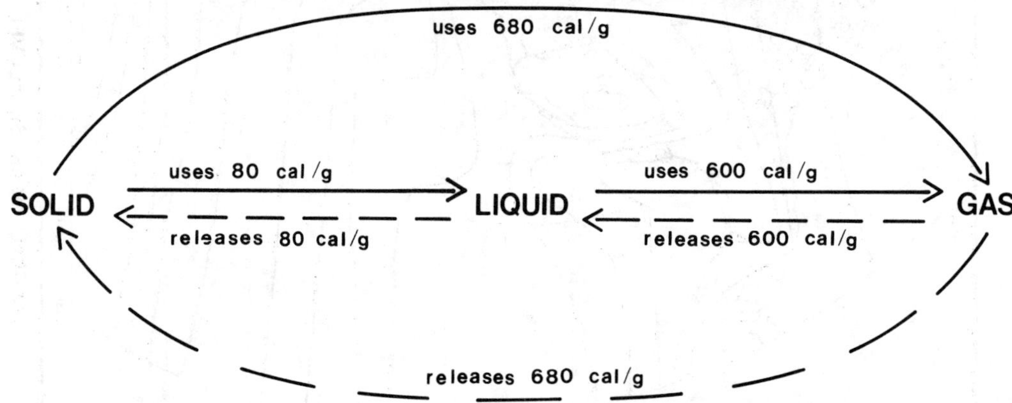

Figure 7-5

Name____________________ Date____________

Instructor____________________ Section____________

7.7 Assume that a relative humidity of 70% is observed at four locations. Each location is also experiencing air temperatures as follows — Location A: 22° C; Location B: 22° F; Location C: 3° C; Location D: —2° C.

(a) Which of these locations has the most humid air in terms of actual vapor content?

____

(b) Which of these locations has the driest air in terms of actual vapor content?

____

(c) If the relative humidity for location D rose to 75%, would its **actual** vapor content now be higher than that for location A? (Hint: reference to Fig. 7-1 may be helpful.)

____

7.8 The "dew point" temperature (T_d) is the temperature to which a parcel of air must be cooled before condensation (i.e RH = 100%) will occur (assuming the actual vapor content remains constant). State whether each of the following situations is saturated or not saturated.

(a)	T = 14° C	RH = 97%	____________
(b)	T = —6° C	RH = 100%	____________
(c)	T = 14° F	RH = 97%	____________
(d)	T =—6° C	T_d = —10° C	____________
(e)	T = —6° C	T_d = —6° C	____________

7.9 The surface energy balance, which expresses how the available net radiation can be used, can be expressed as:

$$Q^* = Q_E + Q_H + Q_G$$

Refer to question #2 in Exercise 6 where you calculated Q* for June 25 and August 14, 1973 for a dwarf apple orchard near Simcoe, Ontario. Assume that Q_H and Q_G were also measured with values as tabulated below (units Wm^{-2}):

Solar Time

		06	**08**	**10**	**12**	**14**	**16**	**18**
June 25:	Q_H	40	140	200	210	190	120	0
	Q_G	10	30	40	50	60	50	20
August 14:	Q_H	0	100	110	110	110	90	—60
	Q_G	0	20	30	40	50	30	0

(a) Complete the following table by using the Q* values calculated in Exercise 6 and by determining Q_E from the equation above.

		06	08	10	12	14	16	18
June 25:	Q*							
	Q_E							
August 14:	Q*							
	Q_E							

(b) Noting that a negative value of Q_E implies that, rather than using energy for evaporation (i.e. Q_E positive) energy is being released. Considering the calculated value for Q_E at 06 hours on August 14, what atmospheric process is occurring?

(c) The calculated values of Q_E should indicate much larger values on June 25 and a much larger percentage of the available Q* is used by Q_E through most of the day. What does this indicate about the general relative humidity of the air on June 25 compared to August 14? Does this make sense in light of the weather conditions you determined for these two days in Exercise 6?

EXERCISE **8**

VERTICAL AIR MOTION, CLOUDS AND PRECIPITATION

Atmospheric pressure is a term that refers to the force exerted by the mass of air above a given point. The movement of air occurs in both vertical and horizontal directions, the latter as a response to varying values of atmospheric pressure (high or low) from place to place. These horizontal motions or "winds" will be examined in detail in Exercise 9. High and low pressure areas at the surface can be related to the direction of vertical air motion where upward movement implies low surface atmospheric pressure and descending air motion implies high surface pressure. Regions where air is ascending (low surface pressure) are generally associated with cloudy or rainy conditions whereas, where air is descending in high pressure areas, weather conditions are usually clear and sunny.

Vertically in the atmosphere, pressure is inversely related to altitude — i.e. higher altitudes have lower pressure values. Therefore, any parcel of air which is caused to ascend through the atmosphere will be exposed to progressively lower environmental pressure. Thus, as the air ascends, it is able to expand its volume, and as its volume expands its temperature decreases. Conversely, any parcel of air which descends through the atmosphere will be exposed to progressively larger environmental pressure. It will become steadily compressed into a smaller volume by the increasing pressure, and (because of the work done by the surrounding air in compressing the parcel's volume) will experience a steady increase in temperature. These temperature changes are referred to as "adiabatic" since they are caused by the expansion and compression of the air, and do not occur as a result of the loss of heat energy from the parcel to any outside agency or as a result of the gain of heat energy from any outside source. An adiabatic process is defined as a process in which a parcel of air undergoes changes in pressure, volume or temperature without any transfer of energy into or out of the parcel.

An individual parcel of air forced to rise will generally behave adiabatically since the event will happen too fast for energy exchange with the surrounding environment to occur. For unsaturated air, the rate of adiabatic cooling or warming accompanying upward or downward air movements has been found constant — a value of −1°C/100m. (The minus sign refers to cooling as one ascends.) This value is referred to as the "dry adiabatic lapse rate" (DALR) since it applies to unsaturated air. If air which is fully saturated with water vapor is caused to ascend, upward through the atmosphere, it will expand and cool. As it cools its capacity to hold water in the vapor form will steadily decrease, and the air will experience condensation or sublimation — giving water-cloud droplets or ice-cloud crystals within the air. The rate of cooling for the saturated air ascending through the cloud will not be as large as the DALR. This rate is referred to as the "saturated or wet adiabatic lapse rate" (SALR) since it applies to saturated air.

The "encouragement" or "discouragement" of vertical motion for a parcel of air in an atmospheric layer depends on how the actual temperature structure of the atmosphere (called the "environmental lapse rate" — ELR) compares to the DALR (assuming the vertical column is not saturated). If the ELR cools faster than the DALR with height, the atmosphere is said to be **unstable** and a parcel of air will be **encouraged** to move upwards. For example, an ELR might be −2° C/100 m. If the surface temperature is 10° C, then the actual temperature at 100m elevation is 8° C. A parcel of air (initially 10° C at the surface) forced to rise would cool adiabatically and be 9° C at 100m elevation. Since it is warmer than its surroundings (which have a temperature of 8° C), it is less dense and would be encouraged to continue to rise.

Conversely, if the ELR is such that the actual temperature lapse rate cools slower than the DALR or gets warmer as one ascends (the latter situation is called a temperature inversion), then the atmosphere is **stable** and vertical air movement will be **discouraged.** For example, if the ELR is −0.5° C/100m and the initial surface temperature is 10° C, then the actual atmospheric temperature at 100m elevation is 9.5° C. Again a parcel of air forced to rise would cool adiabatically and again be 9° C at 100m elevation. This time the parcel finds itself cooler than its new surroundings and therefore more dense. Thus, it will fall back to its original position having been discouraged to rise.

If the ELR is exactly −1° C/100m, then vertical motion will neither be encouraged or discouraged and the atmospheric stability is called **neutral.**

If the atmosphere is saturated (i.e. in a cloud) then the SALR must be used rather than the DALR for making stability determinations.

When the earth is observed from space it is seen that about half of the surface is covered by clouds. These clouds are found, distributed in systematic arrangements, at various altitudes within the troposphere. They are formed by the condensation or sublimation of water vapor around very small dust-particle nuclei. It is estimated that several thousand cloud particles may sometimes exist within a single cubic centimeter of cloud. Clouds develop whenever the air is chilled to or below its dew point as often occurs with adiabatic cooling.

If cloud droplets grow sufficiently, they will produce precipitation. The type and characteristics of the precipitation will depend on the cloud type and other environmental factors. The importance of the precipitation that falls depends on the thermal and radiative regimes for any given location. The concept of "effective precipitation" refers to the availability of precipitation for plant growth. If precipitation that falls, evaporates because of intense radiation and high temperatures, then it is not "effective." Thus, precipitation amounts should always be considered in conjunction with the temperature or radiative values for the location.

Name______________________________ Date______________

Instructor__________________________ Section____________

QUESTIONS

8.1 Given that pressures of 29.92 inches of mercury, 76 cm of mercury, and 1013.25 millibars are equivalent atmospheric pressures (in fact, these values are standard sea-level pressure), complete the following table:

inches mercury	cm mercury	millibars (mb)
____________	____________	950
30.15	____________	____________
____________	78	____________

8.2 In this manual, pressure will be expressed in millibars (mb). The typical range of atmospheric pressures experienced at sea-level is from a low of 980 mb (lower in the center of a tornado or hurricane) to a high of 1030 mb. Low pressure areas represent regions where air is converging at the surface and rising in the center of the low pressure region. Conversely, high pressure areas represent places where air is sinking towards the surface and then diverging at the surface. Fig. 8-1 represents a vertical cross-sectional view of the atmosphere extending across an area where there are sea-level high and low pressure areas. Complete the diagram by drawing arrows depicting the circulation of air.

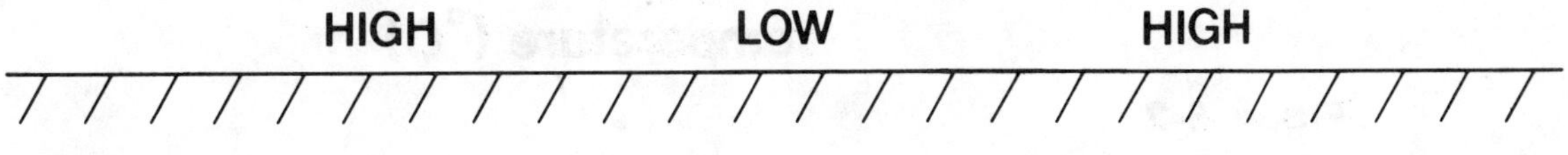

HIGH LOW HIGH

Figure 8-1

8.3 The actual temperature lapse rate of a column of air is known as the "environmental lapse rate." The lapse rates calculated in Exercise 7, question 2 were environmental lapse rates. Fig. 8-2 shows a temperature profile for a location measured by an ascending balloon device.

(a) Calculate the environmental lapse rate for the atmospheric layers numbered 1, 2 and 3.

layer 1: ____________________

layer 2: ____________________

layer 3: ____________________

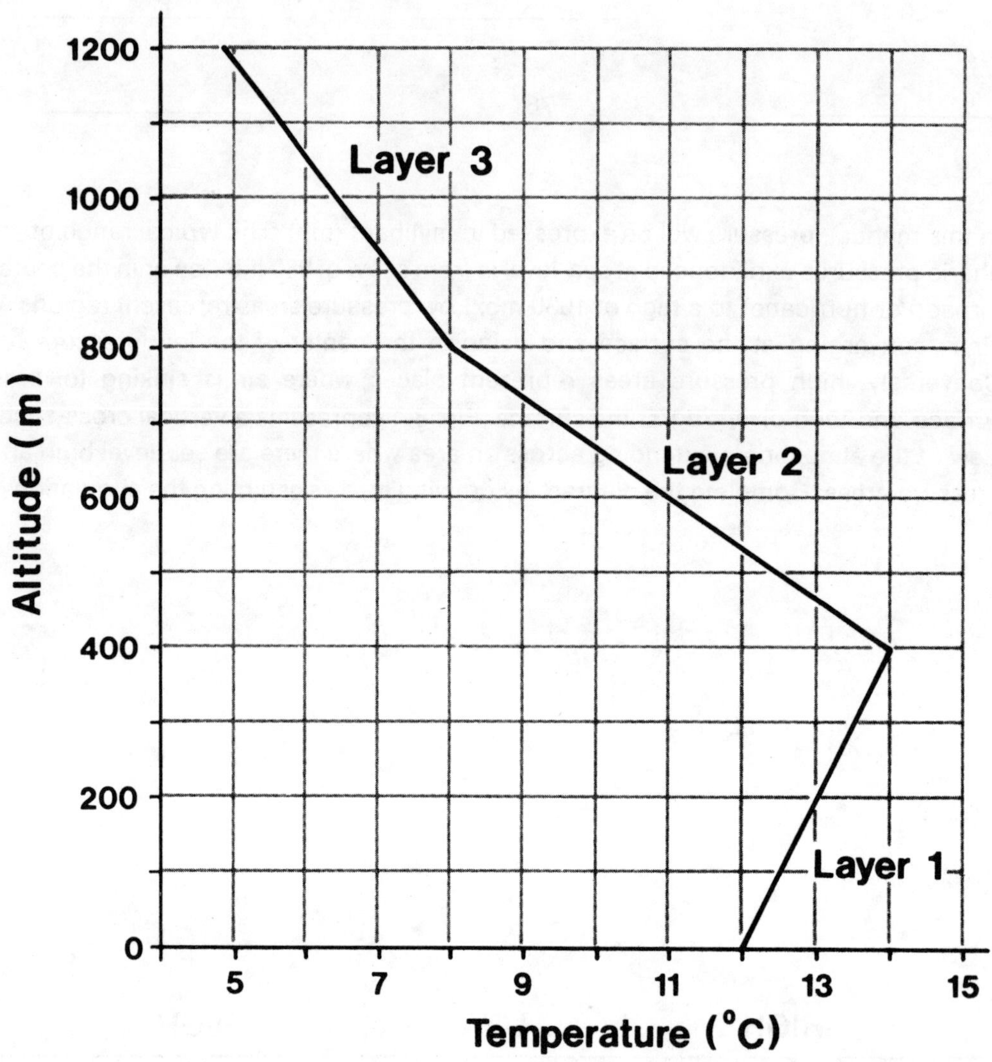

Figure 8-2

Name________________________________ Date________________

Instructor________________________________ Section________________

(b) Determine the atmospheric stability for the three atmospheric layers in Fig. 8-2:

If the atmosphere is unsaturated throughout:

layer 1 —

layer 2 —

layer 3 —

If the atmosphere is saturated throughout and the SALR is −0.5°C/100 m:

layer 1 —

layer 2 —

layer 3 —

8.4 Fig. 8-3 shows an onshore wind blowing against a coastal mountain range (ridge elevation, 2000m). Forced to cross the mountains, the air cools adiabatically on the western (left) side (known as the windward side) and warms adiabatically on the eastern side (known as the leeward side). It descends to a plateau region at 200m elevation. Cooling with height on the windward side produces cloud cover, the base of which occurs at 500m. On the leeward side, cloud base is found at 1500m. If the air blowing off the ocean at point A has a temperature of 10°C, and if the SALR is —0.5°C/100m, calculate the following:

(a) The temperature at B, the windward side cloud base ________________ °C

(b) The temperature at C, the top of the mountain ________________ °C

(c) The temperature at D, the cloud base on the lee slope ________________ °C

(d) The temperature at E, the interior plateau level ________________ °C

(e) Assuming that winds are generally blowing from the west off the ocean, in descriptive terms compare the annual precipitation totals for locations B and E.

What term is used to describe the situation at E? ________________

(f) Your calculations above indicate that the wind blowing down the leeward slope is relatively warm and is sometimes known as a "chinook." These are significant in winter when they abruptly bring warm conditions to previously cool or cold areas. Where in North America are chinooks most frequent?

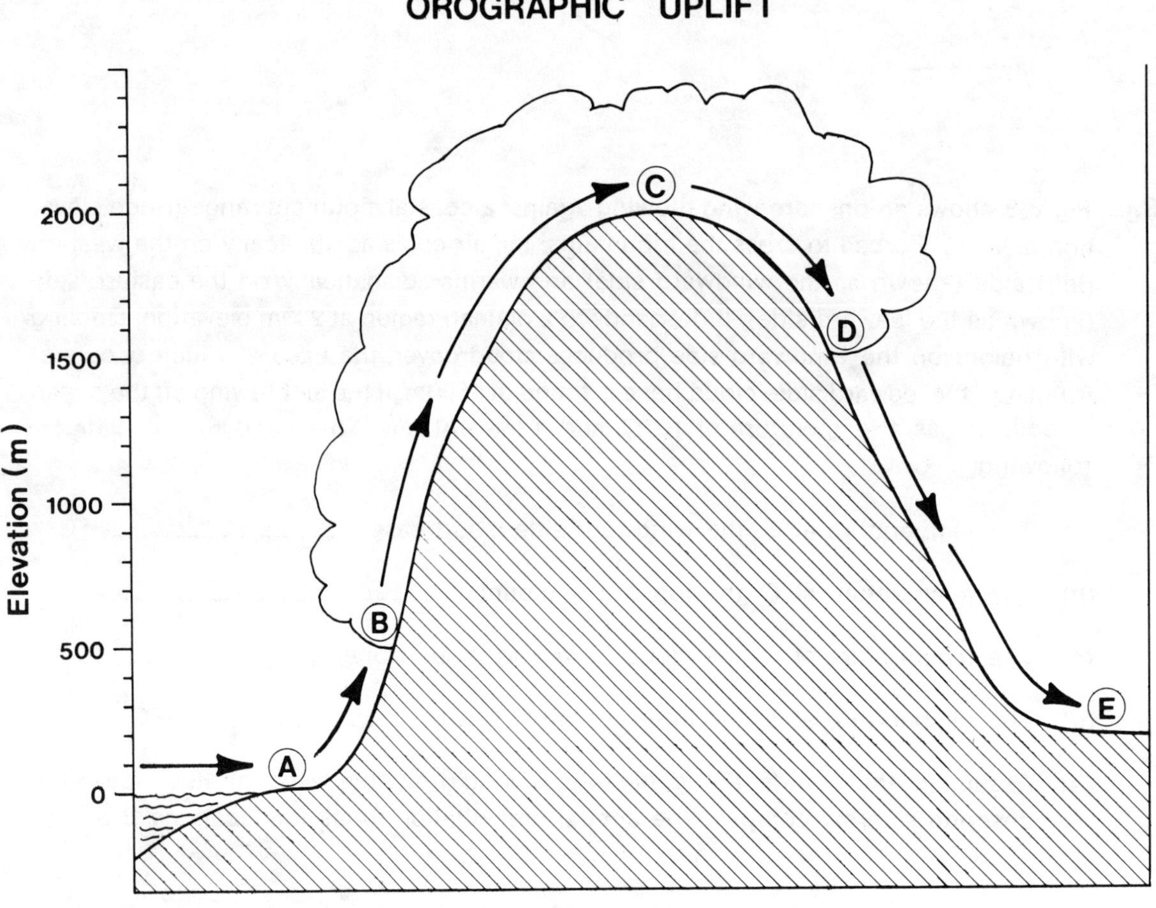

Figure 8-3

Name____________________ Date__________

Instructor____________________ Section__________

8.5 Name the atmospheric stability for the following situations:

(a) vertical air motion is encouraged or enhanced: ____________________

(b) vertical air motion is discouraged or retarded: ____________________

(c) vertical air motion is neither encouraged or discouraged: ____________________

(d) the best atmospheric stability for dispersing air pollution near the surface: ____________________

(e) the worst atmospheric stability for dispersing air pollution near the surface: ____________________

(f) for a calm clear 24 hour period, the atmospheric stability near the surface that you would expect to experience shortly before sunrise: ____________________

(g) for a calm clear 24 hour period, the atmospheric stability near the surface that you would expect to experience in mid-afternoon: ____________________

8.6 By referring to appropriate pictures and descriptions of different types of clouds, draw a schematic cross-section view in Fig. 8-4 depicting at appropriate altitudes the shapes of each of the cloud types listed below. Label each cloud type with its proper letter abbreviation.

Cloud types: Stratus (St)
Stratocumulus (Sc)
Cumulus (Cu)
Cumulonimbus (Cn)
Nimbostratus (Ns)
Altostratus (As)
Altocumulus (Ac)
Cirrus (Ci)
Cirrostratus (Cs)
Cirrocumulus (Cc)

Figure 8-4

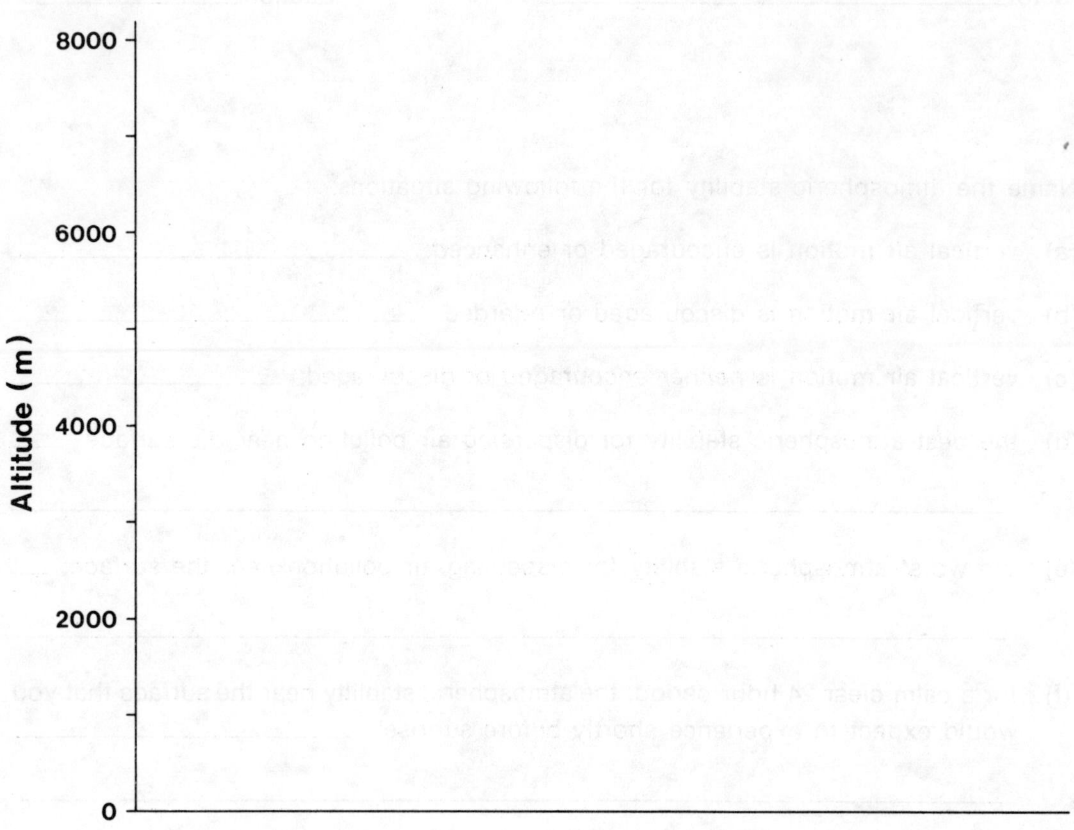

8.7 While some of the cloudiness in the atmosphere is caused by non-adiabatic cooling of the air, most occurrences of extensive cloudiness are the result of adiabatic cooling accompanying the upward movement of air. The large-scale lifting of air necessary to initiate extensive cloudiness and precipitation can result from a number of processes, acting singly or in combination. The three main causes of vertical uplift are: rising air due to intense surface heating warming air near the surface thus reducing its density and causing it to rise (called **convectional** uplift); air forced to rise when two contrasting air masses (warm and cold) meet (zone of meeting is called a front) with the less dense warmer air rising over the cold (called **frontal** uplift); air forced to rise over a physical barrier such as a mountain range (called **orographic** uplift).

(a) Which of these three uplift mechanisms is dominant in the tropics? __________

Which is dominant in the mid-latitudes? __________

(b) Uplift can be either gradually rising while undergoing considerable horizontal movement (such as movement over a gradually sloping mountain or front) or uplift can be more directly vertical (such as movement over a sharp cliff, vertically shaped front or associated with most convectional uplift). Which types of clouds with low bases are associated with gradual uplift?

Which types of clouds with low bases are associated with the more vertical uplift?

Name________________________________ Date________________

Instructor______________________________ Section______________

(c) Considering clouds with low level bases, which types would you generally associate with the following atmospheric stabilities:

unstable: ____

stable: ____

(d) Nimbostratus (Ns) and cumulonimbus (Cn) clouds are precipitation clouds (the term nimbo is from the Greek for rain). Other clouds occasionally produce precipitation but only of a light nature. Compare the intensity and duration of precipitation that would generally be associated with Ns versus Cn clouds. [Hint: Note that the Cn cloud is often associated with thunderstorms.]

8.8 Cloud droplets are very small and must therefore grow by large amounts before falling as precipitation. Two theories of cloud droplet growth are: the Bergeron or ice-crystal process and the coalescence or collision process.

Illustrate each of these two theories of droplet growth by drawing well labelled diagrams to explain how they occur.

Bergeron (ice-crystal):

Coalescence (collision):

Is the Bergeron or Coalescence process believed to be dominant in the tropics? ________

________ Which is believed to be dominant in the mid-latitudes? ________

____

8.9 The mechanisms of vertical uplift for cloud formation in question 7 are often used to name three genetic types of precipitation known as convectional, frontal and orographic precipitation. For the locations listed below and for the seasons indicated, name the dominant genetic type of precipitation:

Location	**Summer**	**Winter**
Athens, Georgia	________	________
Edinburgh, Scotland	________	________
Sitka, Alaska	________	________
Nairobi, Kenya	________	________

8.10 The western coastal region of Oregon receives among the highest annual total rainfalls in the United States, yet inland in eastern Oregon, the climate is almost like a desert. What factor accounts for this radical change in precipitation regime?

8.11 The concept of "effective precipitation" refers to the availability of precipitation for plant growth. If precipitation that falls, evaporates quickly because of intense radiation and high temperatures, then it is not "effective" since it was never available to the plants. One way of elucidating the concept of effective precipitation is by determining a parameter called "potential evapotranspiration." Potential evapotranspiration refers to the maximum possible amount of water that can be evaporated by the available energy if water supply is not limited. Thus, the magnitude of potential evapotranspiration (PET) depends primarily upon net radiation. Net radiation can be roughly related to temperature allowing temperature to be used to estimate PET. With a knowledge of PET and precipitation (P) for a location, a water budget can be established. In developing a water budget or water balance diagram, storage of water in the soil must be considered. For months with PET less than P, a water surplus exists. If PET becomes greater than P, then soil moisture utilization must occur. Thornthwaite established a system whereby 100 mm of stored soil water is assumed available to plants. If PET continues to be greater than P, a water deficit is established. If PET subsequently becomes less than P again, then soil moisture recharge occurs (to 100 mm of storage) until a water surplus is again established. Such a sequence of events is illustrated in the hypothetical example shown in Fig. 8-5.

Name______________________ Date______________

Instructor______________________ Section______________

Figure 8-5

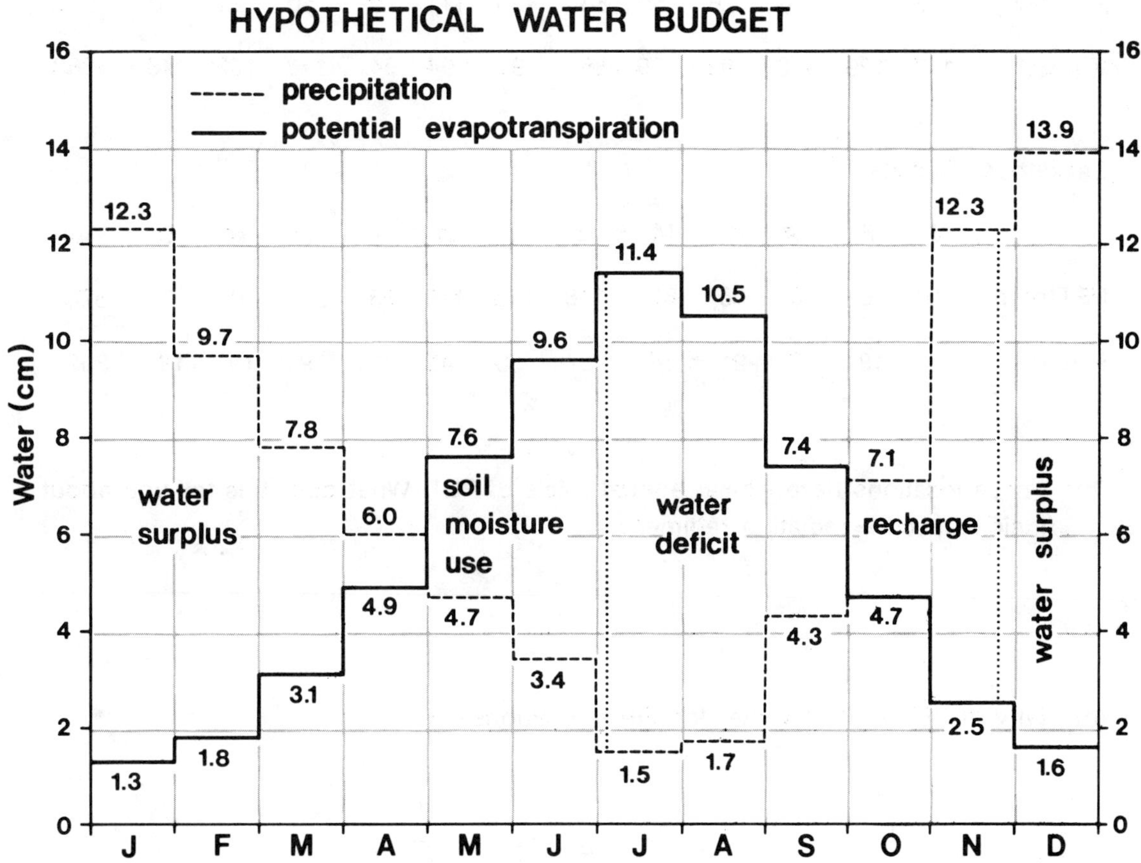

(a) Listed below are PET and P data for Halifax and Saskatoon, Canada. Halifax is located on the eastern coast of Canada while Saskatoon is on the western interior plains of Canada approximately 500km east of the Rocky Mountains. On Fig. 8-6, draw water budget diagrams for both of these locations using the same format shown in Fig. 8-5. Label the times of water surplus, soil moisture utilization, water deficit and soil moisture recharge.

Halifax, Canada:

	J	F	M	A	M	J	J	A	S	O	N	D	Year
PET(mm)	0	0	0	26	58	88	120	111	79	50	24	0	556
P (mm)	147	129	112	105	109	85	92	94	94	113	152	148	1380

Saskatoon, Canada:

	J	F	M	A	M	J	J	A	S	O	N	D	Year
PET(mm)	0	0	0	26	82	116	138	115	65	28	0	0	570
P (mm)	18	18	17	21	34	57	53	45	33	19	19	18	352

(b) These locations have similar annual totals of PET. What does this tell you about their respective radiation regimes?

(c) Why is PET zero in winter for these locations?

(d) When and how long does Halifax have a water deficit during the year?

____

When and how long does Saskatoon have a water deficit during the year?

____

In light of the very different water deficit regimes for these two locations, describe the differences in natural vegetation types that would be expected in the vicinity of these cities.

Name________________________ Date______________

Instructor________________________ Section______________

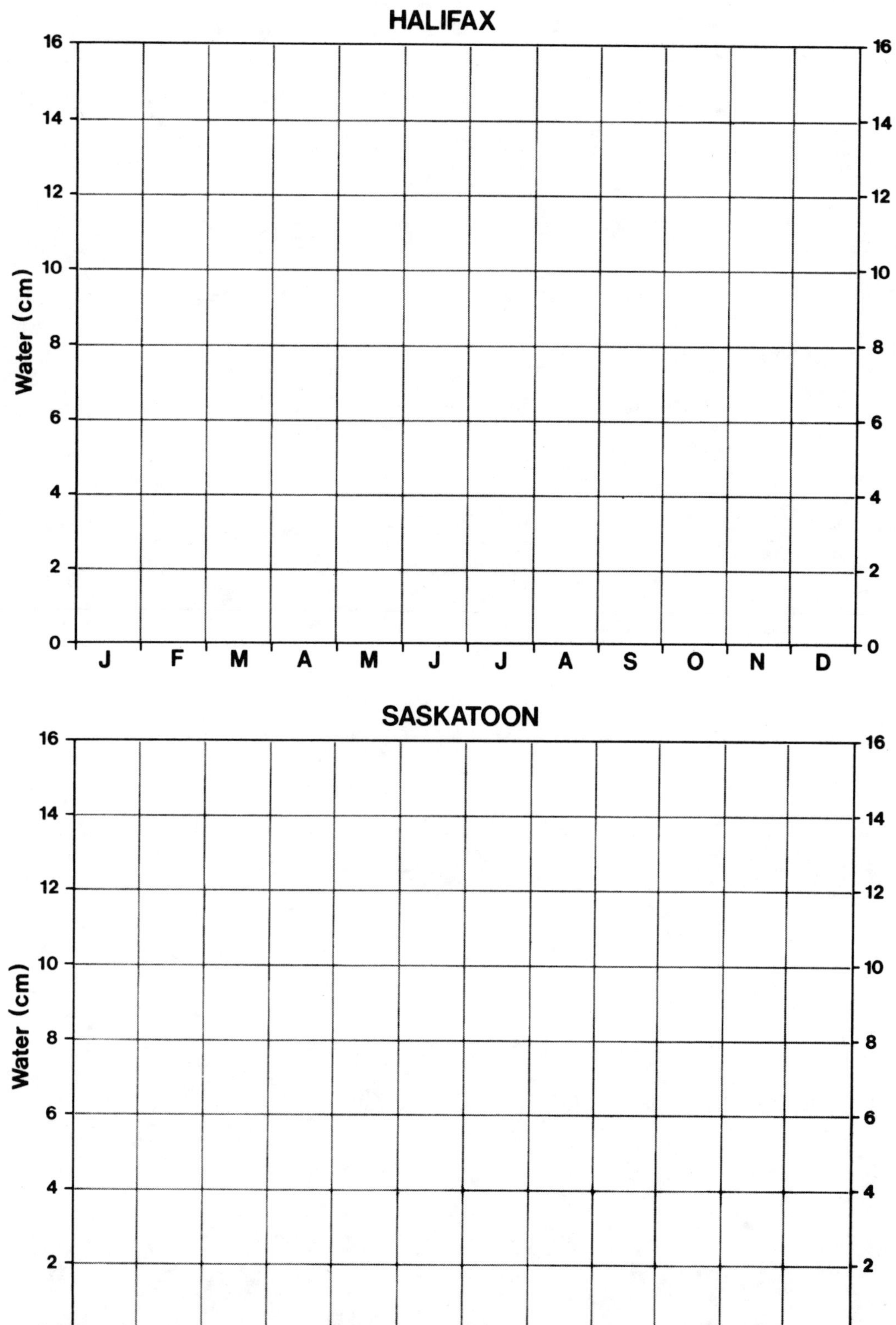

Figure 8-6

EXERCISE 9

HORIZONTAL AIR MOTION AND THE GENERAL CIRCULATION OF THE ATMOSPHERE

Horizontal air motion or advection is more commonly known as wind. Advection occurs when there is a pressure difference along a horizontal surface or at some constant level in the atmosphere. This pressure difference initiates a pressure force which maintains wind movement with the direction of the force acting from the high to the low pressure area. The strength of the force and hence the windspeed is a function of the pressure differential between the high and low pressure zones. This pressure difference is called the pressure gradient (i.e. the change in pressure with the change in horizontal distance, $\Delta P/\Delta D$) and the higher or "steeper" the pressure gradient the greater the windspeed.

Isolines drawn to illustrate barometric pressure are called isobars and are usually drawn at 4 mb intervals. Study of isobaric maps provides useful information for weather analysis. For example, maps with isobars close together have strong or steep pressure gradients and, therefore, strong winds. Conversely, areas with isobars more widely spaced are areas with weaker pressure gradients and weaker winds.

The pressure force acts perpendicular to the isobars on a map, from high pressure directly towards low pressure. However, winds do not go directly from high to low pressure areas because other forces must be considered. As a consequence of the spinning of the earth on its axis, an apparent deflection of winds occurs. The force causing this is known as the Coriolis force and acts to deflect winds to the right in the northern hemisphere and to the left in the southern hemisphere (Coriolis force is zero on the equator). As a result of the interaction of the pressure and Coriolis forces, winds blow **parallel** to the isobars instead of across them. This is illustrated in Fig. 9-1. Note that in the northern hemisphere, the low pressure zone is on the left whereas it is on the right in the southern hemisphere. Upper level winds generally behave in this fashion since only these two forces (pressure force; Coriolis force) are in existence. Such winds are known as "geostrophic winds."

Closer to the surface (lowest 1 km), however, friction must also be considered. Frictional force is caused by the drag of the earth's surface upon moving air causing winds to become weaker as one nears the surface. In this case, the Coriolis force, whose strength depends partially on windspeed, is weakened. Hence, wind direction is across isobars at an angle from high to low pressure areas (see Fig. 9-2).

After applying the above concepts to the horizontal movement of air associated with high and low pressure regions in the first questions of the exercise, vertical and horizontal motions will be combined in an analysis of thermal circulations initially on a local scale and then on a global scale. Finally, global pressure belts, wind patterns and precipitation distributions will be assessed.

Figure 9-1

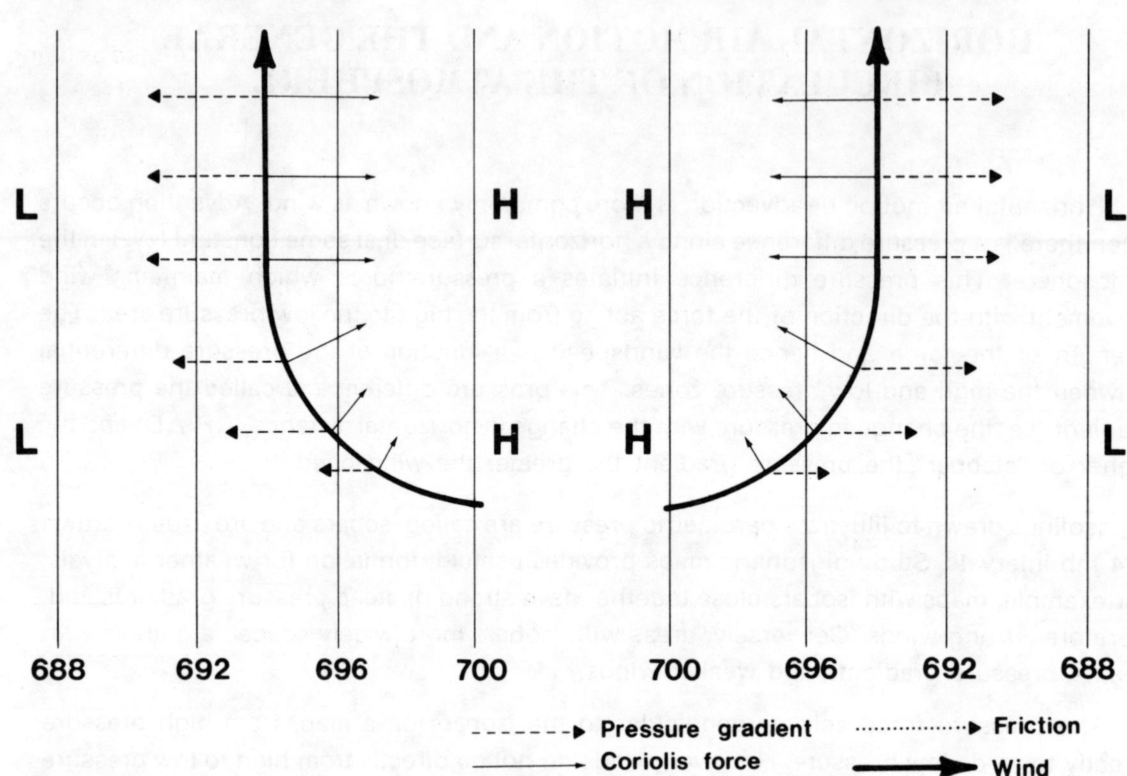
GEOSTROPHIC
(a) Northern Hemisphere
(b) Southern Hemisphere
L
H
H
L
L
H
H
L
688
692
696
700
700
696
692
688
Pressure gradient
Coriolis force
Friction
Wind

Figure 9-2

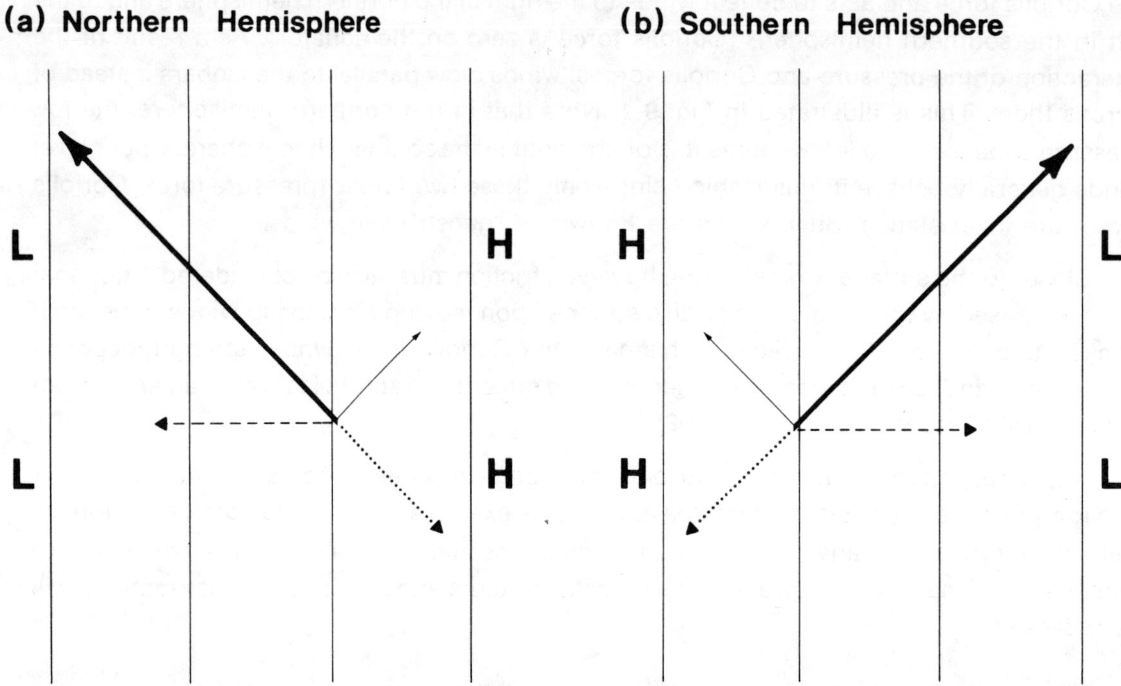
SURFACE
(a) Northern Hemisphere
(b) Southern Hemisphere
L
H
H
L
L
H
H
L
1000
1004
1008
1012
1012
1008
1004
1000

Name________________________________ Date________________

Instructor____________________________ Section______________

QUESTIONS

9.1 Wind velocity refers to both speed **and** direction. The direction of wind is defined as being the horizontal direction from which the wind is coming. Fig. 9-3 shows maps of pressure distributions for several situations. Use arrows to indicate the appropriate wind directions corresponding to the isobaric patterns on each map.

Figure 9-3

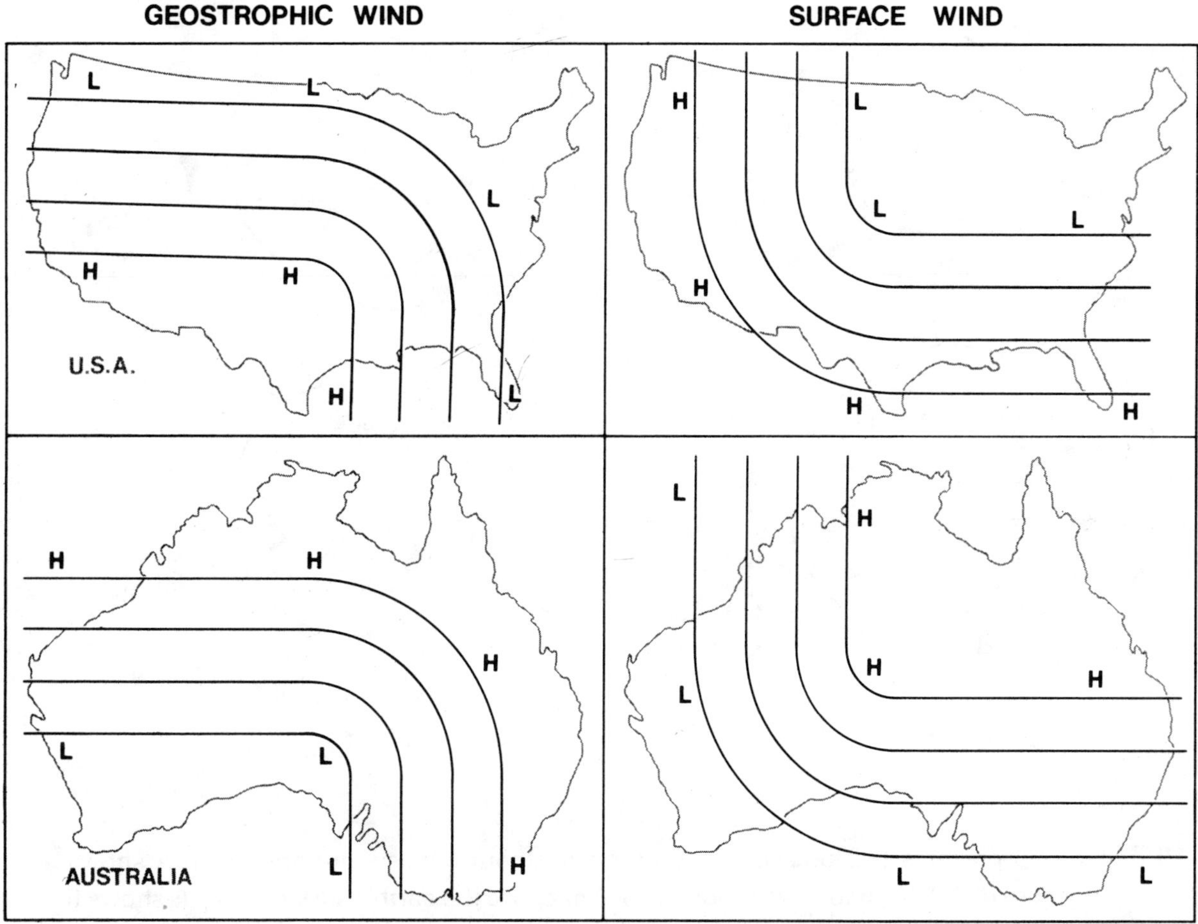

9.2 Fig. 9-4 presents four diagrams showing the wind pattern around high and low pressure areas. Fill in the following table:

	High or Low Pressure	N or S Hemisphere
Diagram A		
Diagram B		
Diagram C		
Diagram D		

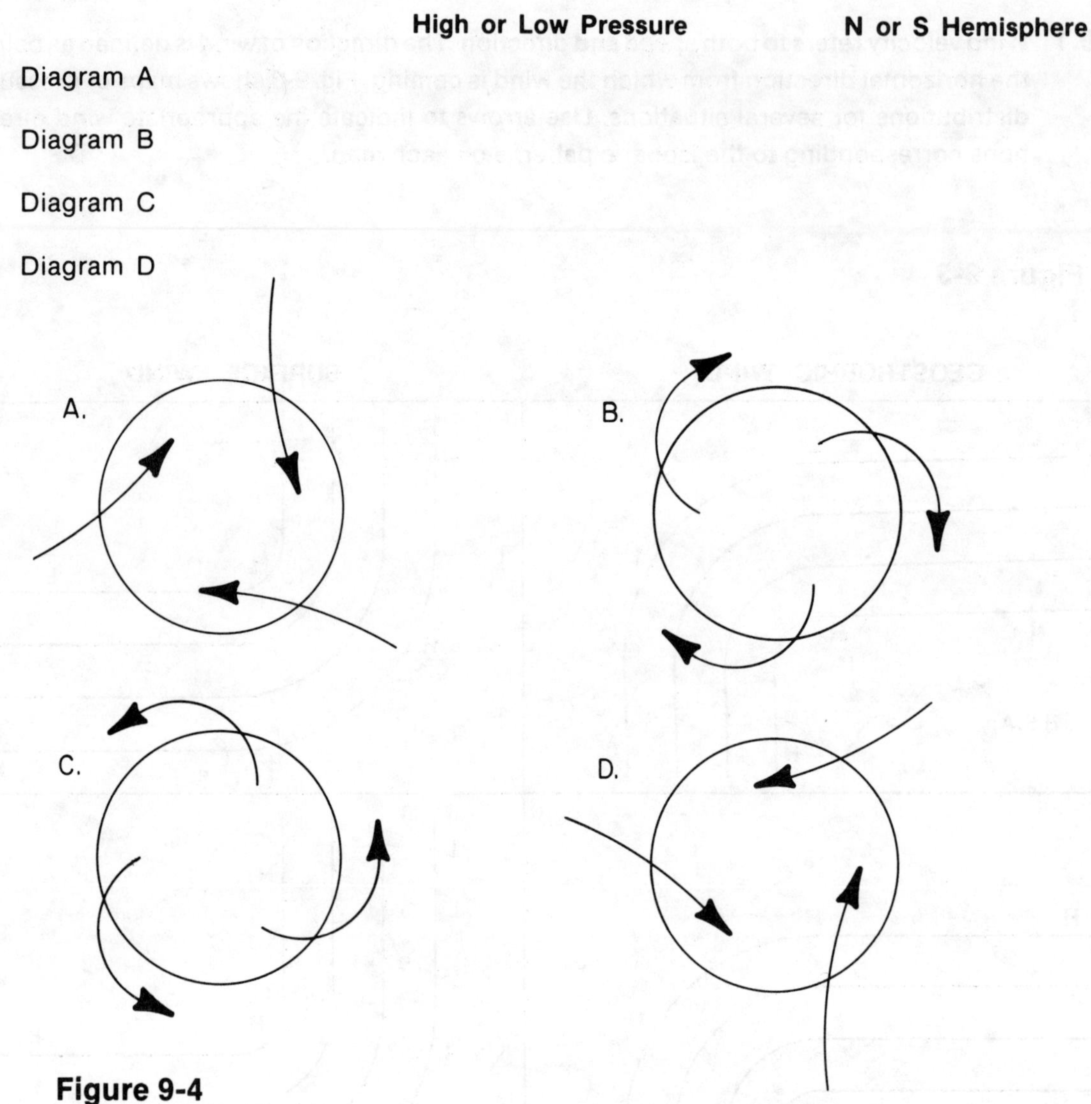

Figure 9-4

9.3 Fig. 9-5 presents two pressure maps for a northern hemisphere area. Map A shows a pressure distribution for an altitude of 2500 m above the earth's surface. Map B shows the surface pressure pattern. In each case, sketch in the wind pattern you would expect to observe. Make the length of your arrows approximately proportional to wind speed.

Name________________________ Date______________

Instructor________________________ Section______________

MAP A: ALOFT PRESSURE

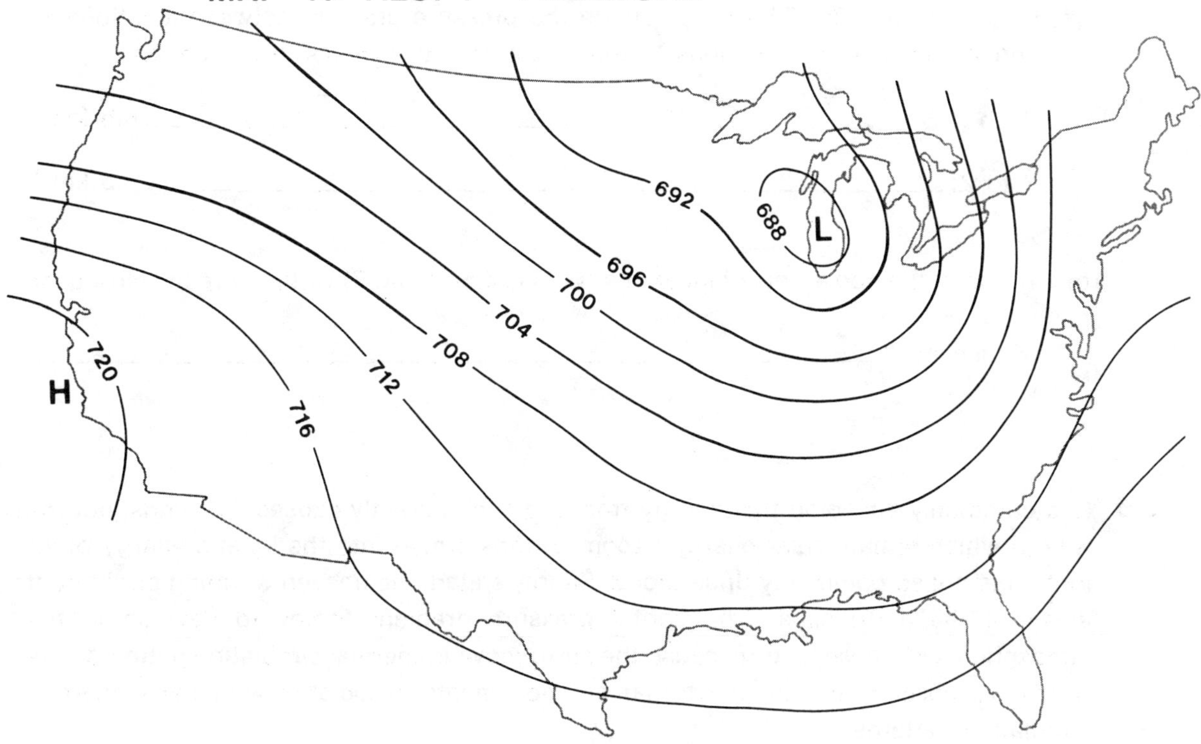

MAP B: SURFACE PRESSURE

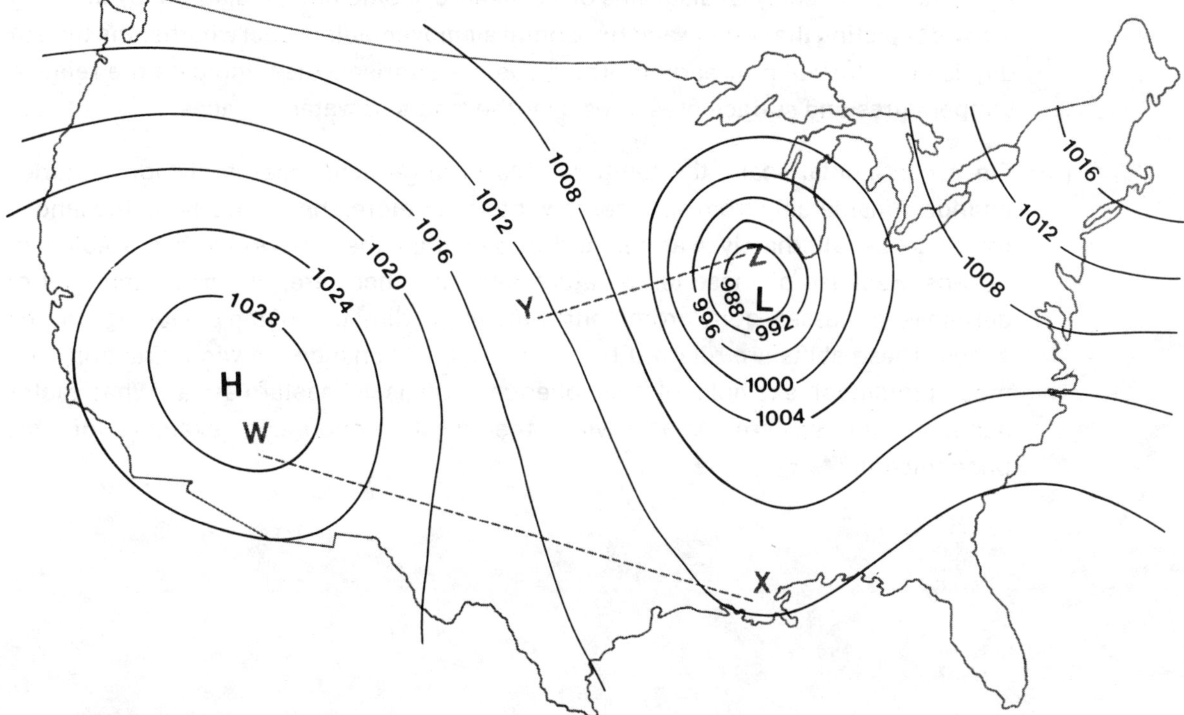

Figure 9-5

9.4 Also shown on Map B of Fig. 9-5 are the locations of four cities.

(a) Assuming that 1 cm on the map is equal to 300 km distance on the earth's surface (a map scale of 1:30,000,000), calculate the pressure gradient between locations W and X and between locations Y and Z. Express the values in mb per km.

W to X:______________________________ mb km^{-1}

Y to Z: ______________________________ mb km^{-1}

(b) Which of the above sets of locations (W and X **or** Y and Z) would have the strongest windspeeds?______________________________

9.5 It is commonly accepted that energy from the sun indirectly causes the winds, but the way in which radiant solar energy becomes transformed into the kinetic energy of air motion is not so commonly understood. Simply stated, the uneven warming of different areas of the earth causes horizontal pressure gradient forces to develop in the atmosphere and these, in turn, cause the air to move in thermal circulation patterns. The winds we observe are the horizontal air movements associated with these thermal circulation patterns.

(a) A local example of such a thermal circulation is the land/sea breeze case. Fig. 9-6 presents cross-sectional diagrams of a seashore. Complete the diagram by drawing arrows depicting the vertical and horizontal air movements. Label whether the time of day is mid-afternoon or early morning (before sunrise). Also, indicate the relative temperatures and surface pressures over the land and water surfaces.

(b) On a continental scale, the temperatures of large land masses at high latitudes change considerably from summer to winter; therefore, the air overlying the land is found to be alternately warmer and cooler than the air overlying the adjacent oceans. Because of these temperature reversals there are, along the margins of certain land masses, quite prominent shifts in the direction of the prevailing surface winds. These shifts are referred to as "monsoonal" changes in wind direction. The most prominent example of this phenomenon is in eastern Asia. What factor accounts for eastern Asia having the most pronounced example of this phenomenon?

Name______________________________ Date______________

Instructor____________________________ Section______________

Figure 9-6

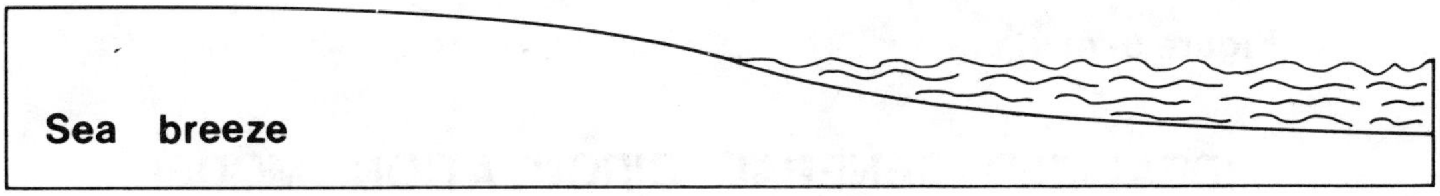

Sea breeze

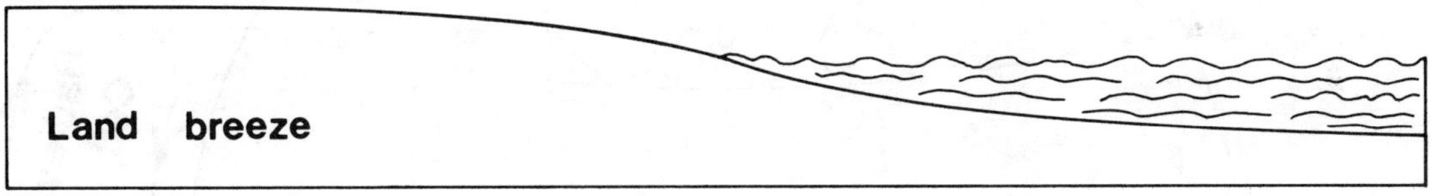

Land breeze

9.6 Fig. 9-7 presents a simplified "3-cell" model of the general circulation of the earth's atmosphere. The diagram shows only the vertical cross-sectional pattern along the right-hand horizon with the Hadley, Ferrel and Polar cells labelled. Complete the diagram by indicating the pressure distribution on the global map and then draw in the surface wind belts using arrows to indicate the sense of wind movement. As a guide, the following is a list of pressure features and wind belts that should be labelled on the map.

Intertropical Convergence Zone (ITCZ)
Subtropical High Pressure (STHP) — both hemispheres
Subpolar Low Pressure (SPLP) — both hemispheres
Polar High Pressure (PHP) — both hemispheres
NE trade winds
SE trade winds
Westerlies — both hemispheres
Polar Easterlies — both hemispheres
doldrums
horse latitudes

Figure 9-7

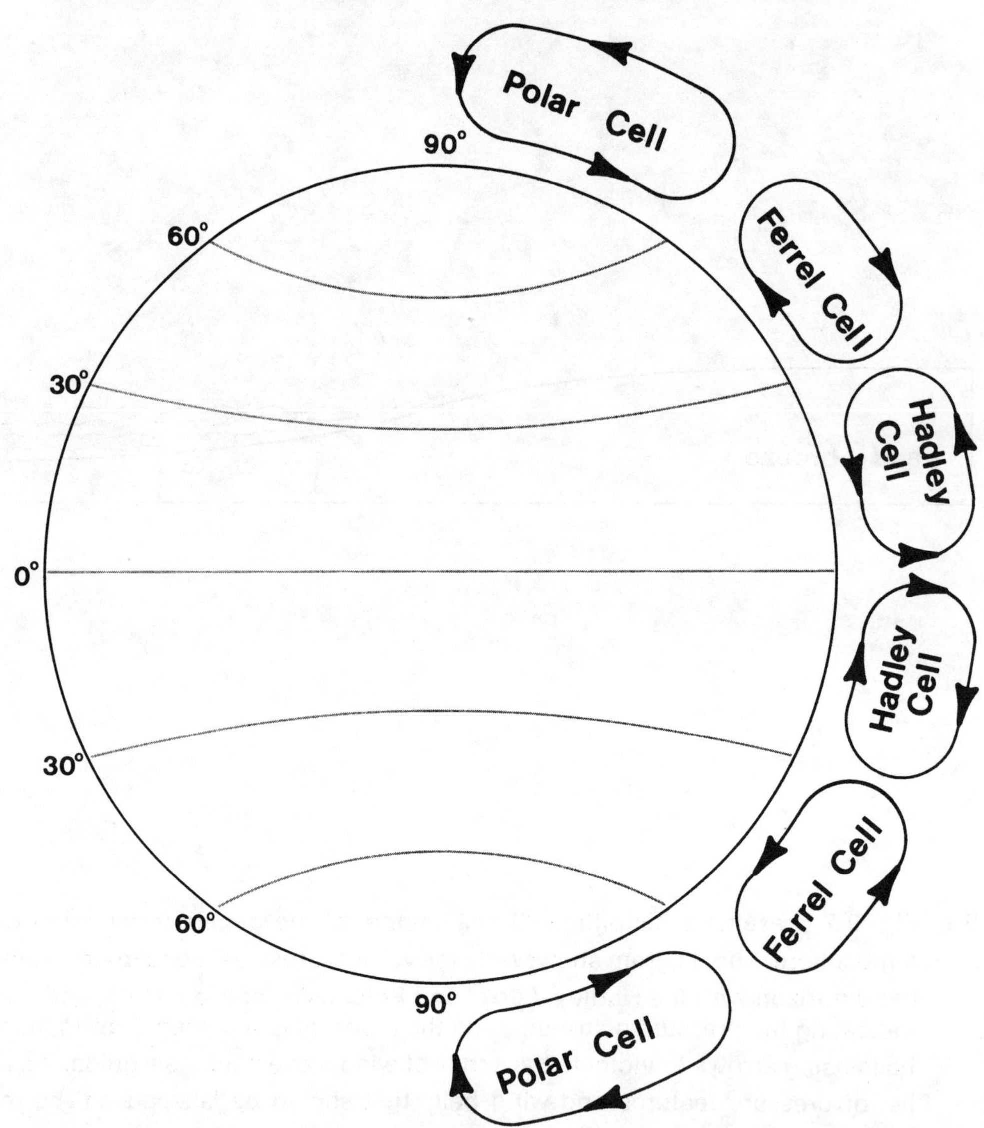

Name______________________ Date______________

Instructor______________________ Section______________

9.7 Briefly, describe factors that will influence or modify the general patterns depicted by the simplified model in question 6. What parts of the world will conform to the patterns in question 6 most closely?

9.8 Fig. 9-8 depicts maps of the global pressure distribution for January and July. With reference to these maps, answer the following:

(a) Near the equator, solar heating varies little during the year. Consequently, atmospheric pressure varies little. What is the approximate mean value of sea-level pressure along the equator in January?__________ in July?__________

(b) Over the oceans, one can find dominant high pressure areas in the subtropics. These are known as the Subtropical Highs (STHP). During which season do the STHP's intensify and expand areally? In which direction latitudinally are they shifting?

(c) Poleward of the STHP, especially in the northern hemisphere, large low pressure areas are evident particularly in one season. Which season?__________. These are known as the Subpolar Lows (SPLP).

(d) In the northern hemisphere during winter, another very strong high pressure region develops. Where? ______________________

(e) Which land area experiences the largest annual pressure change? __________

______________________ Note that this large annual change can be referred to as a monsoonal reversal of pressure and will be important with respect to climate in this region.

(f) The surface westerlies occur in the mid-latitudes. In which season (summer or winter) will they be most widespread and strongest? __________
With reference to the north-south gradient of net radiation and hence temperature, briefly state why the westerlies are strongest in this season.

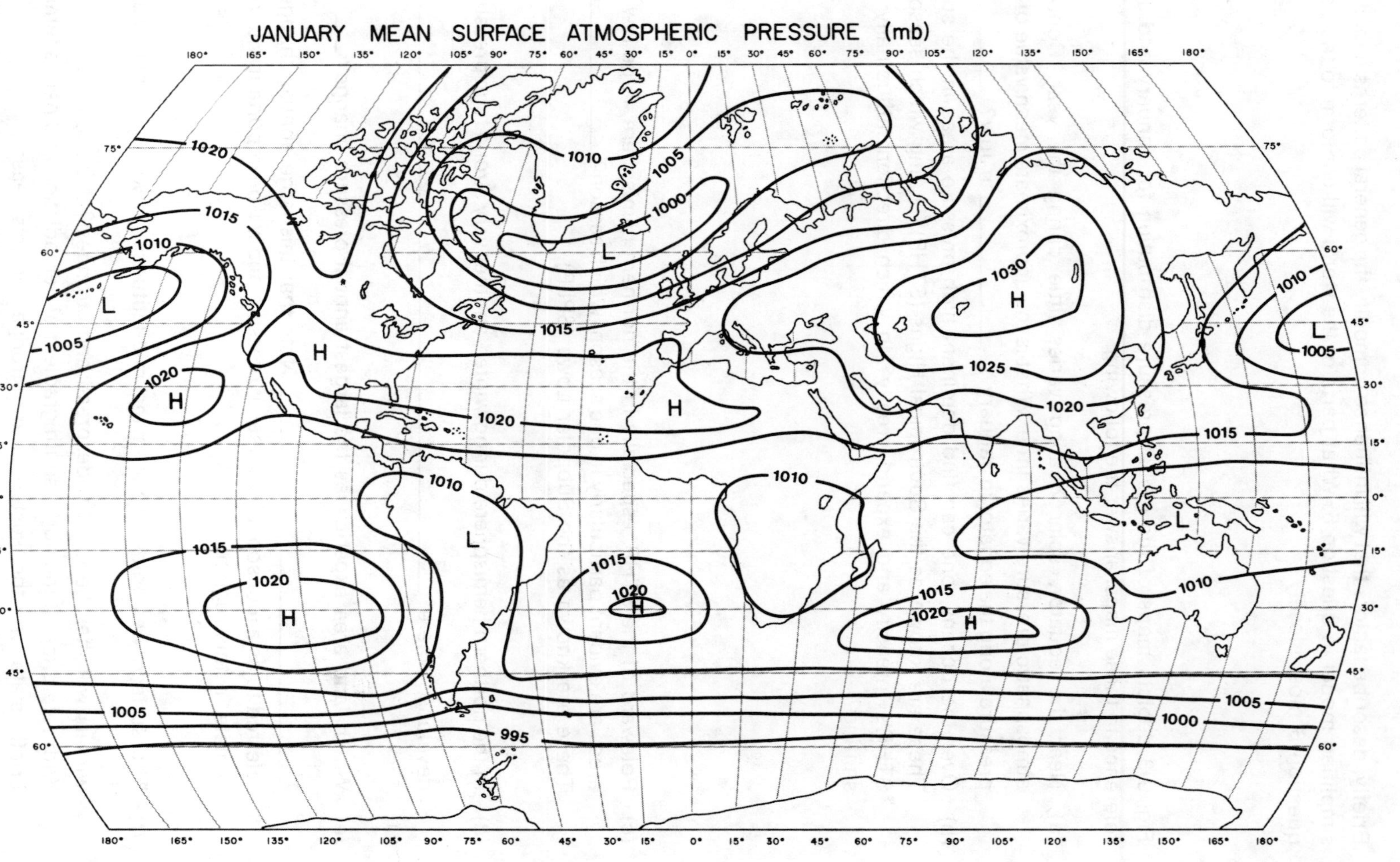

JANUARY MEAN SURFACE ATMOSPHERIC PRESSURE (mb)
1030
1025
1020
1015
1010
1005
1000
995
H
L

Figure 9-8A

Name________________________________ Date____________________

Instructor____________________________ Section__________________

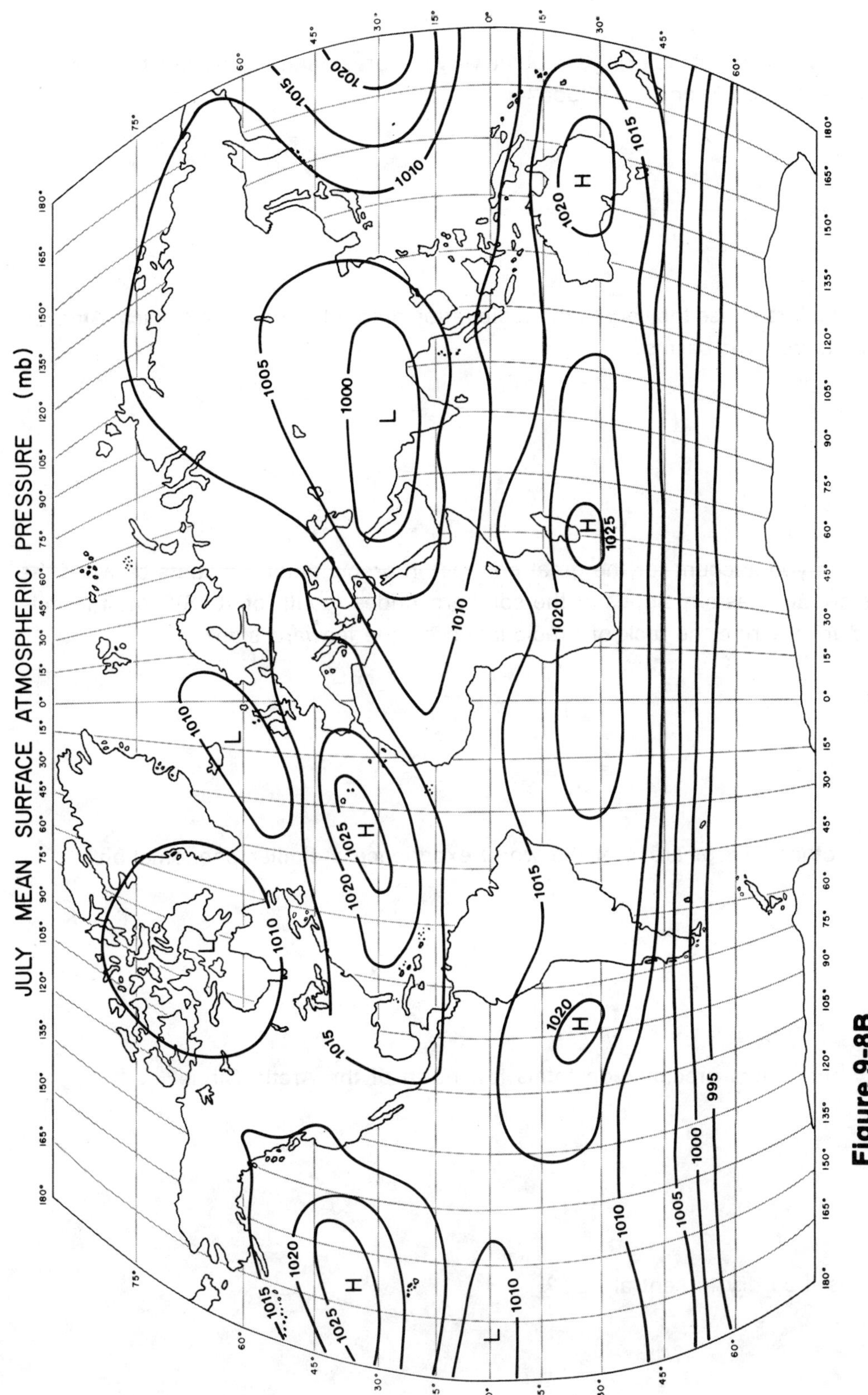

Figure 9-8B

9.9 The global distribution of precipitation is a key factor in determining the types of climates found over the earth. The global distribution of annual precipitation totals is presented in Fig. 9-9.

(a) What factors operate to bring about the very low precipitation amounts for locations around 20°-30° N and S latitude?

(b) What factors cause the very high precipitation amounts within the tropical Amazon and Congo river basins?

(c) How do you account for the great contrast in precipitation amounts between the western and eastern slopes of the southern Andes (south of 40° S)? What term is used to describe the lack of precipitation on the leeward side?

(d) What other area or areas of the world experience a similar phenomenon?

(e) Why are annual precipitation totals low north of the Arctic Circle?

(f) Why is it so dry in central Asia?

Name________________________________ Date________________

Instructor____________________________ Section______________

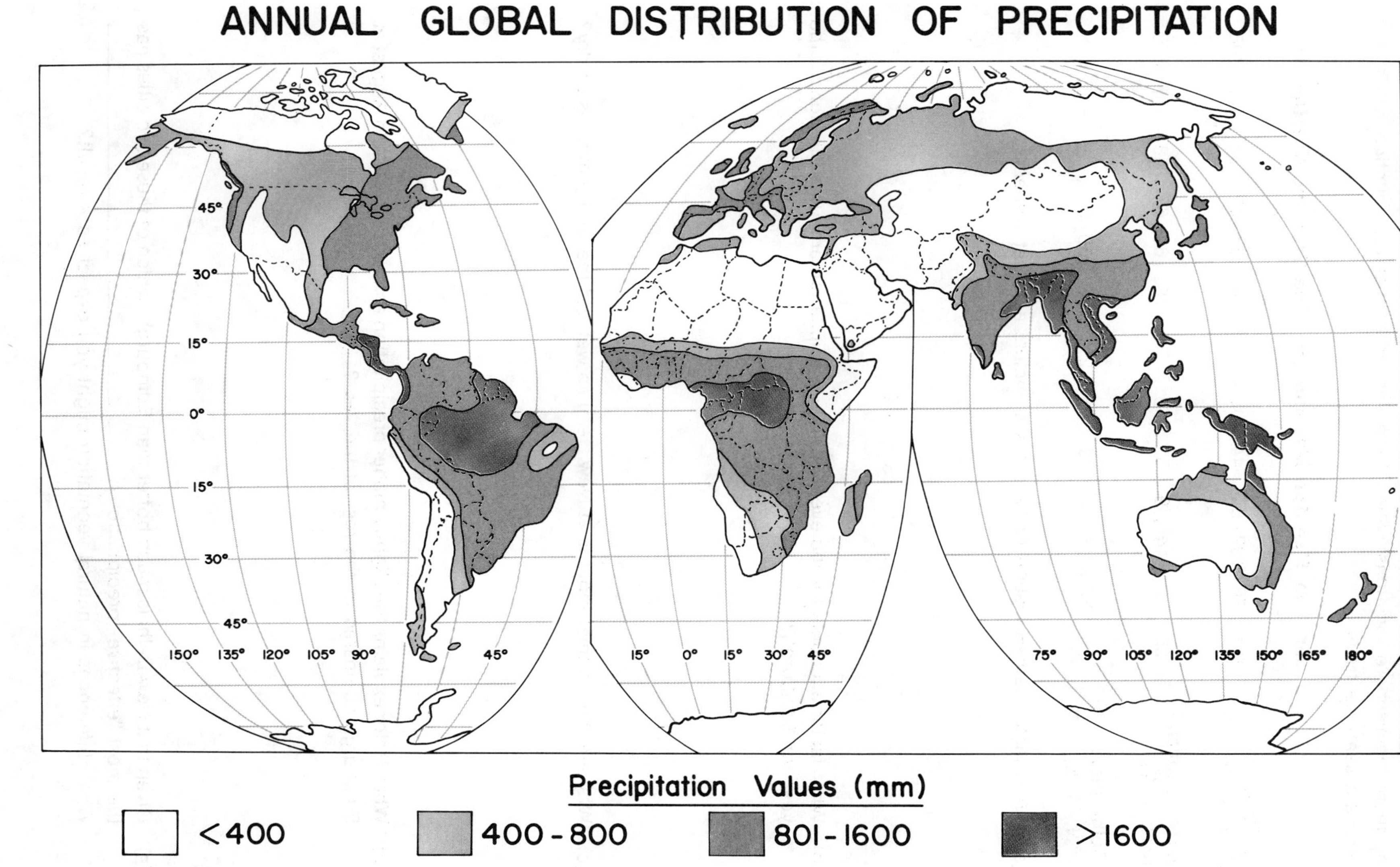

Figure 8-9

9.10 Given below are the mean monthly precipitation amounts, in millimeters, for three northern hemisphere locations which have similar annual precipitation totals, but different seasonal traits.

	Jan	**Feb**	**Mar**	**Apr**	**May**	**Jun**	**Jul**	**Aug**	**Sept**	**Oct**	**Nov**	**Dec**	**Yr**
Lisbon, Portugal (39° N; 9° W)	111	76	109	54	44	16	3	4	33	62	93	103	708
Edinburgh, Scotland (56° N; 3° W)	55	42	44	39	52	53	71	78	60	67	58	57	676
New Delhi, India (22° N; 77° E)	25	22	17	7	8	65	211	173	150	31	1	5	715

(a) What factor causes Lisbon to have dry summers and wet winters?

(b) What factor accounts for the even monthly precipitation amounts throughout the year for Edinburgh?

(c) Why are the summer months at New Delhi so wet and the winter months so dry?

(d) What factor explains the slightly higher precipitation amounts in July and August at Edinburgh compared to the rest of the year?

(e) Given that New Delhi is much hotter than Edinburgh, which of these two cities has the most "effective" precipitation? ______________________________
What differences in natural vegetation might you expect as a result?

EXERCISE **10**

THE MID-LATITUDE CYCLONE

One distinct characteristic of mid-latitude weather is its day-to-day variability. This variability is the result of continually shifting air masses which are large bodies of air with nearly homogeneous characteristics of moisture and temperature. The analysis and prediction of the weather in the mid-latitudes involves recognizing and predicting the paths of these air masses as they move into the mid-latitudes from their respective source regions. Of particular importance is the identification and analysis of weather fronts (areas along which contrasting air masses come into conflict or collide). Frequently a mid-latitude cyclonic storm or series of storms will develop along the front between contrasting air masses. Analyzing and predicting the course these storms will follow and the weather they bring with them is a paramount concern of weather forecasters.

This exercise analyses the concepts of air masses, frontal systems and the formation and characteristics of a typical mid-latitude cyclone. Tornadoes, nature's most violent storms, are a weather phenomenon often associated with cyclonic storms and also will be briefly considered.

Name________________________________ Date____________________

Instructor______________________________ Section__________________

QUESTIONS

10.1 On the map of North America (Fig. 10-1), mark the source regions for the principal air masses which occur over the continent and indicate, with arrows, the major directions of movement of these air masses.

Figure 10-1

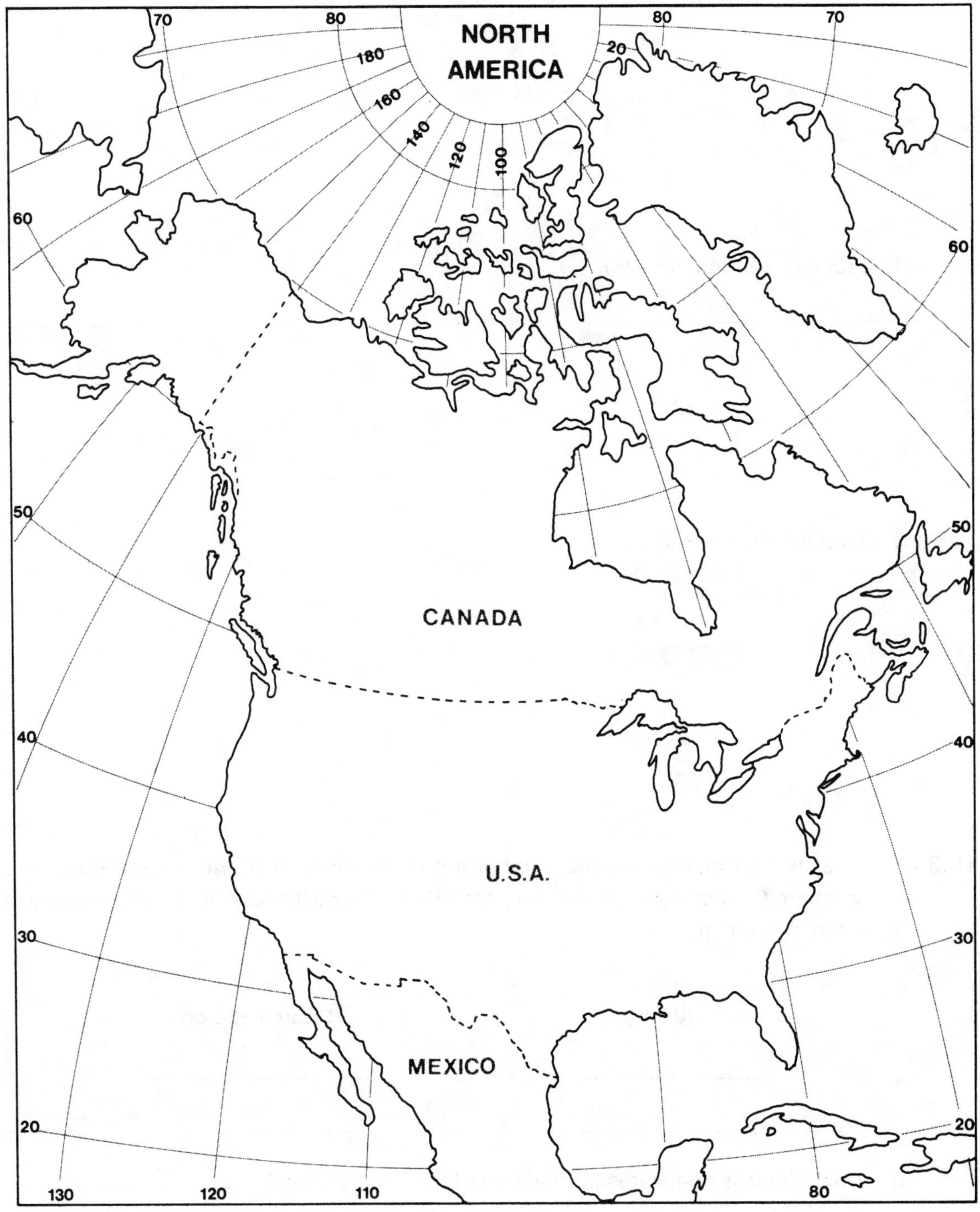

10.2 Describe the major temperature and moisture characteristics, seasonality (if any) and stability of each of the following air masses as they relate to North America:

(a) Polar continental (cP):

(b) Polar maritime (mP):

(c) Tropical continental (cT):

(d) Tropical maritime (mT):

10.3 (a) What two air masses are most important to the weather of the United States and southern Canada east of the Rocky Mountains? Note the specific source regions of these two air masses.

	Air mass	**Source region**
(1)	______________	______________
(2)	______________	______________

(b) What air mass dominates the weather of the Pacific coast? ______________

Name________________ Date________________

Instructor________________ Section________________

(c) Why do mP air masses from the North Atlantic source region seldom affect the weather of the eastern United States?

(d) Identify the source region and air mass that generally provides the greatest amount of moisture to the United States east of the Rocky Mountains?

____

(e) Identify the source region and air mass that would normally be associated with a summer drought in the southern Great Plains?

____

10.4 In Fig. 10-2 partially completed cross-sectional diagrams of a cold front, a warm front (warm air stable), and a warm front (warm air unstable) are presented. Complete the diagrams by labelling areas of warm and cold air, sketch and name cloud types and label the zones of precipitation. Indicate whether these zones of precipitation would contain moderate or heavy precipitation intensity.

(a) In which of these three cases will precipitation be most intense:

____

(b) With which one of these three cases will the duration of precipitation for a locality be longest? ______________________________

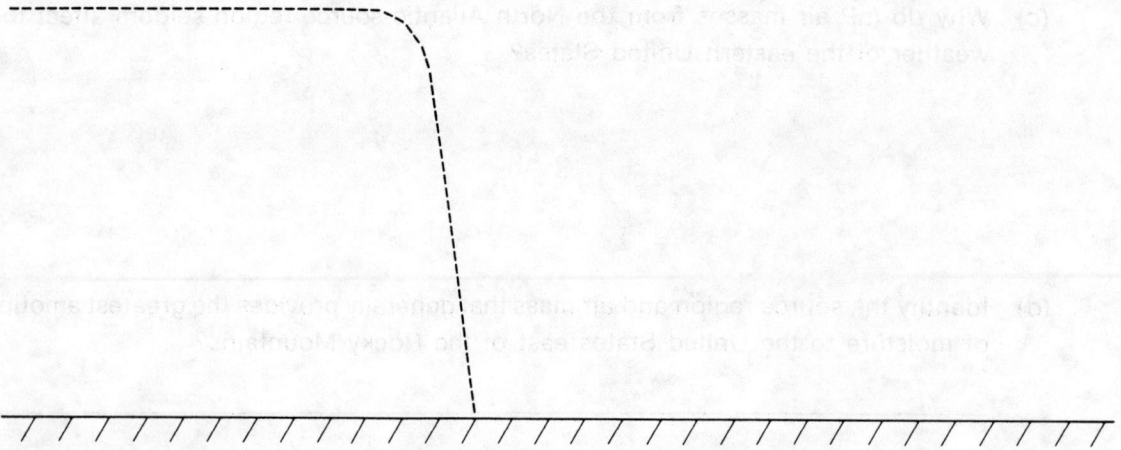

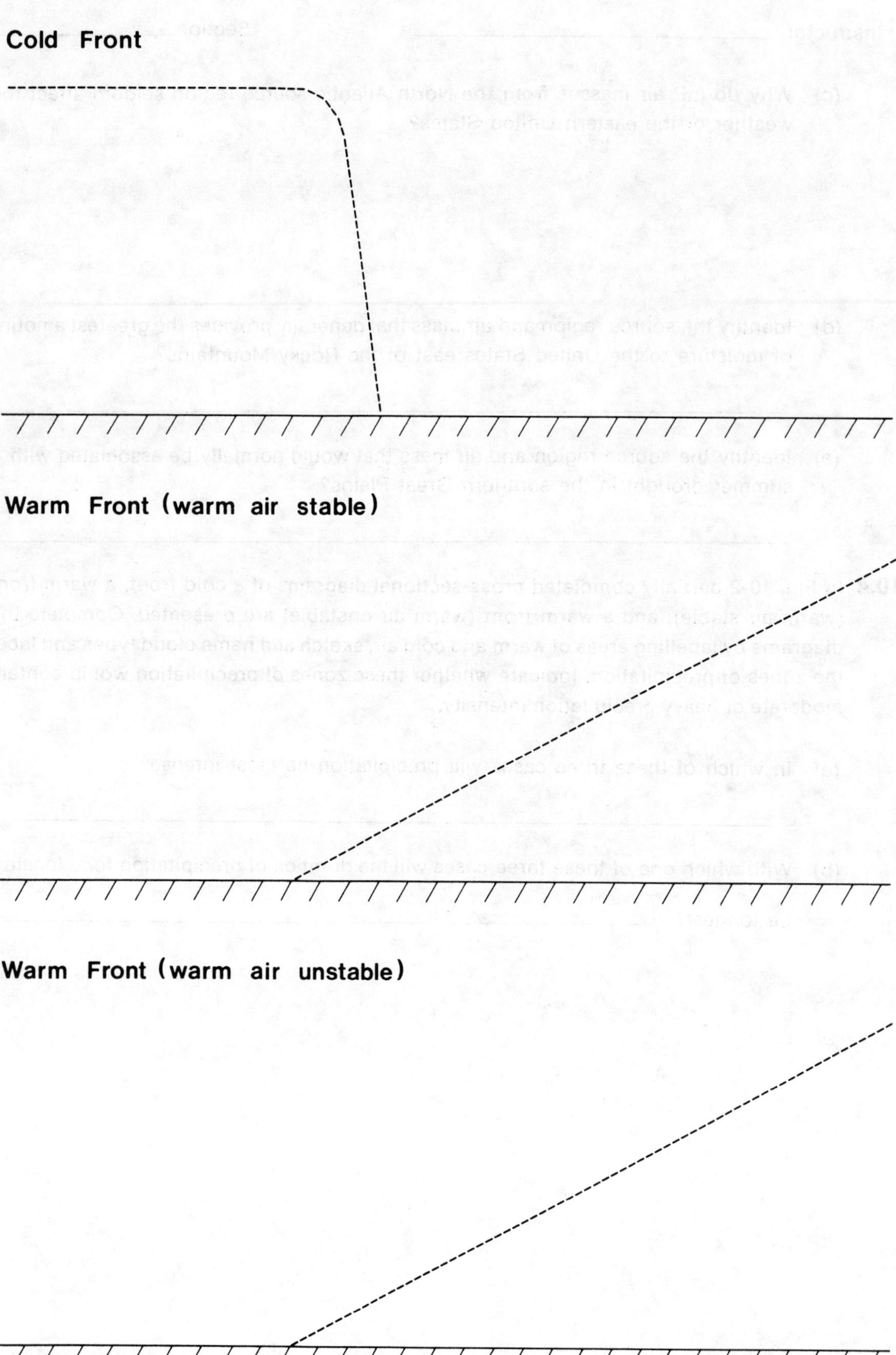
Cold Front
Warm Front (warm air stable)
Warm Front (warm air unstable)

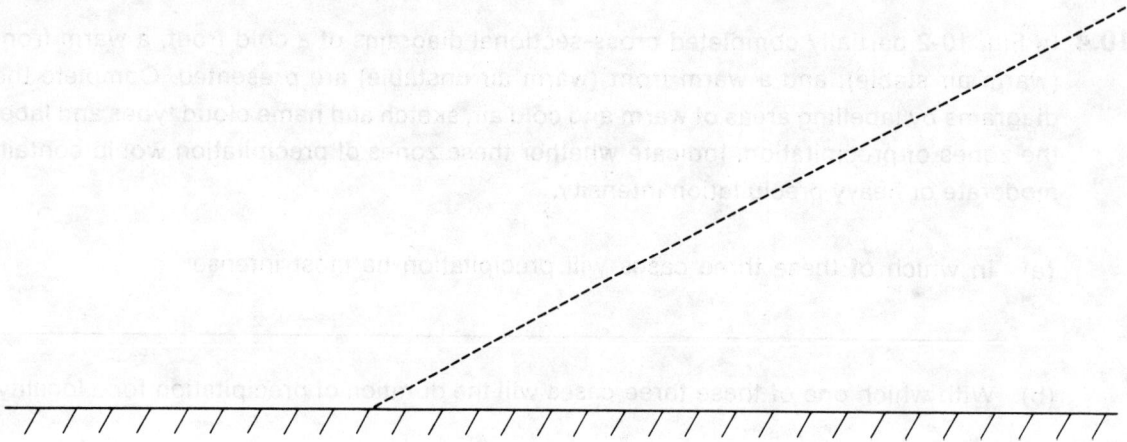

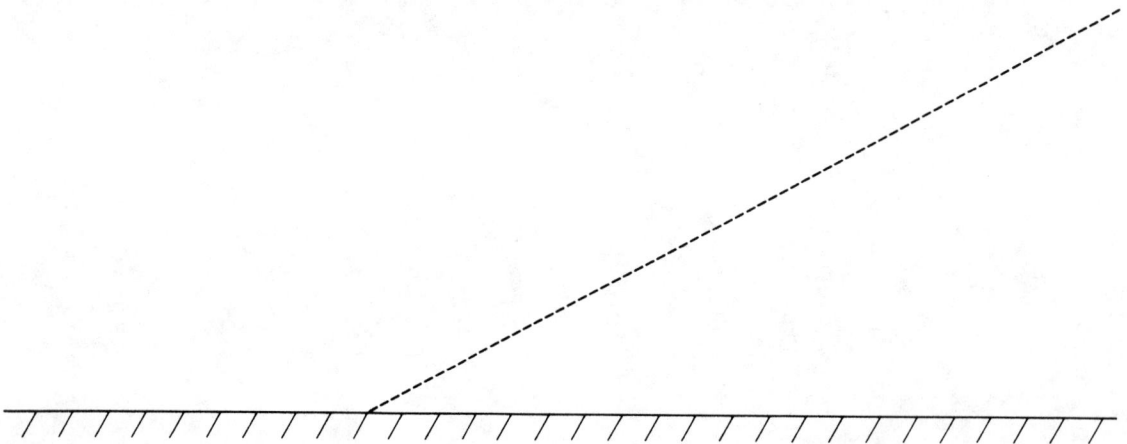

Figure 10-2

Name________________________ Date________________

Instructor____________________ Section______________

10.5 (a) When a traveling cold front "catches up" to a warm front, an occluded front is formed. Draw a cross-sectional diagram similar to those in Fig. 10-2 representing such a front where the air behind the cold front is colder than that preceding the warm. Label areas of warm, cool and cold air. Illustrate and label the cloud and precipitation characteristics.

(b) The above is called a cold-type occluded front and is the most common occluded front to form east of the Rockies. It is possible for the air behind the cold front to be slightly warmer than the air preceding the warm front. These warm-type occluded fronts occur along the Pacific coast where the air behind the cold front is mP rather than cP and therefore not as cold. Draw a cross-sectional diagram below representing this warm-type occluded front. Label areas of warm, cool and cold air, fronts, cloud and precipitation zones.

10.6 The "Polar Front" is established where tropical and polar air masses meet. It is along this macroscale front that mid-latitude cyclones develop with warm and cold fronts. These cyclones are important mechanisms for transferring energy poleward in the mid-latitudes. Fig. 10-3 represents a series of plan or map views of the frontal structure and isobars associated with an idealized life cycle of a mid-latitude cyclone in the northern hemisphere.

(a) On each of the diagrams in Fig. 10-3 indicate the following:

— the center or region of lowest pressure (label L)

— an arrow depicting the general direction the storm is moving

— the warm front (diagrams b, c, d only)

— the cold front (diagrams b, c, d only)

— the occluded front (diagram d only)

— the warm sector or areas of warm air

— the areas of cold air

— the surface wind directions (diagrams c and d. Label with arrows like those given in diagrams a and b).

— shade in areas of probable precipitation.

(b) In the space provided in Fig. 10-4, draw cross-sectional views along the lines from point A to B (from diagram c of Fig. 10-3) and point C to D (from diagram d of Fig. 10-3). On your cross-sections indicate and label the appropriate fronts, the pockets of warm and cold air, any vertical air movements, cloud locations and types (assume that the warm air in the warm sector is stable) and areas of precipitation.

What is the length of the occluded front? ____________________ km.

What is the distance from point A to point B? ____________________ km.

Name______________________________ Date________________

Instructor____________________________ Section______________

Figure 10-4

A B

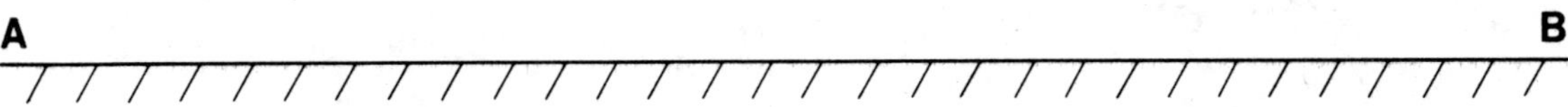

C D

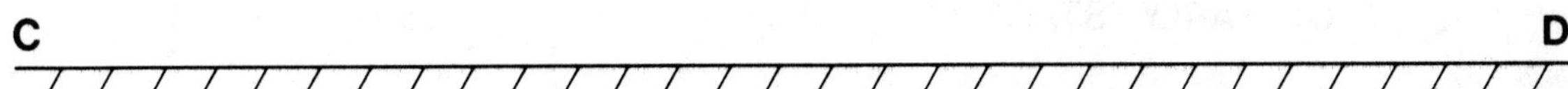

(c) Note that in map (c) of Fig. 10-3, the location of six cities are labelled. Which of these cities is experiencing the lowest atmospheric pressure?______________ Which city probably has the highest air temperature?______________ Which two cities are receiving moderate steady rainfall? ______________ ______________ . Which city has high-level cirrus clouds only in the sky? ______________Which city is experiencing southwesterly winds?______________ ______________ Which city is experiencing northwesterly winds?______________

IDEALIZED LIFE CYCLE OF A MID-LATITUDE CYCLONE

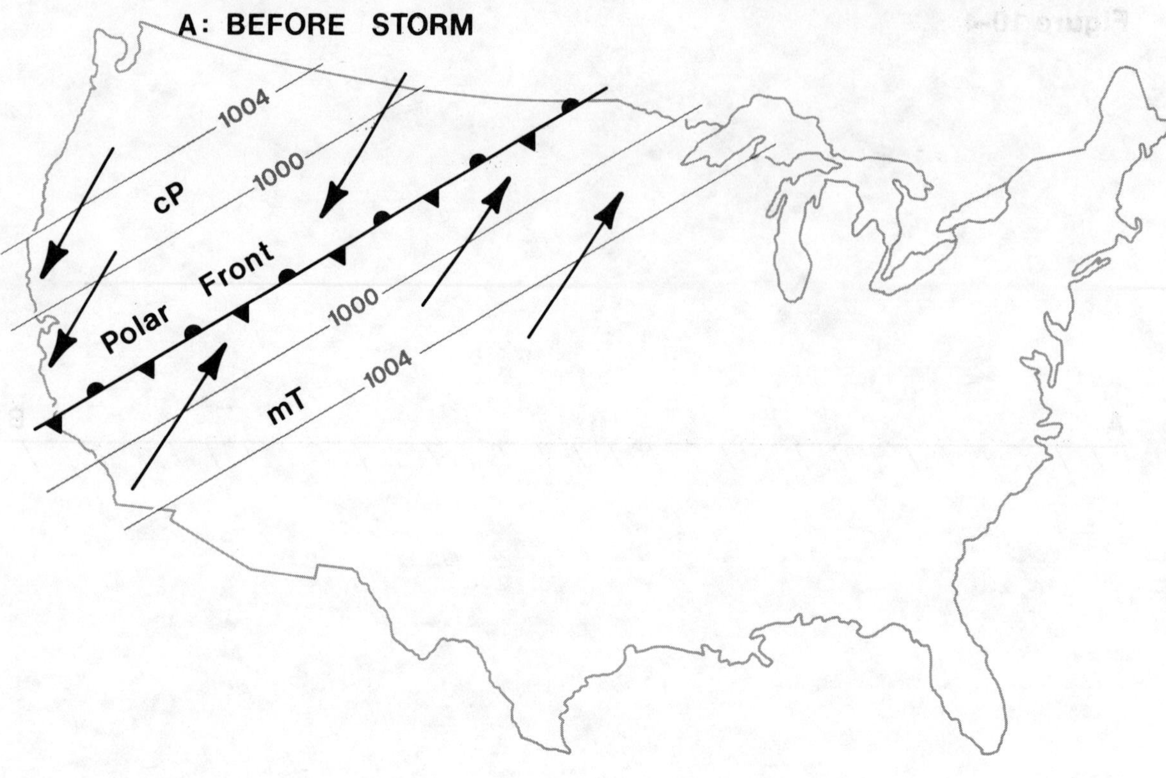

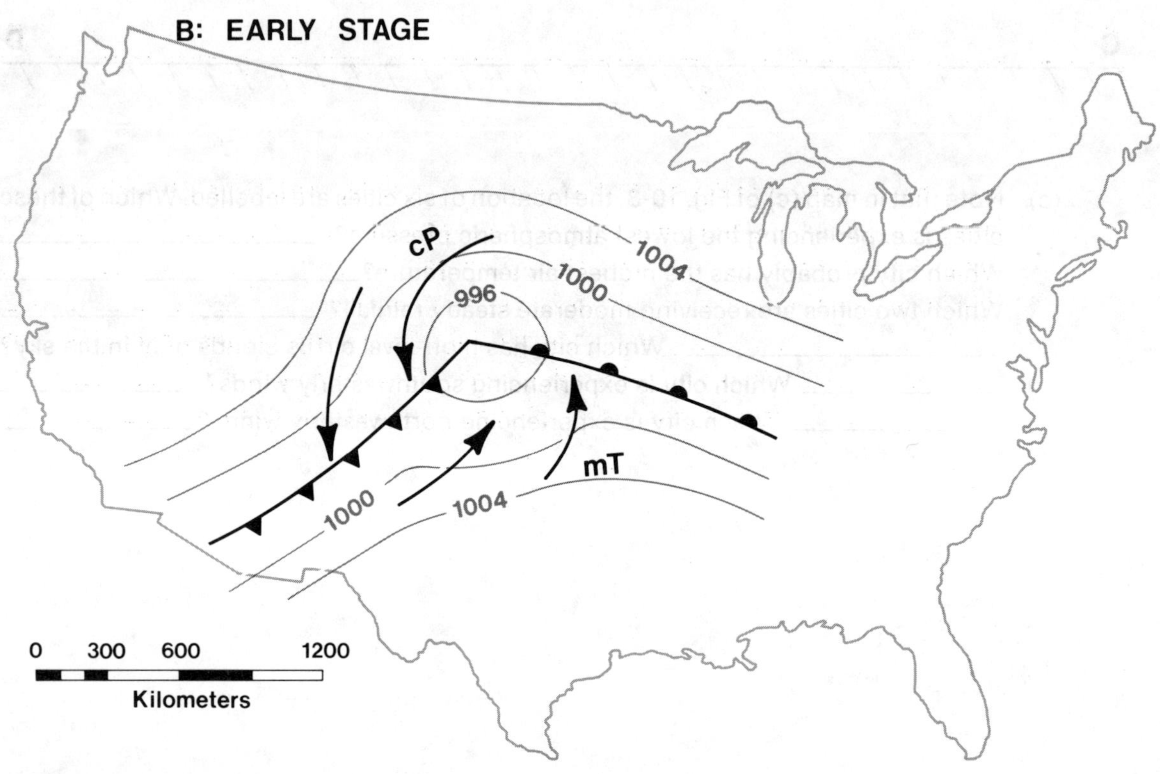

Figure 10-3A & B

Name________________________ Date________________

Instructor____________________ Section______________

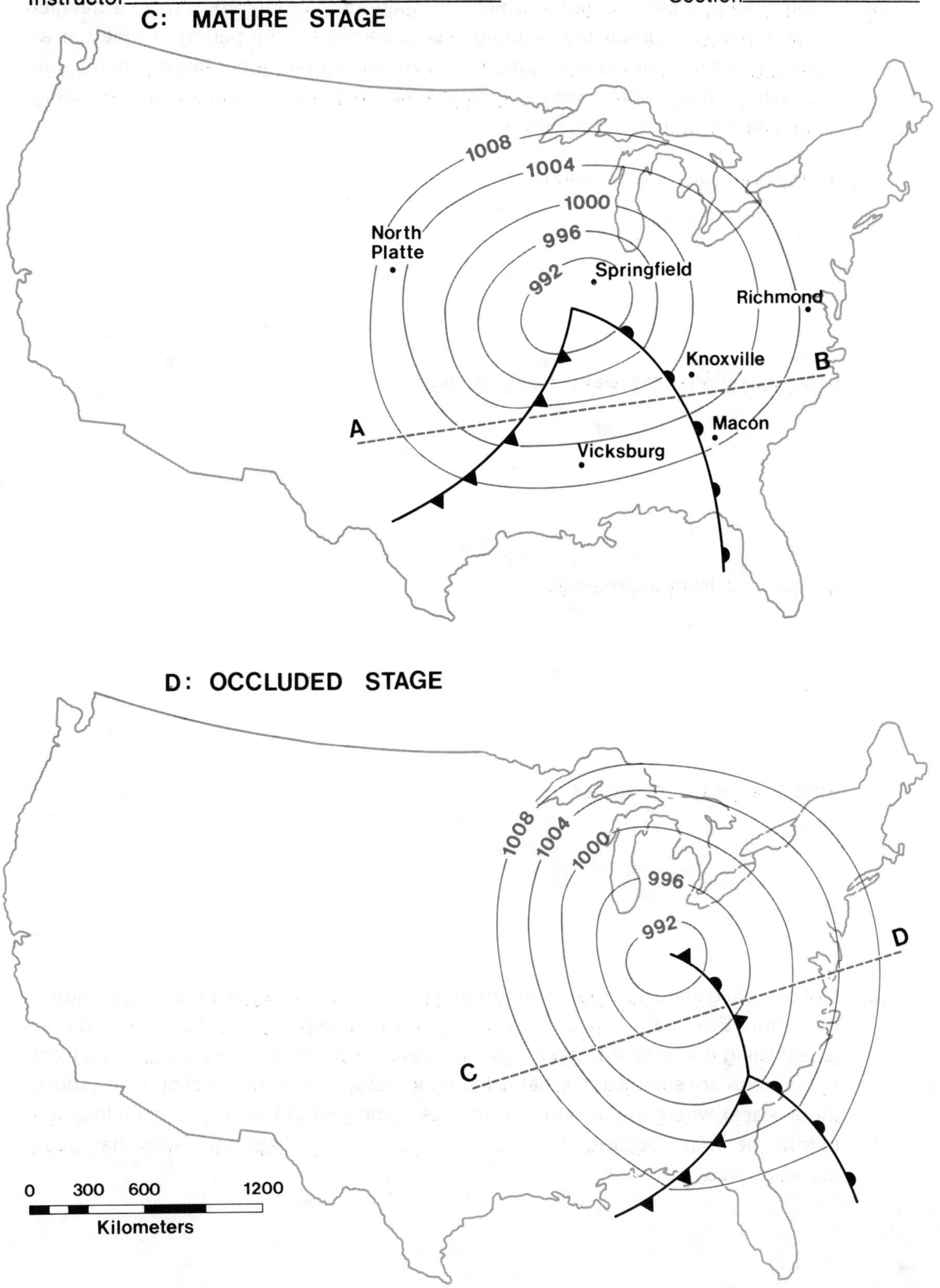

C: MATURE STAGE

1008
1004
1000
996
992
North Platte
Springfield
Richmond
Knoxville
B
A
Macon
Vicksburg

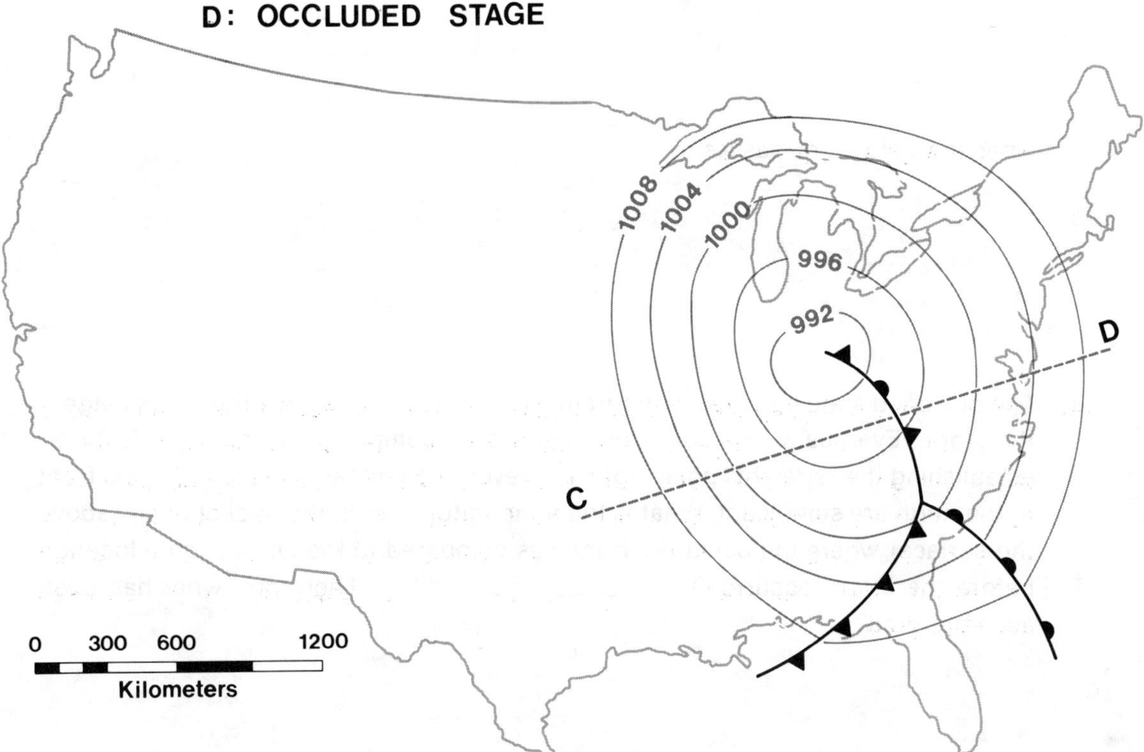

Figure 10-3C & D

(d) If you lived in a city located at point B in diagram c of Fig. 10-3, write a detailed weather forecast (under the headings listed below) for the period of time corresponding to the complete passage of this cyclonic storm (about two days). Include forecasts of temperature, wind direction, pressure behavior, cloudiness (including type) and precipitation conditions.

As the warm front approaches:

Immediately after the warm front passes:

As the cold front approaches:

After the cold front passes:

(e) The occluded state depicted in diagram d of Fig. 10-3 represents the dying stage of the storm. Eventually, the cold front catches up completely to the warm front re-establishing the "straight" polar front. However, the remnants of the occluded front to the north are significant. What is the temperature like in the pocket of air (above the surface) where the occluded front was compared to the air in such a location before the storm occurred? ____________________ Therefore, what has been accomplished?

Name____________________ Date__________

Instructor____________________ Section__________

10.7 Thunderstorms can occur along the cold front but also in organized groups extending along a line preceding the cold front in the warm sector of a mid-latitude cyclone. These thunderstorms would be convectional rather than frontal in origin. Such a line is known as a "squall line" and occurs most often in spring or early summer. Under these conditions, it is possible to get very intense thunderstorms and perhaps tornadoes. Tornadoes are small storms of short duration but of extreme violence. Surface pressures are the lowest found in nature and produce the strongest windspeeds with values up to 650 km h^{-1} in a whirlpool-like structure around the central low. The funnel shaped cloud coming down from a Cn cloud is a trademark of a tornado.

(a) There are approximately 700 tornadoes reported each year in the United States. What is the peak month or season for tornado occurrence? __________
Note that it is during this season that the cool and warm air masses associated with the development of the mid-latitude cyclones are at their greatest contrast.

(b) Below is a table of the number of tornadoes reported for selected states for the 24 year period 1953-1976 according to the National Oceanic and Atmospheric Administration. Given the state areas, complete the table by calculating the **mean annual** number of tornadoes per 10,000 km^2 .

Tornado Incidence for Selected States (1953-1976)

	Area (km^2)	Total Number of Tornadoes	Mean Annual Number per 10,000 km^2
Alaska	1,518,800	1	
California	411,013	68	
Colorado	269,998	348	
Florida	151,670	880	
Georgia	152,488	511	
Illinois	146,075	678	
Indiana	93,993	553	
Iowa	145,790	615	
Kansas	213,063	1122	
Maine	86,026	67	
Massachusetts	21,385	102	
Michigan	150,779	353	
Minnesota	217,735	371	
Missouri	180,486	716	
Montana	381,086	85	
Nebraska	200,017	817	
New York	128,401	80	
Ohio	106,764	323	
Oklahoma	181,089	1326	
Texas	692,405	2775	

(c) Which four states had the highest mean annual incidence of tornadoes per unit area (in order)? 1. ______________ 2. ______________

3. ______________ 4. ______________.

Note that tornadoes can also be associated with hurricanes as well as squall lines. Thus, the high incidence in Florida is partially due to such hurricane-associated tornadoes.

(d) Judging by your calculations in part (b) above, what general regions of North America experience the least number of tornadoes?

(e) Oklahoma has a high incidence of tornadoes yet the number of tornado deaths per unit area is lower than states such as Massachusetts, Indiana or Ohio. Suggest two reasons that would explain this apparent discrepancy.

EXERCISE 11

THE TROPICAL CYCLONE

Tropical cyclones are nonfrontal (i.e. homogeneous air mass) cyclonic storms characterized by a rotary motion and steep pressure gradient generating windspeeds equal to or in excess of 119 km h^{-1}. While mid-latitude cyclones are larger and tornadoes more intense, tropical cyclones, in terms of property damage and loss of life, are the most destructive of all storms. Tropical cyclones that occur over the Atlantic or the eastern North Pacific Ocean are called hurricanes. The same type of storms are called typhoons in the western North Pacific and cyclones in the Indian Ocean. Regardless of the regional name applied, tropical cyclones originate in the warm, moist air of tropical oceanic regions. Fig. 11-1 illustrates the principal regions of tropical cyclone formation and paths of movement.

Figure 11-1

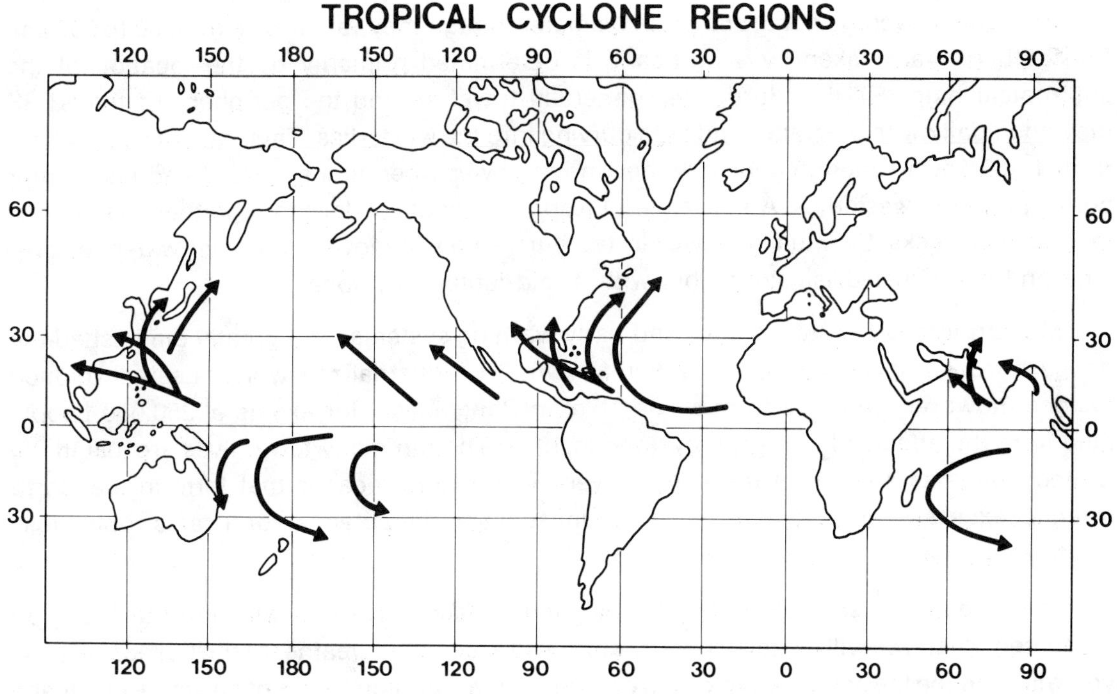

Although tropical cyclone formation is not fully understood scientists have identified several prerequisite conditions for their formation. First, for the rotary motion characteristic of hurricanes to be generated, the Coriolis force must exceed a certain minimum value. Therefore, these storms do not form near the equator and the vast majority originate between 5 and 20 degrees of latitude. A second necessary condition is the presence of a large sea area with surface temperatures over 27°C. The warm moist air overlying the oceanic surface is the hurricane's fuel. Thus, these cyclones cannot persist for long over land or cool oceanic surfaces. The presence of warm, moist air in an area with an appropriate Coriolis force is not, however, sufficient for hurricane formation. In addition, something must initiate the rapid upward movement of air such as a large area of unstable air. In the late summer and early autumn (July through October in the northern hemisphere case) the prevailing winds in the tropics (i.e. the trades) are characterized by "easterly waves." These converging lines of low pressure often contain very unstable air. When an easterly wave comes into contact with the waves of the ITCZ, a sufficiently large core of low pressure may be created to generate a hurricane. For a hurricane to develop, however, the upper divergence (i.e. above the surface low) must exceed the lower level convergence. Only if the outflow greatly exceeds the low level inflow could the extreme low pressure of a hurricane be brought about.

Once formed, the hurricane becomes a vast heat engine generating power from the abundant moist air that surrounds it by forcing the converging air upward where condensation occurs and latent heat energy is released into the upper atmosphere. By this process the hurricane becomes a safety valve, extracting surplus energy from the tropics and transferring it into the upper atmosphere to be moved poleward by the upper air flow.

Hurricanes average about 550 km in diameter though they vary in size from 80 to 900 km. The path or track taken by a hurricane is determined primarily by the location of the subtropical high (STHP). Hurricanes generally travel around the periphery of the STHP moving initially with the trades and then curving into the westerlies. Their rate of travel varies from 16-50 km h^{-1} though generally they move slower when first spawned and faster after curving into the westerlies. A hurricane's lifespan is about six days though they may last as long as two weeks. Eventually, however, the hurricane will move into cooler waters or over land and will either dissipate or join with a mid-latitude cyclone.

Modern forecasting techniques and early warning systems have greatly diminished the dangers from tropical cyclones, particularly in the industrialized world. Less-developed nations, however, continue to be vulnerable. In Bangledesh, for example, 250,000 people died from the effects of a tropical cyclone in 1974. Though many fewer lives are lost in the United States as a result of the annual average of five hurricanes that form in the North Atlantic, excessive property damage is not unusual and the potential for a catastrophic loss of life exists each hurricane season.

In this exercise several questions designed to familiarize you with the hurricane are presented. First, you will examine the structure and associated weather of the hurricane. Next, you will examine the path, life cycle, and environmental characteristics of an actual hurricane. Finally, you will be asked to forecast the weather conditions at a given site along a hurricane track.

Name____________________ Date__________

Instructor____________________ Section__________

QUESTIONS

11.1 Structurally the hurricane is composed of three basic components. A warm central core of low pressure, called the "eye," is characterized by relative calm with light variable winds. The eye which ranges from 8-35 km in diameter is surrounded by dense, dark cumulonimbus clouds. These are called the "collar clouds" and this part of the hurricane contains the most intense winds. Atmospheric conditions are so unstable and the vorticity of the collar so intense that the air converging into the hurricane does not reach the eye of the storm. In addition to intense winds, the collar is characterized by thunder and lightning, heavy rainfall and frequent tornadoes. The average distance through the collar is about 30 km. The third part of the hurricane is composed of stratiform clouds that are collectively called the "spiral bands." Within this section of the hurricane the wind velocity decreases as the distance from the collar clouds increases. Fig. 11-2 illustrates an overhead or plan view and a cross-sectional view of a typical northern hemisphere tropical cyclone. The table at the bottom of this diagram indicates the features associated with each component of the hurricane. Study this diagram and label the following on both the overhead and cross-sectional views: eye, collar clouds, spiral bands. Also indicate wind circulation and direction by drawing short lines with arrows. Complete the table at the bottom of the diagram by providing the missing information.

11.2 The track of Hurricane David (1979) has been plotted as a dark line on Fig. 11-3 with the storm originating around 12° N and 36° W. The data in Table 11-1 and the satellite imagery presented in Fig. 11-4 further illustrate the evolution of this hurricane. These references will aid in completing the following exercises.

(a) On the hurricane tracking chart (Fig. 11-3) label the dates and storm stages (TD, TS or H) of David in a fashion similar to that marked for the first day in the southeastern section of the map. Delimit the days on the north and east side of the track and label the storm stages on the opposite side.

(b) Based on the data presented in Table 11-1, how many hours did it take for David to reach hurricane strength? ____________________

(c) During the course of how many days was David at hurricane strength?

____

(d) What was the greatest windspeed reached by David and at what time and day was this attained? ____________________

(e) While at hurricane strength what was the recorded range of windspeeds?

____

(f) What was the lowest barometric pressure recorded for David and when did it occur? ____________________

HURRICANE STRUCTURE AND ASSOCIATED WEATHER

Figure 11-2

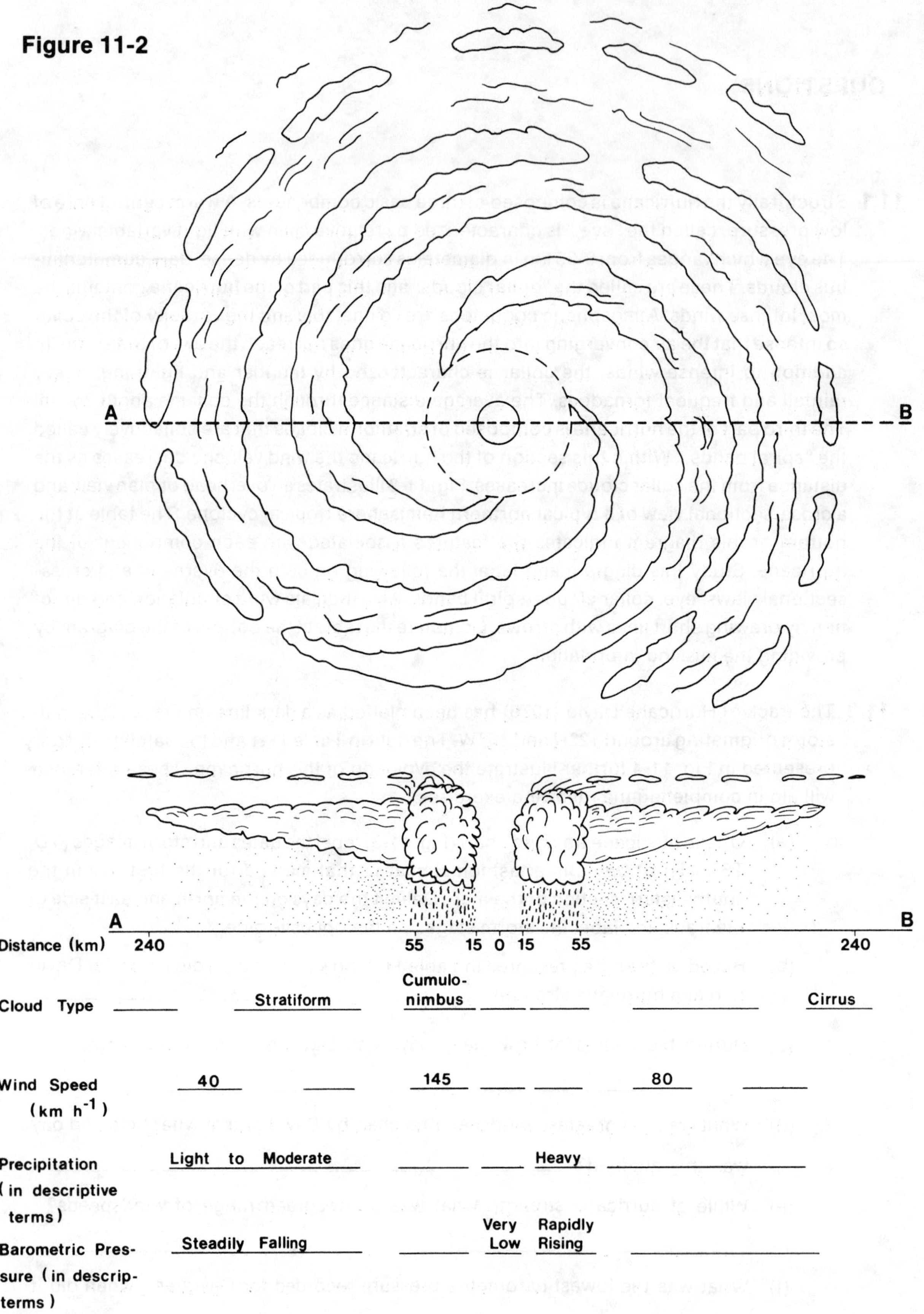

Distance (km)	240		55	15 0 15	55		240
Cloud Type	______	Stratiform	Cumulo-nimbus	______ ______		______	Cirrus
Wind Speed ($km\ h^{-1}$)	40	______	145	______ ______		80	______
Precipitation (in descriptive terms)	Light to Moderate		______	______	Heavy	______	
Barometric Pressure (in descriptive terms)	Steadily Falling		______	Very Low	Rapidly Rising	______	

Name________________________ Date____________

Instructor________________________ Section____________

(g) The satellite imagery for August 31 (see Fig. 11-4) indicates that only the spiral band section of David passed over Puerto Rico, yet portions of that island received more than 450 mm of rainfall as the storm passed. What supplemental mechanism was primarily responsible for generating this large amount of precipitation? ________________________

(h) The satellite photography for 1800 hours on September 1 indicates a change in David's character. What happened to David at that time and what was the cause of this change?

(i) David caused the greatest damage in the Dominican Republic (2000 deaths, 200,000 left homeless, and over a billion dollars of property damage). Cite two factors that explain David's catastrophic effect on the Dominican Republic.

(j) At 1800 hours on September 4, the eye of Hurricane David was over Savannah, Georgia. Use the satellite imagery for that date and estimate the diameter of the storm.________________________

11.3 (a) The tracking chart (Fig. 11-3) contains numerous points denoting surface barometric pressure. Draw an isobaric map by connecting the points of equal value or interpolating between points. Use an isobaric interval of 4 mb with isobars of 1024, 1020, 1016 and so on. Remember that isolines cannot intersect.

(b) What is the name given to the pressure system that you have just drawn?

____

(c) Describe the relationship of this pressure system to the track of hurricane David.

Figure 11-3

HURRICANE DAVID

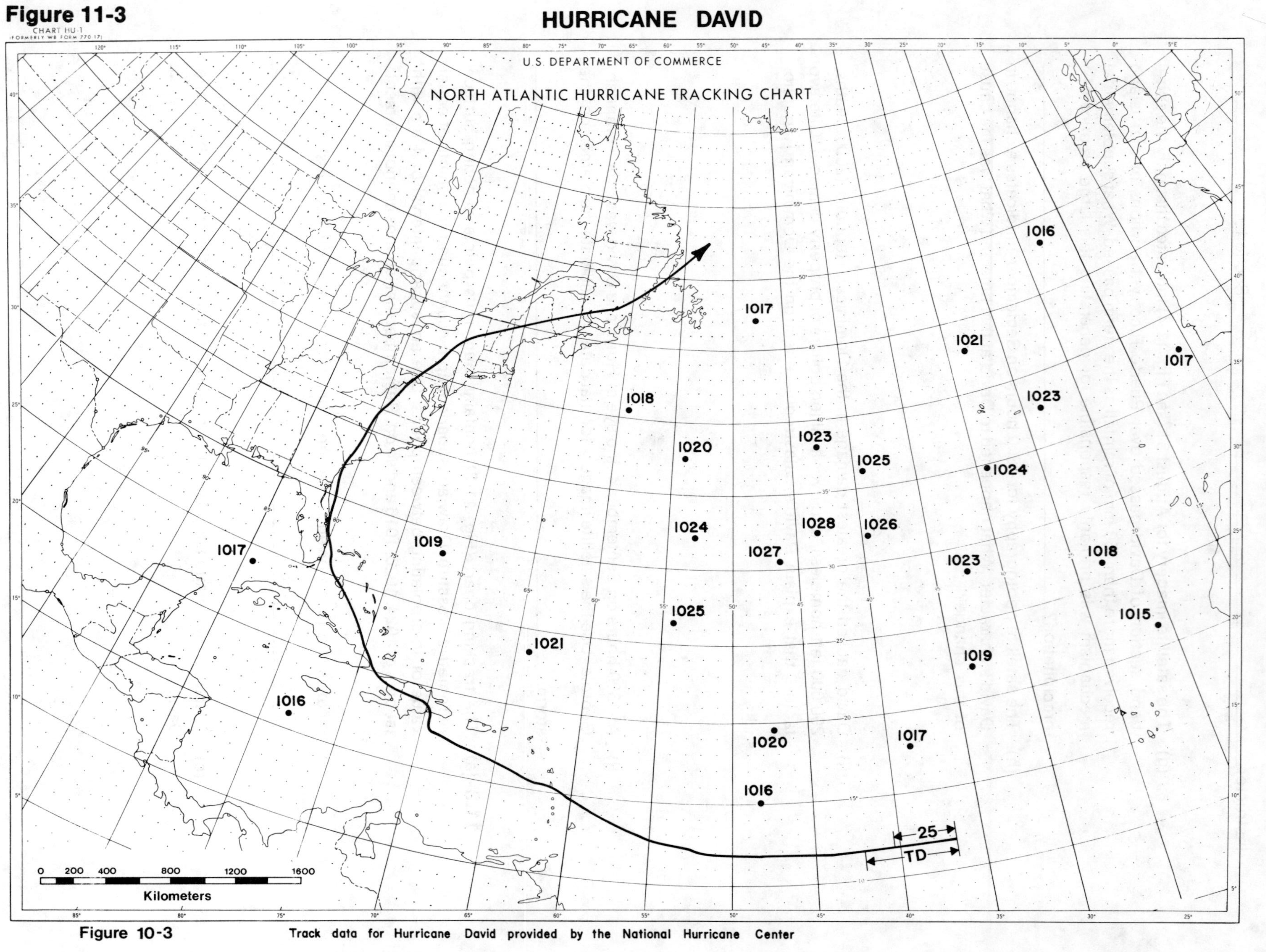

Figure 10-3 Track data for Hurricane David provided by the National Hurricane Center

TABLE 11-1

Generalized Track of Hurricane David
(August 25 to September 8, 1979)

Date	Time (GMT)	Hurricane Center Lat.	Hurricane Center Long.	Pressure (mb)	Wind (km h^{-1})	Storm Stage*
8/25	1200	11.7	36.1	1008	46	Tropical Depression (TD)
	1800	11.7	38.2	1007	46	
8/26	0000	11.7	40.3	1006	56	
	0600	11.6	42.2	1005	65	Tropical Storm (TS)
	1200	11.6	44.0	1003	74	
	1800	11.6	45.5	998	83	
8/27	0000	11.7	47.0	990	102	
	0600	11.8	48.5	980	120	Hurricane (H)
	1200	11.8	50.0	966	148	
	1800	11.9	51.5	954	176	
8/28	0000	12.2	52.9	947	213	
	0600	12.5	54.4	941	231	
	1200	12.8	55.7	938	240	
	1800	13.2	56.9	941	232	
8/29	0000	13.7	58.0	944	222	
	0600	14.2	59.2	942	222	
	1200	14.8	60.3	938	231	
	1800	15.3	61.6	933	231	
8/30	0000	15.6	62.8	929	241	
	0600	16.0	64.2	925	259	
	1200	16.3	65.2	924	268	
	1800	16.6	66.2	924	278	
8/31	0000	16.8	67.3	927	268	
	0600	17.0	68.3	928	268	
	1200	17.2	69.1	927	268	
	1800	17.9	69.7	926	278	
9/01	0000	18.8	70.4	953	240	
	0600	19.3	72.0	978	185	
	1200	19.7	73.7	1002	120	
	1800	20.6	74.6	1002	111	Tropical Storm (TS)
9/02	0000	21.3	75.2	997	120	Hurricane (H)
	0600	21.9	75.5	990	130	
	1200	23.0	76.3	984	130	
	1800	23.9	77.4	979	139	
9/03	0000	24.6	78.3	976	148	
	0600	25.3	79.1	974	148	
	1200	26.3	79.6	973	157	
	1800	27.2	80.2	972	157	
9/04	0000	28.0	80.5	971	157	
	0600	29.1	80.8	970	157	
	1200	30.2	80.9	970	157	
	1800	31.5	81.2	970	148	
9/05	0000	32.5	81.1	972	120	
	0600	33.5	80.9	976	102	Tropical Storm (TS)
	1200	34.9	80.6	980	83	
	1800	36.2	80.1	984	74	
9/06	0000	37.6	79.5	987	74	
	0600	39.2	78.5	989	74	
	1200	41.5	76.3	991	74	
	1800	43.3	73.7	992	74	Mid-latitude cyclone (M)
9/07	0000	45.0	70.0	991	83	
	0600	46.5	66.0	988	92	
	1200	47.5	61.5	987	92	
	1800	50.0	57.0	986	88	
9/08	0000	52.5	52.5	985	111	

*Storm stages are defined in accordence with the following criteria: Tropical Depression (TD) windspeed less than 63 km h^{-1}, Tropical Storm (TS) windspeed between 63-118 km h^{-1}, Hurricane (H) windspeed 119 km h^{-1} or higher.

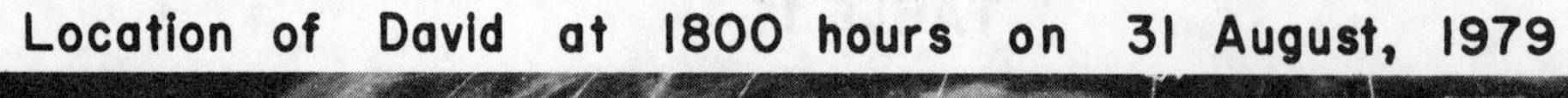
Location of David at 1800 hours on 31 August, 1979

Figure 11-4A

Location of David at 1800 hours on 1 September, 1979

Figure 11-4B

Name______________________________ Date____________________

Instructor____________________________ Section__________________

Location of David at 1900 hours on 2 September, 1979

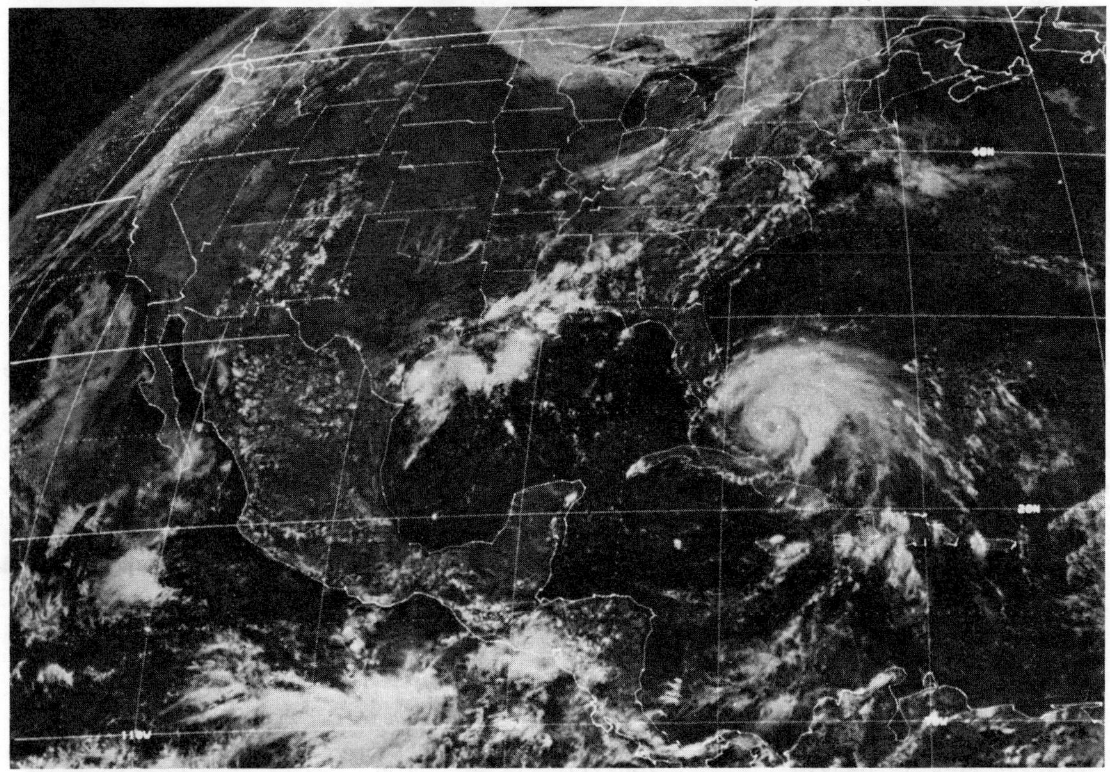

Figure 11-4C

Location of David at 1800 hours on 4 September, 1979

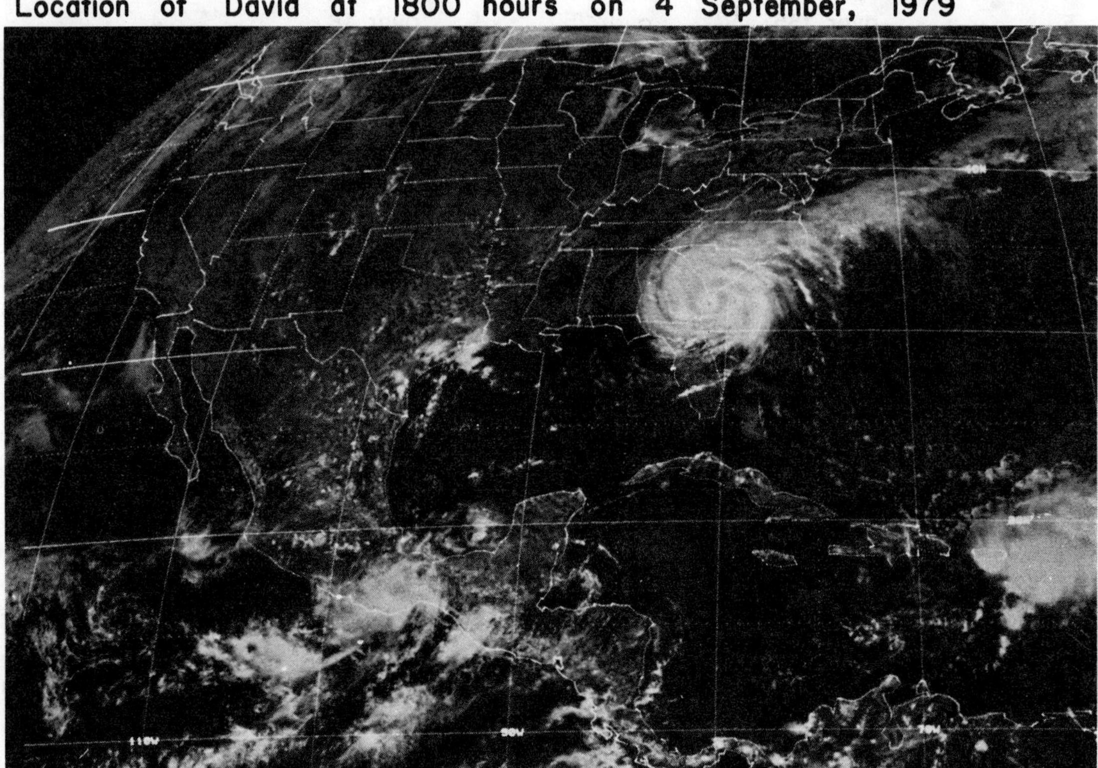

Figure 11-4D

11.4 As a hurricane approaches landfall it moves over the continental shelf and progressively shallower water. Aided by the low pressure of the storm, a tremendous amount of water builds up in front of the hurricane and as the eye approaches landfall the storm literally lifts a wall of water and propells it against the coastline. This wall of water is called the "storm surge" and although it is limited to the center or core of the storm (generally around 80 km wide), it is often the most dangerous aspect of a hurricane. Depending on the strength of the storm and the coastal topography, the storm surge may penetrate several kilometers inland. The vast majority of lives lost in hurricanes are a result of the storm surge and from flooding caused by the intense rainfall generated by the storm.

The National Hurricane Center in Florida uses the Hurricane Disaster-Potential Scale (Table 11-2) to classify hurricanes and to indicate the potential danger from each storm. The scale illustrates the relationship between windspeed and potential storm surge intensity.

TABLE 11-2

Definition of the Hurricane Disaster-Potential Scale

Category	Winds (mph)	Winds (km h^{-1})	Storm Surge (ft)	Storm Surge (m)
1	74-95	119-153	4-5	1.2-1.5
2	96-110	154-177	6-8	1.6-2.4
3	111-130	178-209	9-12	2.5-3.6
4	131-155	210-250	13-18	3.7-5.4
5	over 155	over 250	over 18	5.4

(a) On the Hurricane Disaster-Potential Scale, what category was Hurricane David when the eye reached Savannah and what was the potential storm surge intensity?

____

(b) What category was David when it reached the Dominican Republic and what size storm surge was generated by the storm?____________________

____

11.5 At 2:00 A.M. on July 30, a hypothetical category 5 hurricane (illustrated in Fig. 11-5) is situated off the Georgia coast with the edge of the hurricane 20 km away from Savannah. The hurricane is 480 km in diameter with an eye 30 km in diameter and the distance through the collar clouds is 30 km.

Name____ Date____________________

Instructor____________________________________ Section__________________

(a) Assuming that the hurricane travels at a constant rate of 20 km h^{-1} and follows the track indicated in Fig. 11-5, forecast the weather conditions for the times indicated and place your answers at the bottom of Fig. 11-5 in the space provided. Refer to Fig. 11-2 and Table 11-2 for assistance with this question.

(b) At what time did the highest intensity winds first strike Savannah? ____________

____

(c) How high was the storm surge from this hurricane and at approximately what time did it occur? ____

(d) How long did it take for the eye to pass over Savannah and what weather conditions were associated with the passage of the eye? ______________________________

____

Figure 11-5

HURRICANE WEATHER FORECASTING

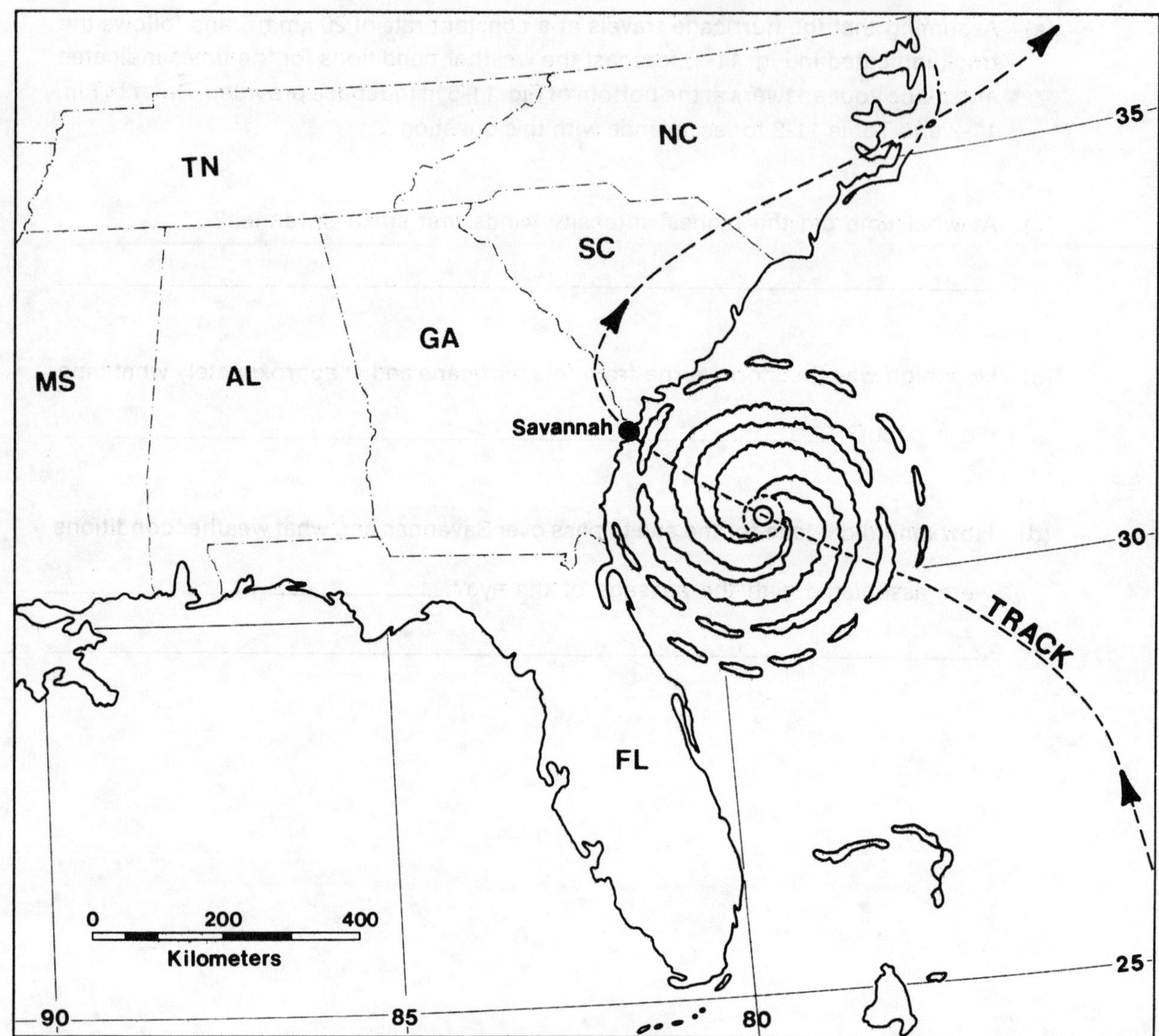

FORECAST CHART

Time	2:00 AM	8:00 AM	2:00 PM	8:00 PM
Component of hurricane over Savannah				
Cloud type				
Precipitation (in descriptive terms)				
Barometric pressure				
Wind direction				
Wind speed				

EXERCISE **12**

WEATHER MAP ANALYSIS

Since individual air masses cover extensive areas and the pressure and wind systems which cause air masses to move and interact are also large-scale features, it is necessary to view atmospheric conditions simultaneously over a wide geographic area in order to understand logically the causes for weather events taking place in any one part of the area. This is called a synoptic approach and weather maps showing the development and movement of weather systems are therefore primary tools used by the weather forecaster.

To prepare the surface weather map, weather observations from many places are collected and transcribed on a map. These weather reports plotted on the map enable the weather analyst or meteorologist to analyze air masses, wind flow patterns, fronts, cyclones and other atmospheric phenomena. From this analysis and by using supplementary maps such as maps of upper air patterns and with computer models, a forecast can be prepared.

In this exercise, a surface weather map will be studied in detail. In order to expedite data presentation and analysis, international codes for plotting weather information on surface maps have been devised. The location of each reporting station is indicated by a small circle which is called the station circle. A specific arrangement of symbols and numbers is placed around this circle, which is then called a station model. For our purposes, a simplified version of the station model will be used. Fig. 12-1 presents an annotated station model and illustrates an actual example. The appropriate weather map symbols used are outlined in Fig. 12-2. As illustrated in Fig. 12-1(a), cloud cover is represented by the amount of shading in the station circle (see Fig. 12-2). Wind direction is indicated by a line drawn in the direction **from** which the wind is coming (labelled dd). Windspeed is indicated by a series of tails (whole and half) at the end of the wind direction line. Values represented by the tails are illustrated in Fig. 12-2. Numbers located in the spaces labelled TT and T_dT_d represent the air temperature and dew-point temperature respectively. Between these numbers (labelled ww) a space is reserved for a symbol representing any special "present weather" conditions such as rain or snow (see Fig. 12-2 for the various symbols used). If no special weather is occurring, the space is left blank. Sea-level pressure is represented by a 3-digit number located in the space labelled PPP in Fig. 12-1(a). Units of pressure are millibars to the nearest tenth. Only the last 3 digits of the pressure value are given. A value of 9 or 10 must be placed in front of the 3 digits given since typical sea-level pressures are in the 960.0 to 1050.0 mb range. Actually, values seldom reach these low and high extremes. If the first digit given is greater than 5, place a 9 in front and if the first digit is less than or equal to 5, place 10 in front. For example, 956 becomes 995.6 mb and 317 becomes 1031.7 mb. In Fig. 12-1(b), an example is given with the interpreted values listed.

Figure 12-1

SIMPLIFIED WEATHER STATION MODEL

(a) Symbolic Station Model	(b) Example
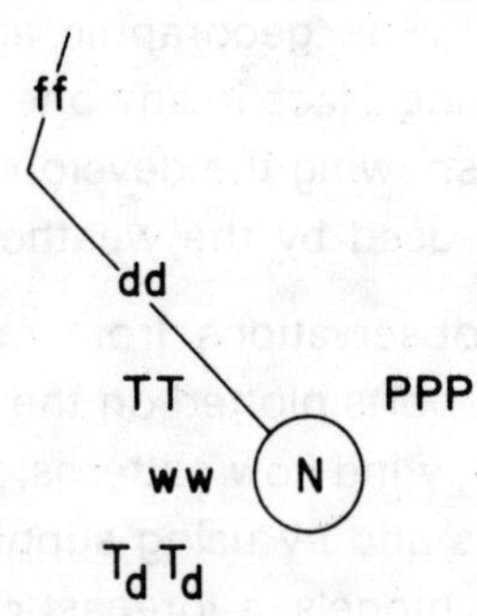	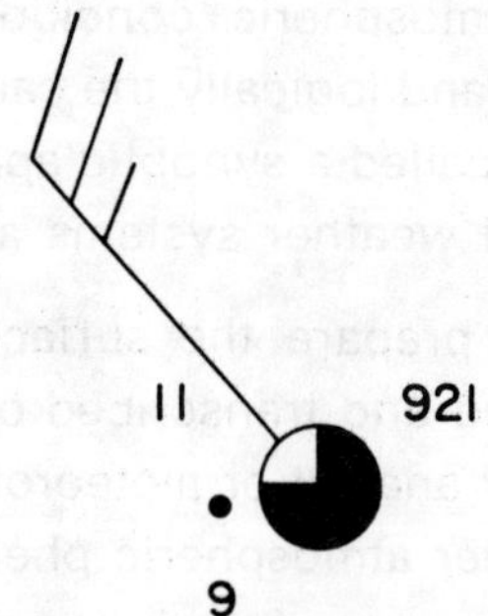
N = Sky coverage (see Fig. 11-2)	7/10 or 8/10
dd = Wind direction (from)	Wind from NW
ff = Windspeed (see Fig. 11-2)	23-27 knots
TT = Air temperature (°C)	11°C
$T_d T_d$ = Dew-point temperature (°C)	9°C
ww = Present weather (see Fig. 11-2)	Intermittent rain
PPP = Sea-level pressure (tenths of mb) (if first digit > 5, add 9 in front; if first digit ≤ 5, add 10 in front)	992.1 mb

Figure 12-2

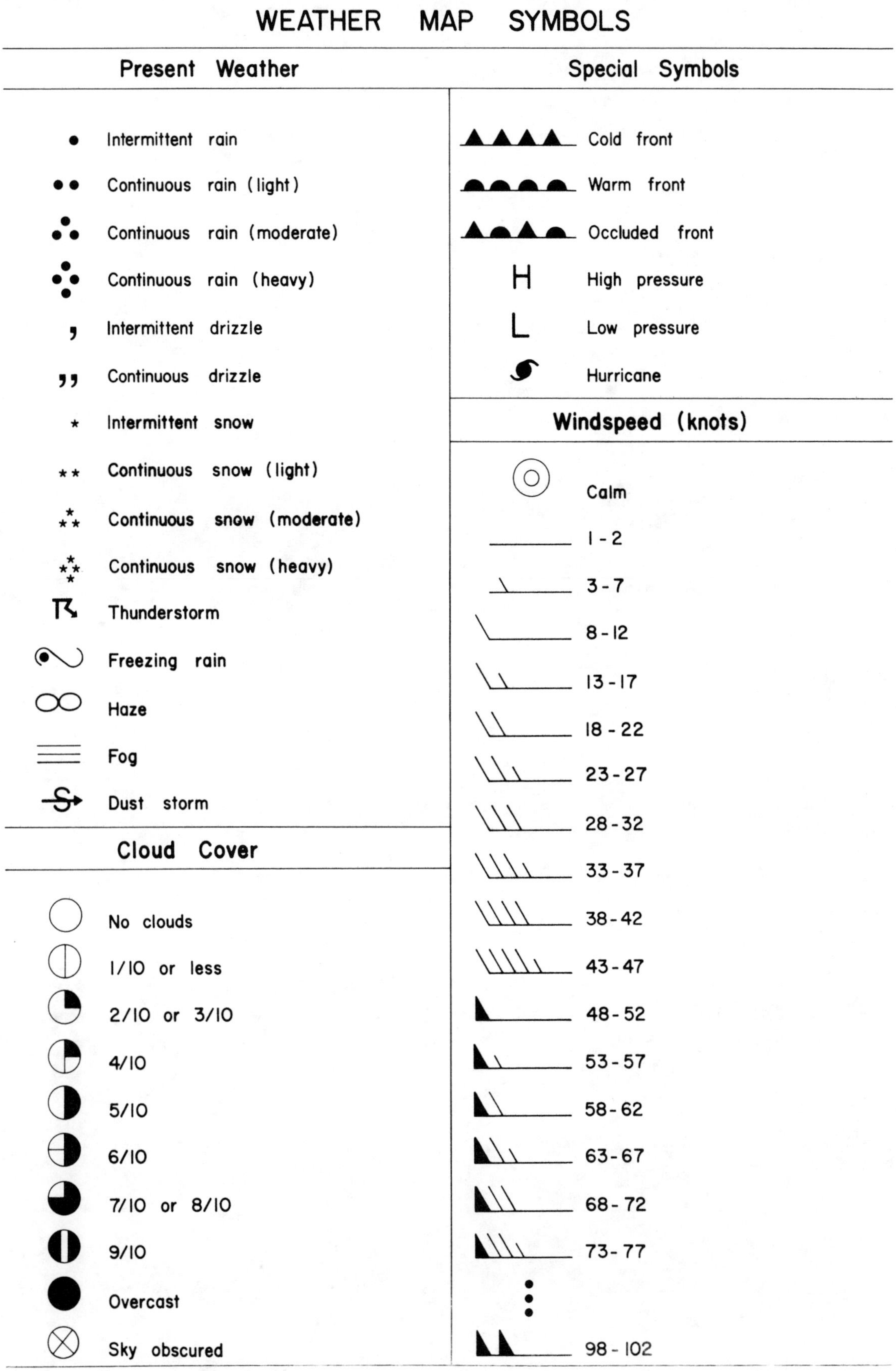

WEATHER MAP SYMBOLS

Present Weather

- Intermittent rain
- Continuous rain (light)
- Continuous rain (moderate)
- Continuous rain (heavy)
- Intermittent drizzle
- Continuous drizzle
- Intermittent snow
- Continuous snow (light)
- Continuous snow (moderate)
- Continuous snow (heavy)
- Thunderstorm
- Freezing rain
- Haze
- Fog
- Dust storm

Cloud Cover

- No clouds
- 1/10 or less
- 2/10 or 3/10
- 4/10
- 5/10
- 6/10
- 7/10 or 8/10
- 9/10
- Overcast
- Sky obscured

Special Symbols

- Cold front
- Warm front
- Occluded front
- H High pressure
- L Low pressure
- Hurricane

Windspeed (knots)

- Calm
- 1-2
- 3-7
- 8-12
- 13-17
- 18-22
- 23-27
- 28-32
- 33-37
- 38-42
- 43-47
- 48-52
- 53-57
- 58-62
- 63-67
- 68-72
- 73-77
- ⋮
- 98-102

TABLE 12-1

Weather Observations for an Autumn Day at Selected Stations

Location	Cloud Cover (tenths)	Wind Direction	Windspeed (knots)	Temperature (°C)	Dewpoint (°C)	Pressure (mb)	Present Weather
Albuquerque, NM	1	SE	5	19	6	1010.5	
Amarillo, TX	0	SE	9	13	7	1012.9	
Atlanta, GA	10	NE	14	19	18	1012.0	
Billings, MT	7	NE	2	7	4	1024.0	
Boise, ID	0	SE	2	8	−3	1019.7	
Boston, MA	0	N	8	12	8	1022.0	
Brownsville, TX	1	NW	4	22	21	1008.5	
Buffalo, NY	4	SE	8	11	8	1020.6	
Buoy (27°N, 87°W)	10	S	53	27	27	997.0	Continuous Heavy Rain
Buoy (27°N, 90°W)	10	NW	34	28	28	998.5	Continuous Moderate Rain
Calgary, Alta.	0	—	0	3	1	1027.1	
Chicago, IL	0	S	3	18	16	1017.3	
Columbus, OH	0	SW	2	16	15	1016.1	
Denver, CO	3	E	10	12	10	1015.3	
Des Moines, IA	10	SW	10	17	17	1016.3	Continuous Heavy Rain
Detroit, MI	10	S	9	11	10	1018.0	Continuous Moderate Rain
El Paso, TX	1	S	4	23	10	1009.0	
Fargo, ND	0	NW	9	8	5	1017.2	
Grand Rapids, MI	0	SW	8	16	16	1017.1	
Houston, TX	10	NE	14	22	20	1010.0	Continuous Light Rain
Huntington, WV	0	SW	3	16	14	1014.9	
Kansas City, MO	10	SW	19	17	16	1017.0	Thunderstorm
Little Rock, AR	6	NW	5	18	17	1014.3	
Los Angeles, CA	X	NE	5	21	19	1006.7	Haze
Louisville, KY	0	—	0	19	18	1015.2	
Miami, FL	10	SE	13	26	24	1009.4	
Minneapolis, MN	4	NW	13	10	8	1013.7	
Montgomery, AL	9	NE	18	23	18	1007.2	Intermittent Rain
Montreal, Que.	2	W	2	10	8	1024.2	
Nashville, TN	0	—	0	16	16	1015.0	
New Orleans, LA	10	NE	26	26	20	1006.6	Continuous Heavy Rain

New York, NY	1	E	2	15	9	1020.2	
Oklahoma City, OK	8	SW	5	18	16	1014.3	Intermittent Rain
Omaha, NE	2	NW	5	10	8	1017.1	
Phoenix, AZ	0	E	3	30	16	1004.5	
Pittsburgh, PA	8	S	6	13	12	1017.7	Intermittent Drizzle
Portland, OR	0	NW	2	14	10	1017.8	
Raleigh, NC	X	E	4	18	17	1014.5	Fog
Rapid City, SD	3	N	4	8	7	1020.4	
Salt Lake City, UT	0	SE	7	8	3	1016.5	
San Francisco, CA	0	—	0	19	9	1010.0	
Sault Ste. Marie, MI	10	E	8	12	11	1017.8	Continuous Drizzle
Savannah, GA	10	E	15	24	23	1011.0	Continuous Light Rain
St. Louis, MO	3	SW	3	18	16	1016.4	
Tampa, FL	10	SE	18	26	24	1007.5	Intermittent Rain
Thunder Bay, Ont.	10	SE	8	10	10	1013.0	Continuous Moderate Rain
Washington, DC	5	SE	9	14	11	1016.9	
Wausau, WI	3	SW	9	15	14	1014.0	
Wichita, KS	0	W	8	12	10	1016.5	
Winnipeg, Man.	2	N	7	7	6	1017.3	

Note: X in the cloud cover column means "sky obscured."

Name________________________________ Date________________

Instructor______________________________ Section______________

QUESTIONS

12.1 Table 12-1 presents weather observations for an autumn day at 2200 h GMT (1700 h EST) at 50 selected locations in North America. These locations correspond to the station circles on the weather map given in Appendix E. Also given on the map is the isobaric pattern with high and low pressure areas indicated. Using the symbols shown in Fig. 12-2, plot the data given in Table 12-1 onto the weather map (Appendix E) by completing the station models for each of the 50 sites. The use of an atlas or place name map of North America will assist in locating the various cities.

12.2 What is the weather system located at about 25°N latitude and 87°W longitude? ____

____

12.3 The low pressure area located at about 48°N latitude and 91°W longitude is the center of a mid-latitude cyclone. By carefully considering the weather information you have plotted, draw the fronts on the map that are associated with this weather system. Use the appropriate front symbols illustrated in Fig. 12-2.

12.4 The low pressure area located in the southwestern corner of the map is **not** a mid-latitude cyclone. What causes a low pressure area to exist in this location at this time?

12.5 Consider windspeed values for locations near the high pressure areas around Montana and New England. Are they generally higher or lower than those for locations near the mid-latitude cyclone? ________________________________

12.6 Considering the expected flow around a high pressure area, what wind direction **should** exist at Portland? ________________________________
However, Portland's recorded wind direction is from the NW. What local event could cause this? ________________________________

12.7 Detroit is experiencing continuous moderate rainfall while Pittsburgh is experiencing only intermittent drizzle. Why is Detroit's rainfall heavier? ________________
In terms of precipitation, what can Pittsburgh expect in the next few hours? ________
________________________________ . What type of precipitation can Detroit expect in the next few hours? ________________________________

12.8 Chicago is presently experiencing pleasant clear sky conditions. What should the weather forecast indicate about Chicago's weather later in the evening? __________

____

12.9 The cloud cover report for Los Angeles and Raleigh is "sky obscured." What is causing this at Los Angeles? ____

at Raleigh? ____

12.10 What weather forecast should be issued for the coastal regions of Louisiana, Mississippi, Alabama and northwest Florida? ____

____

EXERCISE 13

CLIMATE DATA AND CLIMATIC REGIMES

A statement of the climate of a place is in essence a composite or generalized statement of the variety of day-to-day weather conditions which occur at that place. Climate is usually expressed in terms of the magnitude of mean values for selected weather elements and the mean frequency of occurrence of selected weather phenomena, and variations which occur about those mean values. Because atmospheric conditions tend to vary in well-defined annual and daily cycles, it is customary to provide more than just the annual means when describing the climates of places. Such supplemental data may include descriptions of the climates for individual seasons or months, for individual days of the year, or for other selected time periods.

When considering climate, usually an emphasis is placed upon temperature and precipitation. Although mean wind direction, percent of possible sunshine and other climatological parameters are important for certain specific applications, temperature and precipitation are the two most important for both natural and man influenced concerns. In this context, the following terms are often used:

average daily temperature — the average of the day's maximum and minimum temperatures.

average monthly temperature — the average of the average daily temperatures in the month.

annual temperature — the average of the 12 average monthly temperatures.

diurnal temperatures range — the difference between the day's maximum and minimum temperatures.

annual temperature range — the difference between the warmest average monthly temperature and the coolest average monthly temperature in the year. (Recall that the annual temperature range is effected by latitude by generally increasing with increasing latitude and is effected by continentality being greater for locations inland compared to coastal or maritime sites.)

monthly precipitation — the total of precipitation falling during the month.

annual precipitation — the total of precipitation falling during the entire year.

Note that precipitation values include rain plus melted snow (i.e. about 10 cm of snow would be reported as 1 cm of precipitation).

When generalizing the temperature and precipitation data to describe the climate for a location, longterm (30 or more years) means of the monthly and annual temperature and monthly and annual precipitation values are used. These are called "normal" temperature and precipitation values.

One form of presenting climatic data for a location is by means of a table where normal values of average monthly temperature (T) and monthly precipitation (P) are presented as follows:

Washington, D. C. 39°N, 77°W; 20 m(elevation)

T(°C)	3	3	7	13	19	23	26	25	21	15	9	3	14
P(mm)	77	63	82	80	105	82	105	124	97	78	72	71	1036

The first 12 numbers represent the normal average monthly temperature and precipitation values beginning with January. The 13th number is the normal annual temperature value and precipitation total. The top line is always temperature and the bottom line is precipitation. Thus, the terms T(°C) and P(mm) at the beginning are usually omitted. Usually the station's latitude, longitude and elevation are given as in this example.

The monthly values can be plotted on a diagram called a climograph as illustrated in Fig. 13-1. Temperature values are joined by a smooth curve while monthly precipitation totals are expressed using a bar graph.

Figure 13-1

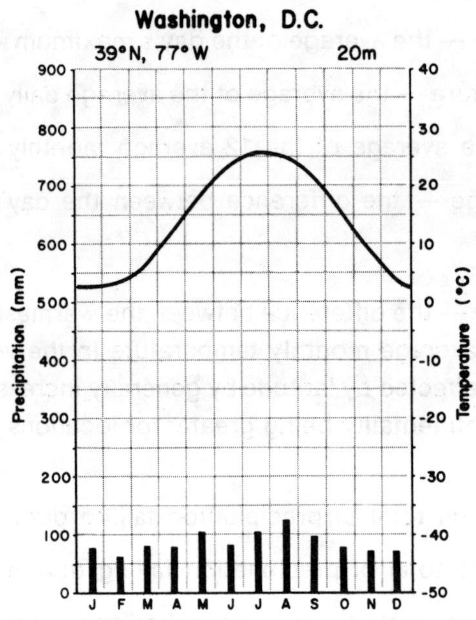

Name________________________________ Date________________

Instructor____________________________ Section______________

QUESTIONS

13.1 Listed below are the maximum (T_{max}) and minimum (T_{min}) temperatures that were recorded for the days during May 1976 at Toronto International Airport (from *Monthly Record,* Environment Canada).

Date (May 1976)	T_{max} (°C)	T_{min} (°C)	T_{avg} (°C)	Date (May 1976)	T_{max} (°C)	T_{min} (°C)	T_{avg} (°C)
1	14	5		17	19	6	
2	18	6		18	10	2	
3	8	1		19	11	1	
4	11	1		20	21	7	
5	22	6		21	16	4	
6	13	3		22	14	3	
7	4	0		23	15	2	
8	12	1		24	14	3	
9	18	5		25	14	7	
10	23	3		26	19	6	
11	12	4		27	23	5	
12	13	2		28	23	8	
13	17	3		29	22	12	
14	21	5		30	19	12	
15	20	7		31	18	13	
16	21	12					

(a) Calculate the **average monthly** temperature for May 1976 at Toronto.

____

(b) Often, temperatures are compared to the "normal" for the date or month. The normal is usually the mean value for a recent 30 year period. The normal average monthly temperature for May at Toronto is 12° C. Calculate the amount by which the value for May 1976 is above or below normal.

____

(c) On May 5, the temperature reached a maximum of 22° C, yet by May 7, temperatures were much cooler with a daytime maximum of only 4° C. What weather event probably caused this dramatic temperature drop for Toronto between May 5 and May 7?

Name________________________________ Date________________

Instructor____________________________ Section______________

13.2 Tabulated below are the average monthly temperatures and monthly precipitation totals for January and July measured at Tulsa, Oklahoma for the period 1941-1970. (Adapted from *Local Climatological Data 1979,* National Climatic Center, Asheville, NC.)

Year	Average Temperature (°C)		Precipitation Total (mm)	
	January	July	January	July
1941	4.7	27.8	41	44
1942	2.6	26.9	5	61
1943	2.3	29.2	0	7
1944	3.7	27.2	37	48
1945	3.4	25.9	16	67
1946	3.9	28.6	87	4
1947	3.5	26.0	22	20
1948	-0.4	26.8	10	92
1949	0.0	27.5	169	57
1950	3.7	24.4	47	229
1951	3.7	27.6	53	54
1952	6.0	28.0	23	46
1953	5.1	26.9	51	132
1954	2.8	32.3	24	1
1955	4.3	29.8	20	112
1956	2.0	29.6	25	35
1957	1.1	29.9	41	13
1958	3.7	27.4	45	85
1959	0.9	25.7	20	250
1960	2.8	26.2	31	229
1961	1.3	25.9	17	276
1962	-1.0	27.5	34	123
1963	-1.9	29.4	25	269
1964	4.9	29.4	16	46
1965	4.3	28.4	40	86
1966	0.4	30.7	18	51
1967	4.2	25.3	38	175
1968	2.2	27.0	83	35
1969	2.7	29.9	41	27
1970	-1.3	28.2	10	3

(a) Calculate the 30-year "normal" values of:

January average temperature (°C) ______________________

July average temperature (°C) ______________________

January precipitation total (mm) ______________________

July precipitation total (mm) ______________________

(b) For 1979, the data tabulated below were collected at Tulsa. Complete the table by calculating the departures (amount above or below) from the normals.

	Departure from normal
January average temperature: -4.9°C	________
July average temperature: 28.6°C	________
January precipitation total: 53 mm	________
July precipitation total: 68 mm	________

13.3 Tabulated below are the normals for average monthly and annual temperature values for seven locations. Plot these temperature data on the graphs given in Fig. 13-2.

Normal Temperatures (°C)

Site	Monthly Values												Annual
1	26	26	27	27	27	27	27	27	27	27	27	27	27
2	4	4	5	7	10	13	15	14	12	9	6	4	8
3	-2	1	5	12	18	24	27	26	21	15	6	1	13
4	-24	-18	-13	-3	8	15	16	13	6	-3	-16	-22	-3
5	-4	-9	-16	-23	-24	-24	-26	-26	-24	-19	-10	-4	-17
6	13	15	19	21	26	31	33	34	31	22	18	14	23
7	16	17	15	12	9	6	5	7	9	11	14	16	11

(a) Which of these sites are in the southern hemisphere?

(b) Sites 3 and 7 are at similar latitudinal distances from the equator. Which is the most continental? ______________________

Name____________________________ Date______________

Instructor_________________________ Section____________

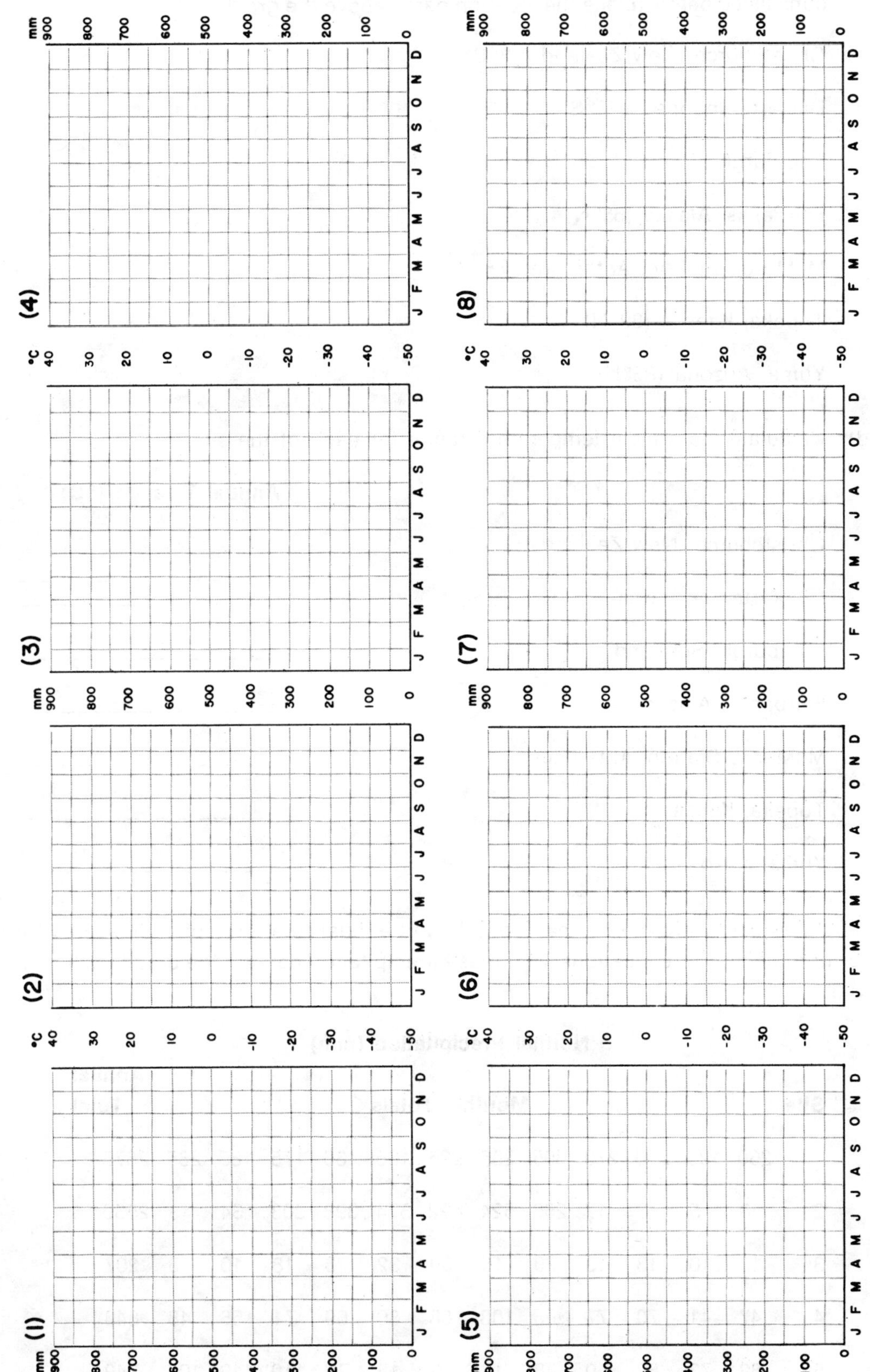

Figure 13-2

(c) Using your knowledge of the effects of continentality and latitude on annual temperature ranges, match the seven temperature graphs drawn in Fig. 13-2 with the locations listed below (place the location name above the graph).

Christchurch, New Zealand (43°S)

Djakarta, Indonesia (6°S)

Edinburgh, Scotland (56°N)

Fairbanks, Alaska (65°N)

McMurdo Station, Antarctica (78°S)

Topeka, Kansas (39°N)

Yuma, Arizona (33°N)

(d) Calculate the annual temperature range for each of the sites:

	Annual T range (°C)
Christchurch, New Zealand	________________
Djarkarta, Indonesia	________________
Edinburgh, Scotland	________________
Fairbanks, Alaska	________________
McMurdo Station, Antarctica	________________
Topeka, Kansas	________________
Yuma, Arizona	________________

13.4 Tabulated below are the normal monthly and annual total precipitation values for six northern hemisphere locations. Plot these precipitation data on the graphs given in Fig. 13-3.

Normal Precipitation (mm)

Site						**Monthly Totals**							**Annual Total**
1	260	195	250	270	305	235	225	185	130	175	185	265	2675
2	8	5	6	17	260	524	492	574	398	208	34	3	2530
3	10	10	13	13	20	15	30	32	23	18	10	13	207
4	47	41	70	77	95	103	86	80	69	71	56	48	843
5	83	73	52	50	48	18	9	18	70	110	113	105	749
6	133	96	83	69	68	56	62	80	87	104	138	150	1126

Name________________________ Date______________

Instructor____________________ Section____________

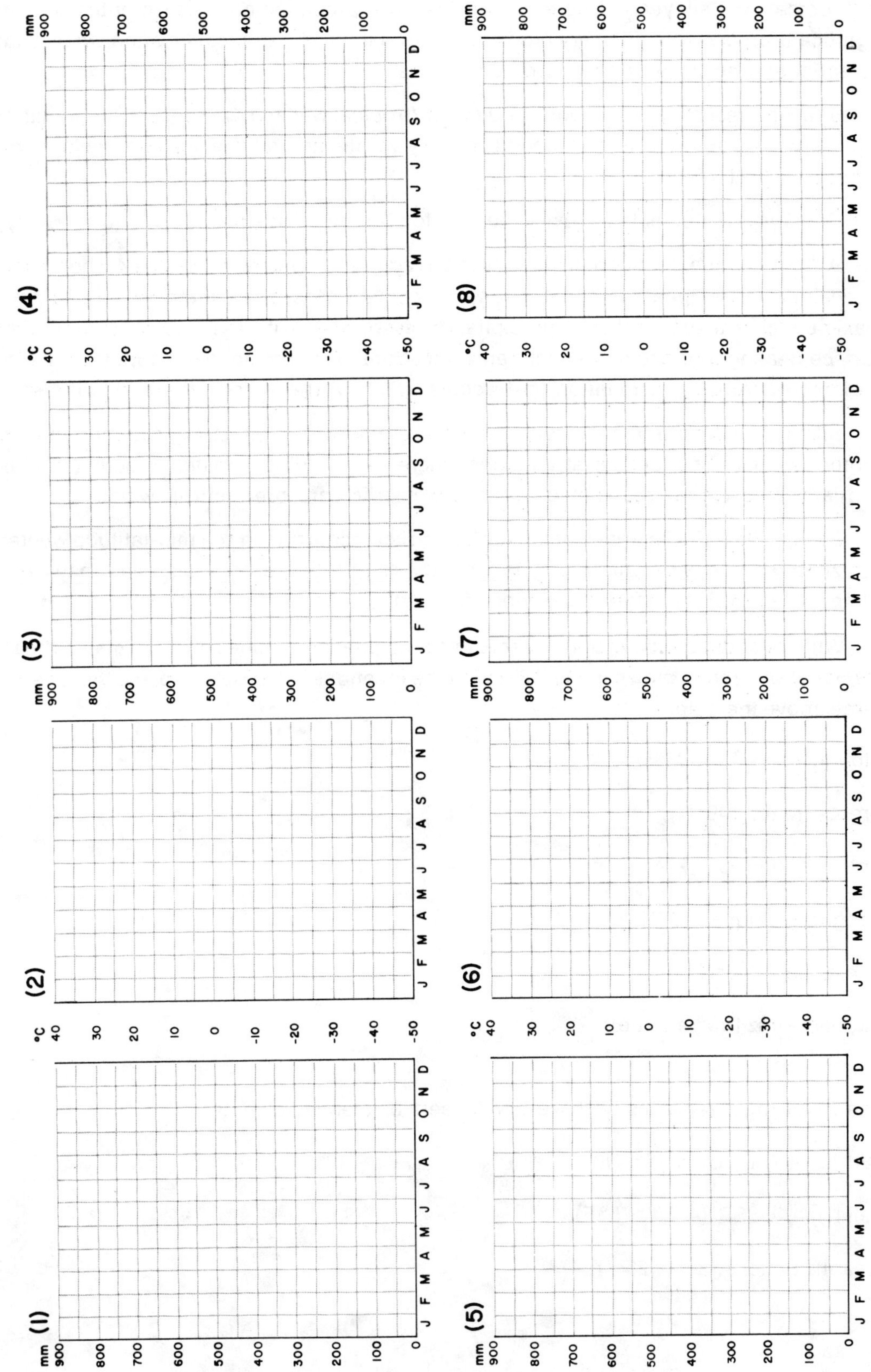

Figure 13-3

Each of these 6 sites can be used to represent a specific precipitation regime. Site 1 is "tropical wet all year." There is considerable precipitation in all months and no significant dry season. The very high annual total implies a tropical location where high potential and actual evapotranspiration occurs.

Site 2 represents a "monsoon" or distinct "tropical wet and dry season" precipitation regime where the dry season is indeed very dry yet the annual total may be similar to the tropical wet all year.

Site 3 is a dry climate all year ("desert" for very dry; "steppe" for not quite as dry).

A "mid-latitude continental" precipitation regime is represented by site 4 where there is much more precipitation than that for a dry climate. Most often, there will be a summer maximum due to convectional precipitation associated with the intense summertime surface heating that occurs in continental locations. This precipitation supplements the cyclonic or frontal precipitation that occurs much of the year in the mid-latitudes.

Site 5 is a "Mediterranean" precipitation regime with distinct dry summer and wet winter seasons. This occurs at latitudes where the STHP dominates in summer while mid-latitude cyclones associated with the westerlies are prevalent in winter.

Lastly, site 6 is a "marine" precipitation regime occurring in the mid-latitude westerlies zone with considerable precipitation year round. Sometimes the peak may be in winter when cyclonic storms are more frequent.

With these descriptions of the respective precipitation regimes in mind, match the six precipitation graphs drawn in Fig. 13-3 with the locations listed below (place the location name above the graph).

Albuquerque, New Mexico (35°N)

Brest, France (48°N)

Chicago, Illinois (42°N)

Rangoon, Burma (17°N)

Rome, Italy (42°N)

Uaupes, Brazil (0° latitude)

13.5 The following three sites all have dry (desert or steppe) climates:

Salt Lake City, Utah (41°N, 112°W)

Balkhash, USSR (47°N, 75°E)

Riyadh, Saudi Arabia (25°N, 47°E)

Name____ Date____________________

Instructor____________________________________ Section__________________

There are three major controls causing dry climates. Each is represented by one of these sites. For each site, name the control causing the dry climate and briefly describe how this control causes the climate to be dry.

Salt Lake City:

Balkhash:

Riyadh:

13.6 San Francisco and Sacramento, California are at about the same latitude (38° N) with San Francisco on the Pacific Coast and Sacramento to the east in an interior valley. Given below are the normal monthly and annual temperature and precipitation values for these two cities. Note that you are not told which is which.

8	10	12	16	19	22	25	24	23	18	12	9	17
81	76	60	36	15	3	0	1	5	20	37	82	414
9	10	12	13	15	16	17	17	18	16	13	10	14
102	88	68	33	12	3	0	1	5	19	40	104	475

Plot these data as climographs (similar to Fig. 13-1) on Fig. 13-4 and select which climograph represents San Francisco and which represents Sacramento (put the appropriate city name above each climograph).

Briefly explain how you decided which data set represented San Francisco and which represented Sacramento.

Figure 13-4

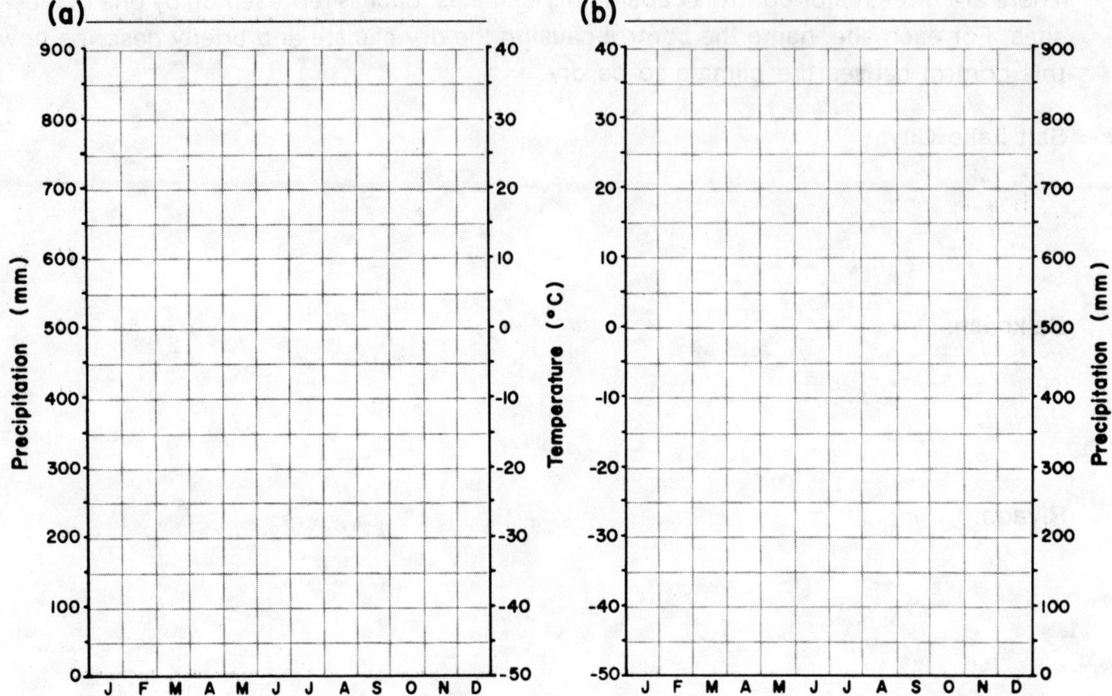

(a)
(b)
Precipitation (mm)
Temperature (°C)
Precipitation (mm)
900
800
700
600
500
400
300
200
100
0
40
30
20
10
0
-10
-20
-30
-40
-50
J F M A M J J A S O N D

EXERCISE 14

CLIMATE CLASSIFICATION

No two places have climates exactly alike, but areas can be usefully defined within which the climates of places are meaningfully similar. Such areas are called climatic regions. A grouping of climatic regions which have characteristics in common constitutes a climatic type. A system which defines and relates different climate types and subtypes is referred to as a climate classification. There are a number of different climate classification systems, developed to meet different needs, and using different criteria for establishing climatic types and subtypes. One of the most commonly used climatic classifications is that first proposed by W. Köppen early this century and subsequently revised by others. There are many useful classifications that have been developed; this one will be used here simply because of its widespread use.

Köppen's classification uses a series of letters to designate a climate. First, there are five basic climate types:

A — tropical humid climates
B — dry climates
C — warm temperate or subtropical climates
D — cool temperate climates
E — polar climates.

Second and third letters are added to the basic type to create subtypes. These second and third letters constitute precipitation and temperature modifications. The criteria for determining the basic climate type and the subtypes associated with the second and third letter are summarized in Table 14-1. Average monthly and annual temperature and precipitation normals are needed in order to proceed with classifying the climate of a location. These will normally be given to you in the tabular form discussed in exercise 13.

To use Koppen's classification, one must determine the first letter by following the order E, B, A, C and then D (i.e. check to see if it is an E climate first; if not, then is it B; and so on). In order to assist with the classification procedure, a flow chart is provided in Fig. 14-1. Also, Fig. 14-2 is provided for determining the second letter for an A climate. The annual precipitation total and the monthly total for the driest month are plotted on the graph to determine Am versus Aw or As. Note that if the driest month exceeds 60 mm, it is automatically Af even if one season seems distinctly drier than another.

The concept of effective precipitation introduced in Exercise 8 is incorporated into Köppen's classification. When determining whether a climate is a dry (B) climate, the precipitation totals are considered in relation to season and temperature. First one needs to determine whether most of the precipitation occurs in summer (high sun season), winter (low sun season) or is evenly distributed through the year. For this purpose, the year is divided into two 6 month periods or seasons (April to September and October to March). If 70% or more of the annual total falls in one of these 6 month periods, then that season is said to have

TABLE 14-1

KÖPPEN SYSTEM OF CLIMATE CLASSIFICATION

Letter Symbol			Description and Criteria
1st	2nd	3rd	
A			**Tropical wet** (Monthly temperatures all greater than or equal to 18°C)
	f		Wet all year (Every month receives 60 mm or more of precipitation)
	m		Monsoon (Short dry season — see Fig. 14-2 for A climate second letter determination)
	w		Dry winter (see Fig. 14-2)
	s		Dry summer (see Fig. 14-2; note that the As climate is rare)
B			**Dry** (Potential evapotranspiration exceeds precipitation — see Fig. 14-3 for B climate versus A, C or D determination)
	W		Desert (Driest version — refer to Fig. 14-3 for second letter determination)
	S		Steppe (Refer to Fig. 14-3)
		h	Tropical or hot (Annual temperature is 18°C or greater)
		k	Mid-latitude or cool (Annual temperature is less than 18°C)
C			**Warm temperate** (Average temperature of coldest month is at least −3°C but less than 18°C; warmest month at least 10°C)
	w		Dry winter (Driest winter month less than or equal to 1/10 of precipitation in wettest summer month)
	s		Dry summer (Driest summer month less than or equal to 1/3 of precipitation in wettest winter month)
	f		Wet all year (Criteria for w or s not met or all months receive at least 30 mm)
		a	Hot summer (Warmest month greater than or equal to 22°C)
		b	Warm long summer (At least 4 months have temperatures of 10°C or above but under 22°C)
		c	Short summer (1 to 3 months have temperatures of 10°C or above but under 22°C)
D			**Cool temperate** (Coldest month less than −3°C; warmest month at least 10°C)
	w		same as for C climate
	s		same as for C climate
	f		same as for C climate
		a	same as for C climate
		b	same as for C climate
		c	same as for C climate
		d	Severe winter (Coldest month less than −38°C)
E			**Polar** (Monthly temperatures all less than 10°C)
	T		Tundra with growing season (Warmest month at least 0°C)
	F		Ice with no growing season (All months less than 0°C)

the precipitation concentration. In Fig. 14-3, three diagrams are given — one for summer precipitation concentration, one for winter concentration and one for precipitation evenly distributed. Since winter precipitation is more effective (less evaporates immediately), the graph is different than the one for summer where much of the rainfall immediately evaporates and is not available to plant growth. Using the appropriate one of these three diagrams, the annual total precipitation and annual temperature values are plotted. If they meet on the right side of the solid boundary line, the climate is not dry and must be an A, C or D climate. If they meet on the left side, the climate is B. The diagram also illustrates whether the climate is BS (steppe) or BW (desert), the latter being the drier. To illustrate how the concept of effective precipitation is incorporated, consider the following two examples.

example 1: Locations 1 and 2 both have annual temperature normals of 10° C and annual precipitation totals of 400 mm. However, location 1 receives a summer concentration of precipitation while location 2 receives a winter concentration. Plotting this information on Fig. 14-3 and using the summer concentration diagram, location 1 is a B climate (specifically BS) whereas, using the winter concentration diagram, location 2 is not a B climate. This is appropriate, since the precipitation for location 2 is more effective by occurring in winter.

example 2: Locations 3 and 4 both receive annual precipitation totals of 500 mm distributed evenly through the year. The normal annual temperature for location 3 is 8° C while the value for location 4 is 19° C. From Fig.14-3 using the evenly distributed diagram, location 4 is found to be a B climate (specifically BS) whereas location 3 is not. This is appropriate since location 3 is much cooler and therefore its 500 mm of precipitation is more effective.

Figure 14-2

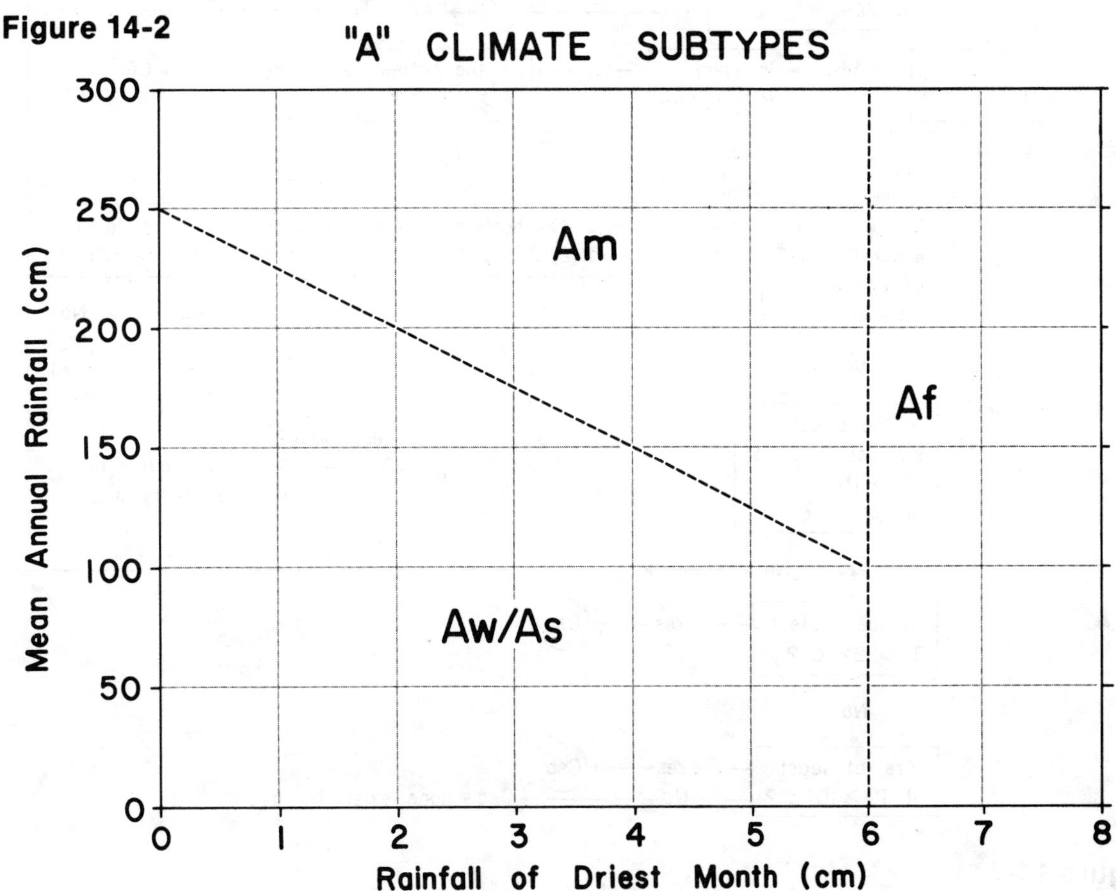

FLOW CHART FOR KÖPPEN'S CLIMATE CLASSIFICATION

Questions are given in rectangular boxes. Final climate subtypes are circled. T and P refer to normal monthly values of temperature and precipitation.

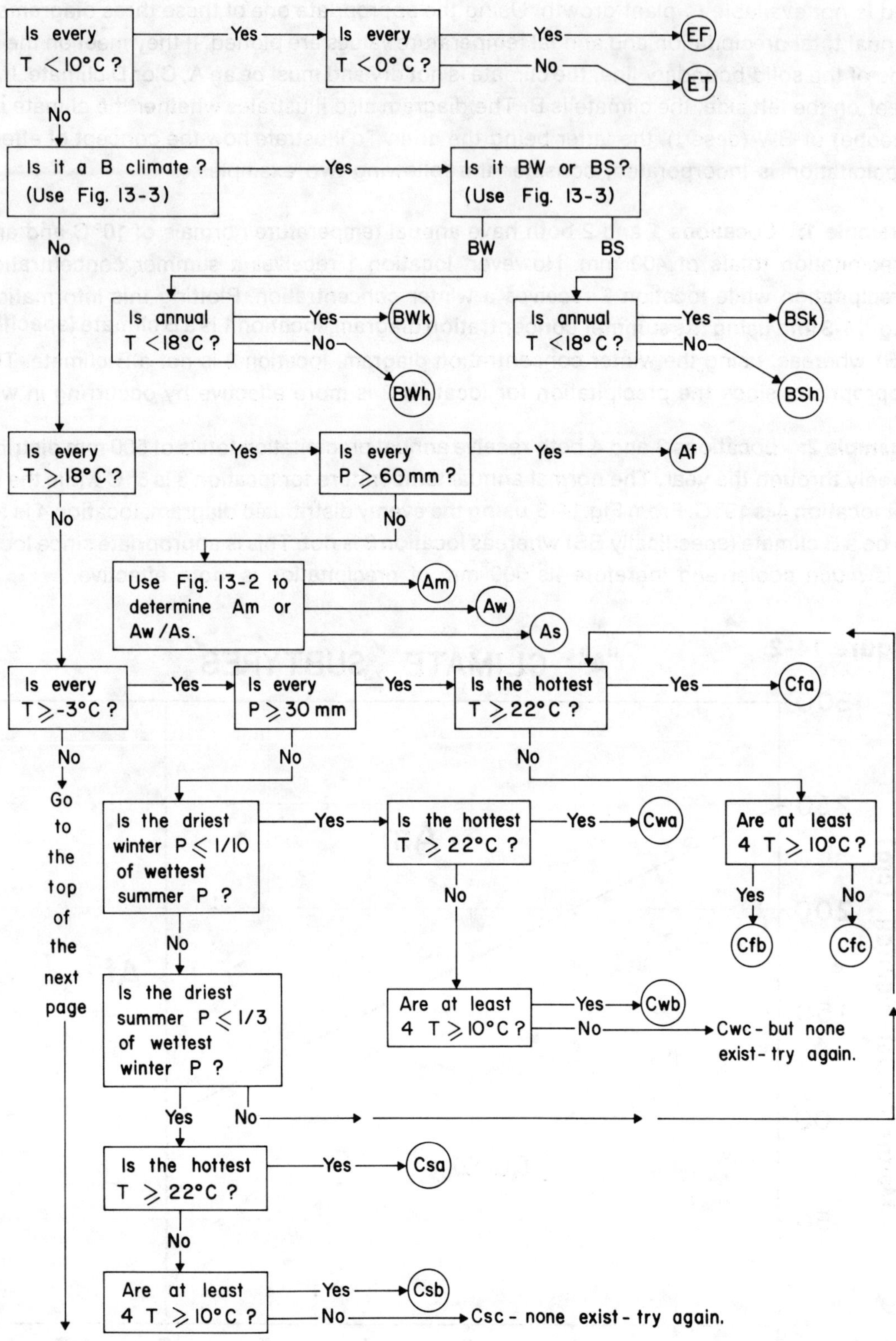

Figure 14-1

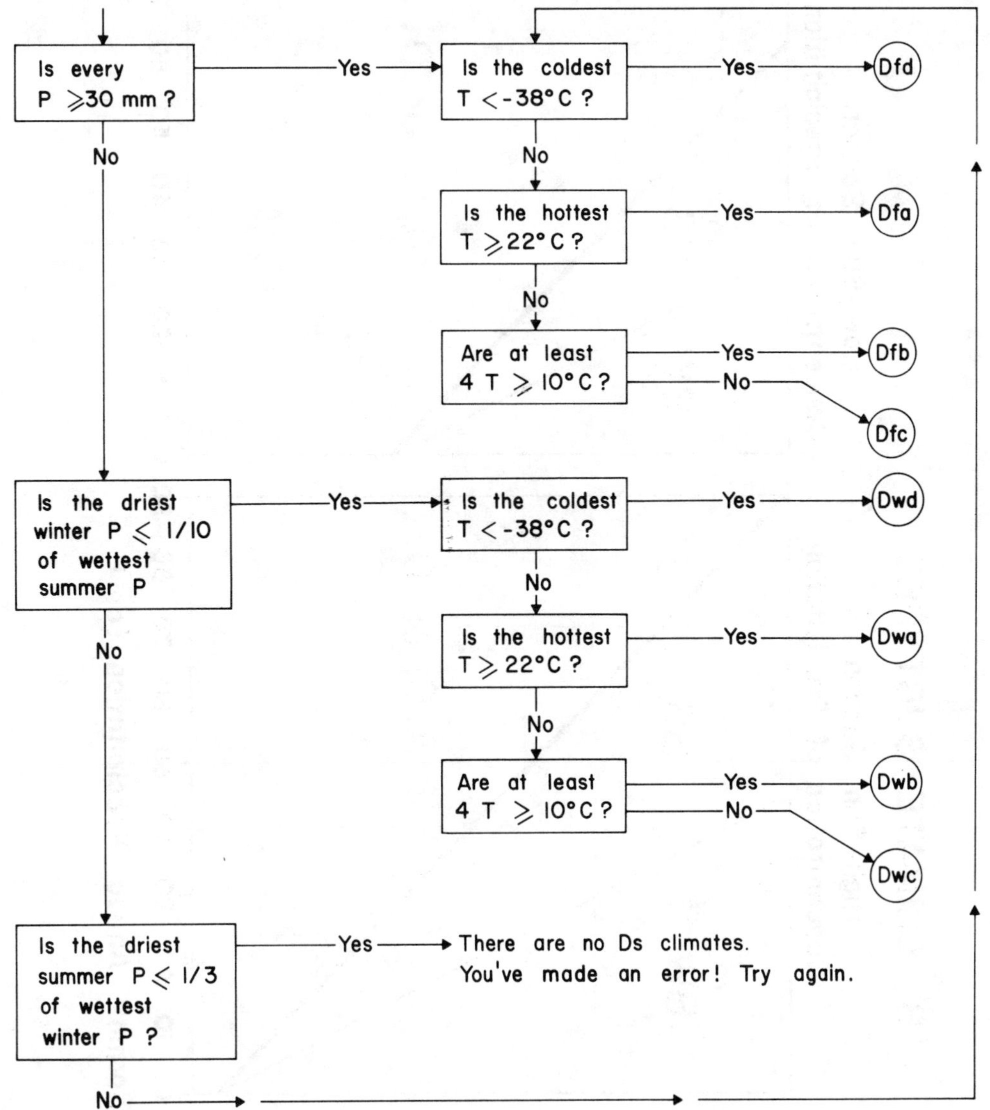

The concept of effective precipitation is also incorporated into Köppen's classification when determining wet and dry seasons for C and D climates (refer to Table 14-1). A summer is considered dry if any summer month receives 1/3 or less of the precipitation occurring in the wettest winter month. However, for the winter to be considered dry, it must indeed be very dry with the one driest month receiving 1/10 or less of the precipitation of the wettest summer month. Again this is appropriate since the little that does fall in winter would be reasonably effective due to the cooler temperatures.

Summarizing Köppen's system, the possible climate subtypes are: Af, Am, Aw, As, BWh, BWk, BSh, BSk, Cwa, Cwb, Cwc, Csa, Csb, Csc, Cfa, Cfb, Cfc, Dwa, Dwb, Dwc, Dwd, Dsa, Dsb, Dsc, Dsd, Dfa, Dfb, Dfc, Dfd, ET and EF. Several of these do not exist in the world (Cwc, Csc and all Ds subtypes). Others are grouped together under one name (e.g. Csa and Csb are both called Mediterranean climates). The climate subtypes, their characteristics and distribution will be examined in detail in Exercise 15.

Figure 14-3

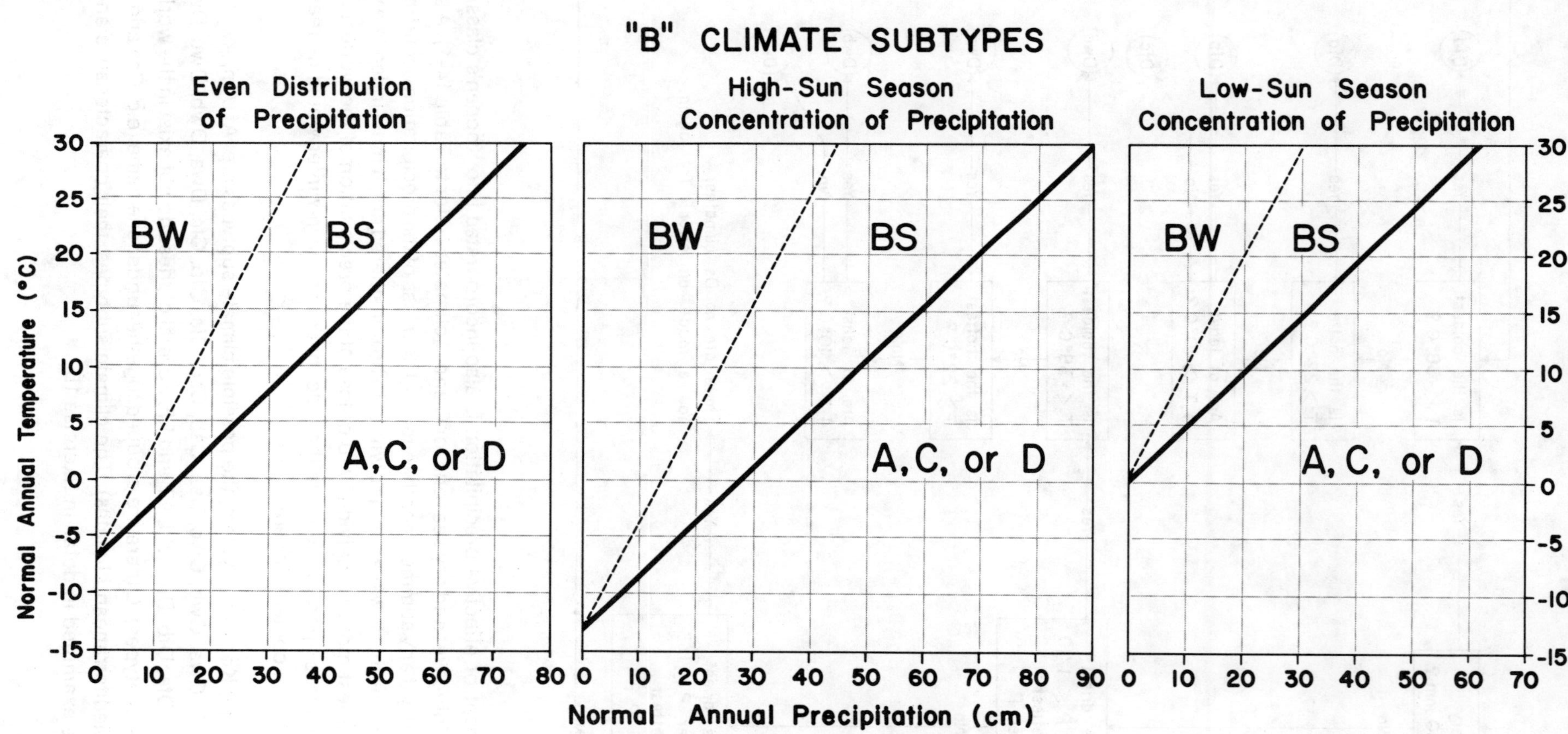
"B" CLIMATE SUBTYPES
Even Distribution of Precipitation
High-Sun Season Concentration of Precipitation
Low-Sun Season Concentration of Precipitation
BW
BS
A,C, or D
Normal Annual Temperature (°C)
Normal Annual Precipitation (cm)
30
25
20
15
10
5
0
-5
-10
-15
0
10
20
30
40
50
60
70
80
90

Name______________________ Date______________

Instructor______________________ Section______________

QUESTIONS

14.1 The climatic data for 20 locations are presented in tabular form. Classify the climate for each of these sites according to the Köppen climate classification.

______________ Albuquerque, New Mexico, USA 35°N, 107°W; 1593 m

2	4	8	13	18	24	26	25	21	15	7	3	14
10	10	13	13	20	15	30	32	23	18	10	13	207

______________ Alice Springs, Australia 24°S, 134°E; 546 m

29	28	25	20	15	12	12	14	18	23	26	28	21
43	33	28	10	15	13	8	8	8	18	30	38	252

______________ Allahabad, India 25°N, 82°E; 98 m

16	18	25	31	34	34	30	29	29	26	20	16	26
23	15	15	5	15	127	320	254	213	59	8	8	1062

______________ Benghazi, Libya 32°N, 20°E; 25 m

13	14	17	19	22	24	26	26	26	24	19	15	20
66	41	20	5	3	0	0	0	3	18	46	66	268

______________ Buenos Aires, Argentina 35°S, 58°W; 27 m

24	23	21	17	14	11	10	12	14	16	20	22	17
104	82	122	90	79	68	61	68	80	100	90	83	1027

______________ Bulawayo, Zimbabwe 20°S, 28°E; 1344 m

21	21	20	19	16	13	14	16	19	22	22	21	19
134	111	65	21	9	3	0	1	5	25	89	125	588

________________ Eismitte, Greenland 71°N, 41°W; 2953 m

−42 −47 −40 −32 −24 −17 −12 −11 −11 −36 −43 −38 −29

15 5 8 5 3 3 3 10 8 13 13 25 111

________________ Fairbanks, Alaska, USA 65°N, 148°W; 134 m

−24 −18 −13 −3 8 15 16 13 6 −3 −16 −22 −3

23 13 18 8 15 33 48 53 33 20 18 15 297

________________ Freetown, Sierra Leone 9°N, 13°W; 28 m

27 27 28 28 27 26 25 25 26 26 26 27 27

10 5 30 79 241 363 742 927 566 361 140 30 3494

________________ Lanchow, China 36°N, 104°E; 1508 m

−6 −2 5 12 17 21 23 21 16 10 2 −5 10

1 3 8 14 34 40 66 92 55 18 4 2 338

________________ London, England 51°N, 0°W; 5 m

4 4 7 9 12 16 18 17 15 11 7 5 10

54 40 37 38 46 46 56 59 50 57 64 48 595

________________ Moscow, USSR 56°N, 38°E; 156 m

−10 −10 −4 5 12 17 19 17 11 5 −2 −7 4

31 28 33 35 52 67 74 74 58 51 36 36 575

________________ Peking, China 40°N, 116°E; 52 m

−5 −2 5 14 20 25 26 25 20 13 4 −3 12

4 5 8 17 35 78 243 141 58 16 10 3 623

________________ Perth, Australia 32°S, 116°E; 60 m

23 24 22 19 16 14 13 14 15 16 19 22 18

7 12 22 52 125 192 183 135 69 54 23 15 889

Name________________________ Date________________

Instructor______________________ Section______________

____

Point Barrow, Alaska, USA 71°N, 157°W; 7 m

−28	−28	−26	−17	−8	0	4	3	−1	−9	−18	−24	−12
5	5	3	3	3	10	20	23	15	13	5	5	110

____

Rio de Janeiro, Brazil 23°S, 43°W; 26 m

26	26	25	24	22	21	21	21	22	22	23	24	23
137	137	143	116	73	43	43	43	53	74	97	127	1086

____

Rome, Italy 42°N, 13°E; 3 m

8	9	11	14	18	22	24	24	22	17	13	10	16
83	73	52	50	48	18	9	18	70	110	113	105	749

____

Singapore 1°N, 104°E; 10 m

26	27	27	28	28	28	27	27	27	27	27	26	27
285	164	154	160	131	177	163	200	122	184	236	306	2282

____

Walvis Bay, Namibia 23°S, 15°E; 7 m

19	19	19	18	17	16	15	14	14	15	17	18	17
0	5	8	3	3	1	1	2	0	0	0	0	23

____

Yakutsk, Siberia, USSR 62°N, 130°E; 163 m

−43	−37	−23	−9	5	15	19	16	6	−9	−29	−41	−11
23	5	10	15	28	54	43	66	30	35	15	23	348

EXERCISE 15

CLIMATES OF THE WORLD

The climatic regions of the world using the Köppen classification will be studied in this chapter. Fifteen climates can be identified:

Köppen Symbols	**Descriptive Climate Name**
Af	Tropical Rainforest
Am	Tropical Monsoon
Aw, As	Tropical Savanna
BWh	Tropical (hot) Desert
BSh	Tropical (hot) Steppe
BWk	Mid-latitude Desert
BSk	Mid-latitude Steppe
Cfa	Humid Subtropical
Csa, Csb	Mediterranean
Cwa, Cwb	Subtropical Monsoon
Cfb, Cfc	Marine
Dfa, Dfb, Dwa, Dwb	Humid Continental
Dfc, Dfd, Dwc, Dwd	Subarctic
ET	Tundra
EF	Ice

The global distribution of these climates is shown in Appendix F. Questions in this exercise will examine the general temperature and precipitation characteristics associated with each of these climate types, the global distribution of each climate and the basic climatic controls operating in each case. Representative climographs are presented for each climate. As a review, you are encouraged to check the classification of each of these climates using the criteria given in exercise 14.

Name____ Date____________________

Instructor___________________________________ Section__________________

QUESTIONS

15.1 Examples of locations having A (tropical wet) climates are presented in climograph format in Fig. 15-1. One example is presented for each of the Tropical Rainforest (Af) and Tropical Monsoon (Am) climates while two examples of the Tropical Savanna (Aw) climate are given. For the latter, a northern hemisphere site and a southern hemisphere location are shown both with distinct dry winters. Note that As is also called Tropical Savanna but is very rare.

(a) Of the 20 locations in exercise 14 for which you classified climate, which were A climates?

Tropical Rainforest (Af): ____

Tropical Monsoon (Am): ____

Tropical Savanna (Aw): ____

(b) Briefly compare the annual temperature range for the three A climate types.

(c) The Tropical Rainforest (Af) climate has ample rainfall all year of at least 6 cm per month. Both the Tropical Monsoon (Am) and Tropical Savanna (Aw) have distinct dry seasons. Differentiate between the precipitation regimes for the Am and Aw climates.

(d) Name the three **main** regions of the world where the Tropical Rainforest (Af) climate occurs: ____________________, ____________________, ____________________.

(e) What climatic control is responsible for the "wet all year" precipitation regime of the Af climate: ____

(f) Most Tropical Monsoon (Am) climates are found along tropical coastlines. Name 4 coastal areas that have this climate.

____________________________, ____________________________,

____________________________, ____________________________.

Figure 15-1

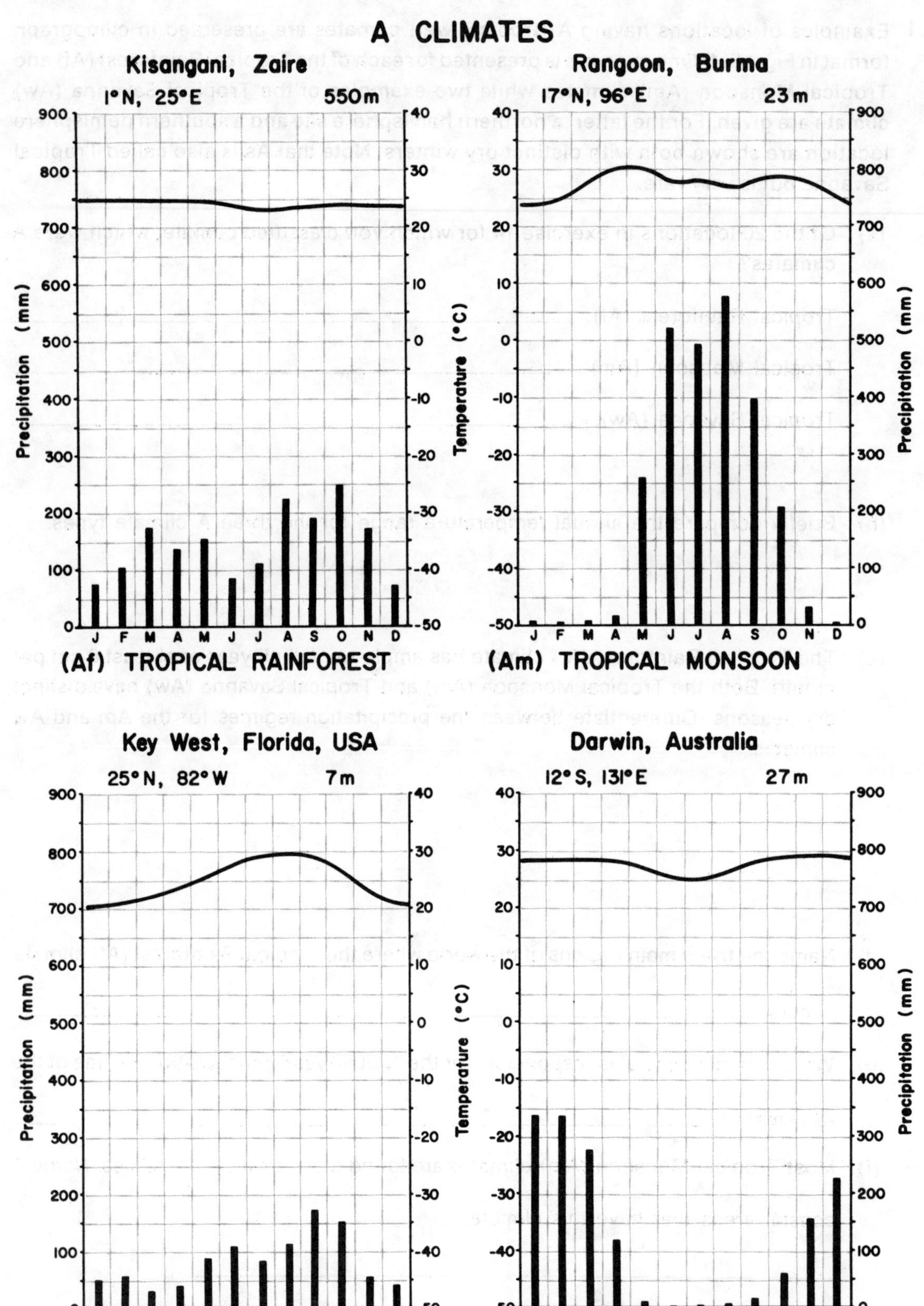
A CLIMATES
Kisangani, Zaire
1°N, 25°E
550m
Rangoon, Burma
17°N, 96°E
23m
Precipitation (mm)
Temperature (°C)
J F M A M J J A S O N D
(Af) TROPICAL RAINFOREST
(Am) TROPICAL MONSOON
Key West, Florida, USA
25°N, 82°W
7m
Darwin, Australia
12°S, 131°E
27m
(Aw) TROPICAL SAVANNA
(Aw) TROPICAL SAVANNA

Name______________________________ Date________________

Instructor____________________________ Section________________

(g) Where in relation to Af or Am climates does one usually find the Tropical Savanna (Aw) climate?

(h) Describe in detail the climatic control operating to produce the distinct wet and dry season for the Aw climate.

(i) Differentiate between the natural vegetation expected for the Aw, Am and Af climates.

15.2 Fig. 15-2 presents climographs of locations having B (dry) climates. One North American example is presented for each of the 4 climates: Tropical (hot) Desert (BWh), Tropical (hot) Steppe (BSh), Mid-latitude Desert (BWk) and Mid-latitude Steppe (BSk). The tropical B climates are sometimes referred to as "subtropical" as they often occur in latitudes corresponding to the STHP.

(a) Which of the 20 locations studied in exercise 14 were B climates?

Tropical Desert (BWh): ______________________________

Tropical Steppe (BSh): ______________________________

Mid-latitude Desert (BWk): ______________________________

Mid-latitude Steppe (BSk): ______________________________

(b) How do the annual temperature ranges for the tropical B climates compare to that for the A climates? If different, state why.

(c) The Sahara Desert in northern Africa is the largest region in the world having the BWh climate. Name three other deserts in each of the northern and southern hemispheres having the BWh climate.

Northern hemisphere: ____________, ____________, ____________.

Southern hemisphere: ____________, ____________, ____________.

(d) What major control causes BWh climates? ______________________________ .

(e) Where are Tropical Steppe (BSh) climates found in relation to BWh climates?

Name________________________ Date______________

Instructor____________________ Section____________

Figure 15-2

B CLIMATES

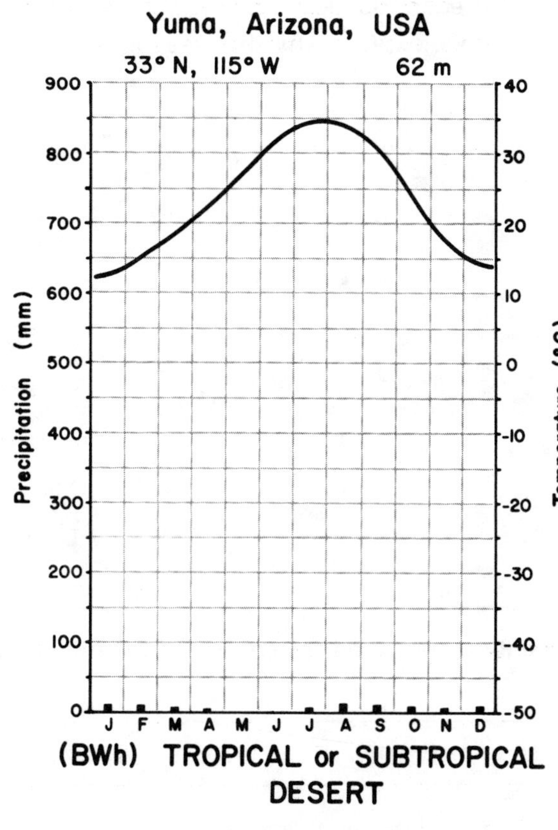

(BWh) TROPICAL or SUBTROPICAL DESERT

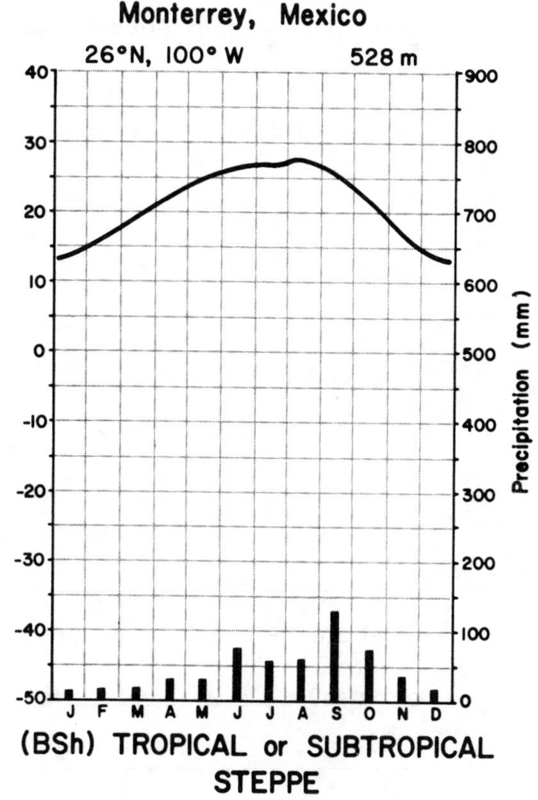

(BSh) TROPICAL or SUBTROPICAL STEPPE

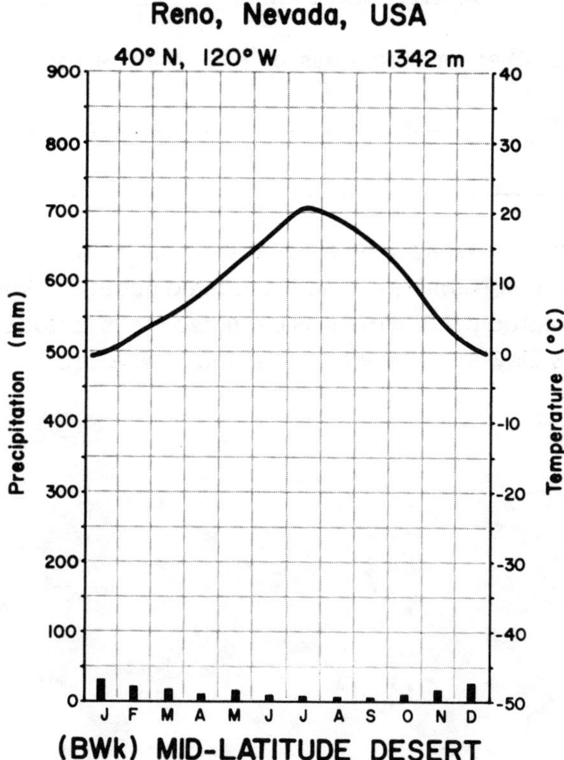

(BWk) MID-LATITUDE DESERT

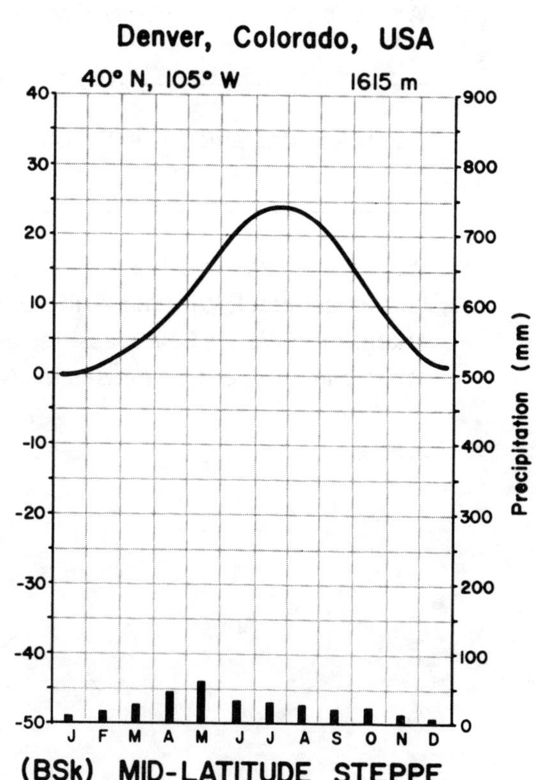

(BSk) MID-LATITUDE STEPPE

(f) Steppe climates are slightly wetter than deserts and can therefore support grassland vegetation. Consider two locations both with BSh climates. One is located in the BSh area just north of the Sahara Desert and the second location is in the BSh zone just south of the Sahara. Both receive most of their precipitation in one season (either summer or winter). For each location, state which season the rain occurs and, in terms of climatic controls, explain why that season is the wet season.

location one —

location two —

(g) What are the two climatic controls **mainly** responsible for causing mid-latitude B climates? ______________________________,

______________________________.

(h) Where is the largest region of Mid-latitude Desert (BWk) climate? ______________

______________. Name two other smaller areas of BWk climate (one in North America and one in South America). ______________________________,

______________________________.

(i) Walvis Bay, Namibia has a BWk climate despite the fact that it is not in the mid-latitudes. In fact, Windhoek (the capital of Namibia), which is located about 270 km inland from Walvis Bay, classifies as a tropical B climate as expected for its latitude. What climatic control causes Walvis Bay to be much cooler than expected?

Name ______________________ Date ______________

Instructor ______________________ Section ______________

(j) Albuquerque, New Mexico and Yuma, Arizona both have desert climates. Although they are located at similar latitudes, Albuquerque is much cooler classifying as BWk compared to Yuma's BWh. What climatic control accounts for the much cooler temperatures at Albuquerque?

(k) Where are the major regions of Mid-latitude Steppe (BSk) climate?

In North America: ______________________

In South America: ______________________

In Europe and Asia: ______________________

15.3 Warm temperate or subtropical (C) climates are represented by the climographs given in Fig. 15-3 with one example for each of Humid Subtropical (Cfa), Mediterranean (Csa, Csb), Subtropical Monsoon (Cwa, Cwb) and Marine (Cfb, Cfc).

(a) Consider the 20 sites classified in exercise 14. Which were C climates?

Humid Subtropical (Cfa): ______________________

Mediterranean (Csa, Csb): ______________________

Subtropical Monsoon (Cwa, Cwb): ______________________

Marine (Cfb, Cfc): ______________________

(b) How does the annual temperature range for the Humid Subtropical (Cfa) climate compare with that for the more tropical A climates? ______________________

____

(c) What genetic type of precipitation (convectional, frontal or orographic) is most prominent for the Cfa climate in summer?

______________________ ; in winter? ______________________.

(d) The Cfa climate is found most often on the east coast of continents between 25° and 40° latitude. List the major areas of this climate in:

North America — ______________________

South America — ______________________

Australia — ______________________

Asia — ______________________

Figure 15-3

C CLIMATES

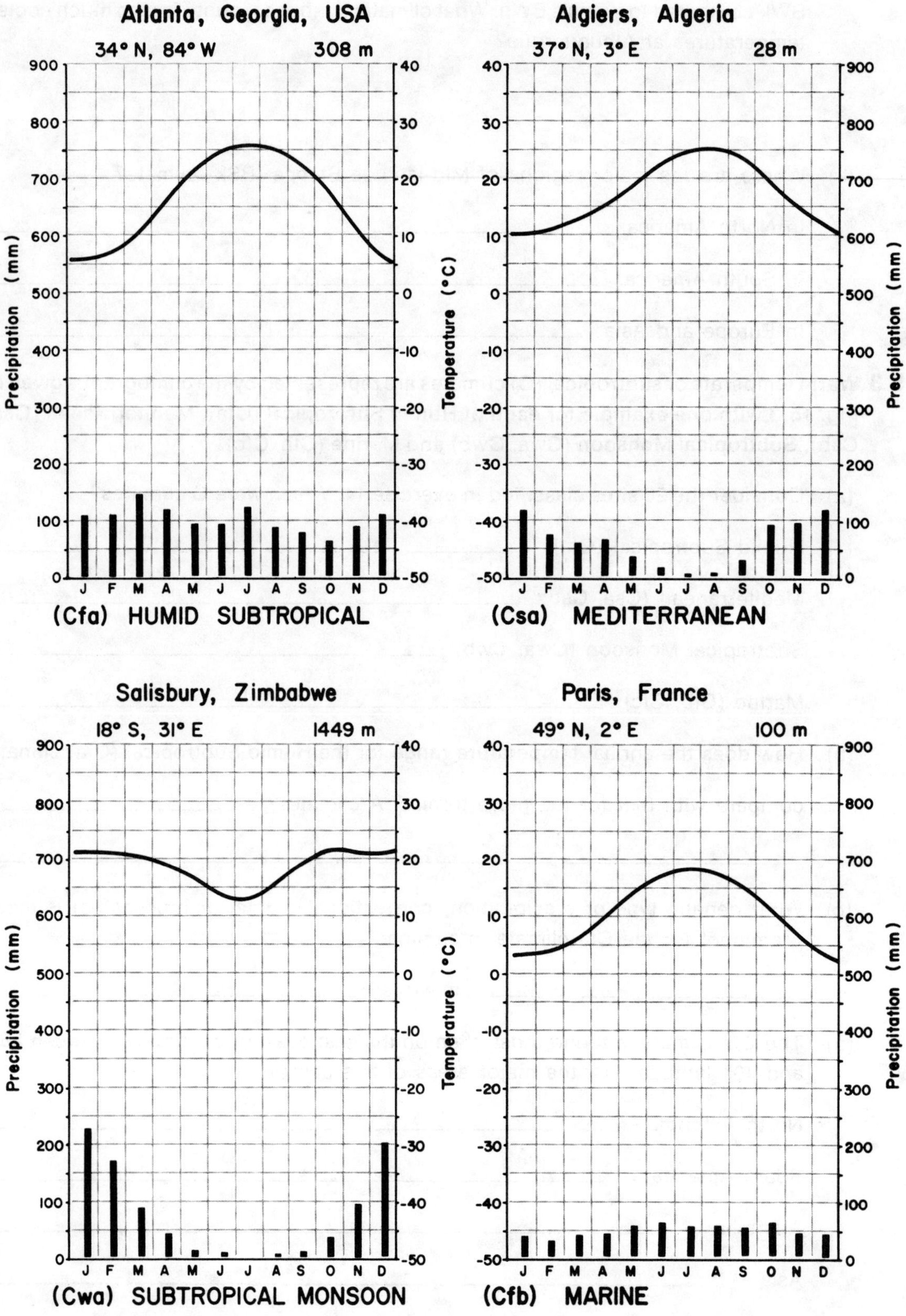

Atlanta, Georgia, USA

34° N, 84° W 308 m

Precipitation (mm): 0 100 200 300 400 500 600 700 800 900

Temperature (°C): -50 -40 -30 -20 -10 0 10 20 30 40

J F M A M J J A S O N D

(Cfa) HUMID SUBTROPICAL

Algiers, Algeria

37° N, 3° E 28 m

Temperature (°C): -50 -40 -30 -20 -10 0 10 20 30 40

Precipitation (mm): 0 100 200 300 400 500 600 700 800 900

J F M A M J J A S O N D

(Csa) MEDITERRANEAN

Salisbury, Zimbabwe

18° S, 31° E 1449 m

Precipitation (mm): 0 100 200 300 400 500 600 700 800 900

Temperature (°C): -50 -40 -30 -20 -10 0 10 20 30 40

J F M A M J J A S O N D

(Cwa) SUBTROPICAL MONSOON

Paris, France

49° N, 2° E 100 m

Temperature (°C): -50 -40 -30 -20 -10 0 10 20 30 40

Precipitation (mm): 0 100 200 300 400 500 600 700 800 900

J F M A M J J A S O N D

(Cfb) MARINE

Name________________________________ Date________________

Instructor______________________________ Section______________

(e) The Mediterranean (Csa, Csb) climate is located on the west coast of continents at similar latitudes to that of the Cfa climate. The major area of Csa or Csb climate virtually surrounds the Mediterranean Sea. List other areas of this climate type in:

North America — ________________________________

South America — ________________________________

South Africa (name the one major city) — ________________________________

Australia (2 areas) — ________________________________ ;

________________________________ .

(f) What control explains the distinct wet winters and dry summers found in Mediterranean climates (describe how this control operates)?

(g) Why don't the Csa, Csb climatic regions receive the summer rainfall experienced by their latitudinal counterparts having Cfa climate?

(h) The Csa climate differs from the Csb climate only in that the Csa has a much hotter summer. Below, climate data are given for San Francisco and Sacramento, California which have Csb and Csa climates respectively:

San Francisco 38° N, 122½° W; 27 m

9	10	12	13	15	16	17	17	18	16	13	10	14
102	88	68	33	12	3	0	1	5	19	40	104	475

Sacramento 39° N, 121½° W, 13 m

8	10	12	16	19	22	25	24	23	18	12	9	17
81	76	60	36	15	3	0	1	5	20	37	82	414

San Francisco is on the coast and therefore modified by the sea. Although only 120 km inland, Sacramento gets a much hotter summer and therefore becomes Csa. What prevents the maritime influence of the sea from penetrating this short distance? ______

The average annual temperature for San Francisco is lower than that for Sacramento. What feature of the ocean accounts for the overall cooler conditions in San Francisco? ______

(i) In relation to the Tropical Savanna (Aw) climate, where are the Subtropical Monsoon (Cwa, Cwb) climates located?

(j) Specifically name regions (portions of countries) having the Subtropical Monsoon climate in:

South America — ______

Africa — ______

Asia — ______

(k) The Cwa, Cwb climates have distinct dry winters like the Aw climates. They are simply cooler (especially in winter) due to increased latitude or altitude. Review the cause for the distinct wet and dry seasons.

(l) The data for Cherrapunji, India (a Cwb climate) are given below:

Cherrapunji, India 26° N, 92° E; 1313 m

12	13	17	19	19	20	20	21	21	19	16	13	17
20	41	179	605	1705	2875	2455	1827	1231	447	47	5	11,437

This site enjoys the distinction of being among the wettest sites in the world. Its July rainfall of almost 2.9 m (9.4 feet) is indeed the wettest single month of normal monthly precipitation. What causes Cherrapunji's summer monsoon season to be so very wet?

______

Name____________________ Date__________

Instructor____________________ Section__________

Why is Cherrapunji's summer cooler than other Cw climates in northern India (resulting in Cherrapunji having Cwb climate rather than the Cwa experienced in much of northern India)? ____________________

____

(m) The Marine (Cfb, Cfc) climate is often located poleward from the Mediterranean climates on the west coast of continents. As a result, it is often called the "Marine West-coast" climate. List the major regions for this climate in:

North America — ____________________

Europe — ____________________

South America — ____________________

Australia and New Zealand — ____________________

(n) Unlike the Mediterranean climate, places having Marine climates experience precipitation year round. What causes the year round precipitation?

(o) Why does the Marine climate region in Europe extend much further inland than the Cfb, Cfc region in North America?

15.4 Fig. 15-4 presents a map of Northern Africa and Western Europe. Consider annual precipitation totals as one moves northward from Zaire to France along the line labelled A to B. These values are illustrated in the graph at the bottom of Fig. 15-4. Notice that precipitation totals are highest near the equator and lowest around 25°N latitude. Labelled on the graph at the bottom of Fig. 15-4 are 7 zones. Name the climatic types according to Köppen's classification for these 7 zones.

Zone 1: ____________________

Zone 2: ____________________

Zone 3: ____________________

Zone 4: ____________________

Zone 5: ____________________

Zone 6: ____________________

Zone 7: ____________________

Figure 15-4

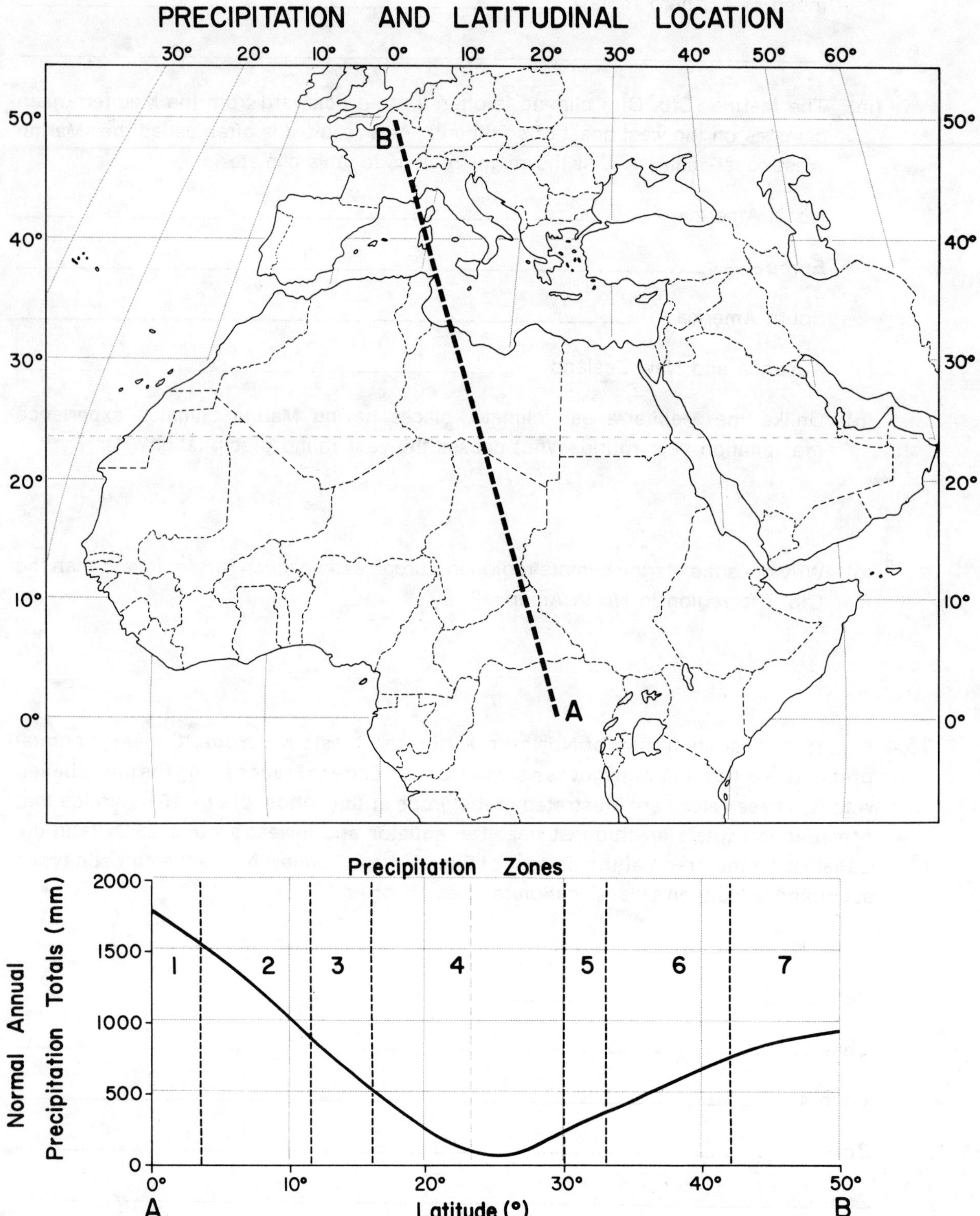
PRECIPITATION AND LATITUDINAL LOCATION
30° 20° 10° 0° 10° 20° 30° 40° 50° 60°
50° 40° 30° 20° 10° 0°
B
A
Precipitation Zones
Normal Annual Precipitation Totals (mm)
2000
1500
1000
500
0
1 2 3 4 5 6 7
0° 10° 20° 30° 40° 50°
A B
Latitude (°)

Name________________________________ Date________________

Instructor______________________________ Section______________

15.5 Examples of locations having D (cool temperate) climates are presented in the climographs shown in Fig. 15-5. Two sites having Humid Continental (Dfa, Dfb, Dwa, Dwb) climate are given: one with hot summer (Detroit — Dfa) and one with warm summer (Winnipeg — Dfb). Two examples of Subarctic (Dfc, Dfd, Dwc, Dwd) are also presented: Churchill (Dfc) and Verkhoyansk (which has the extreme winter case of Dfd).

(a) Which of the locations studied in Exercise 14 have D climates?

Humid Continental (Dfa, Dfb, Dwa, Dwb): ______________________________

Subarctic (Dfc, Dfd, Dwc, Dwd): ______________________________

(b) There are no D climates in the southern hemisphere. Why not?

(c) The Humid Continental climates are usually found poleward of the Humid Subtropical (Cfa) climate or inland from the Marine (Cfb, Cfc) climate at latitudes corresponding to those for the Marine climates. Name the major areas having this climate.

(d) Although the Humid Continental and Marine climates are at similar latitudes, they have very different annual temperature ranges. How do their annual temperature ranges differ and what climatic factor causes this difference?

(e) Subarctic (Dfc, Dfd, Dwc, Dwd) climates are found poleward of the Humid Continental climates and therefore have much shorter and cooler summers. This climate is where the northern coniferous or Boreal forest (sometimes called Taiga) exists. List the two main regions where this climate exists: ______________

____ .

(f) Verkhoyansk is considered the "most continental" climate station in the world. What is meant by this term and why is Verkhoyansk the most continental?

Figure 15-5

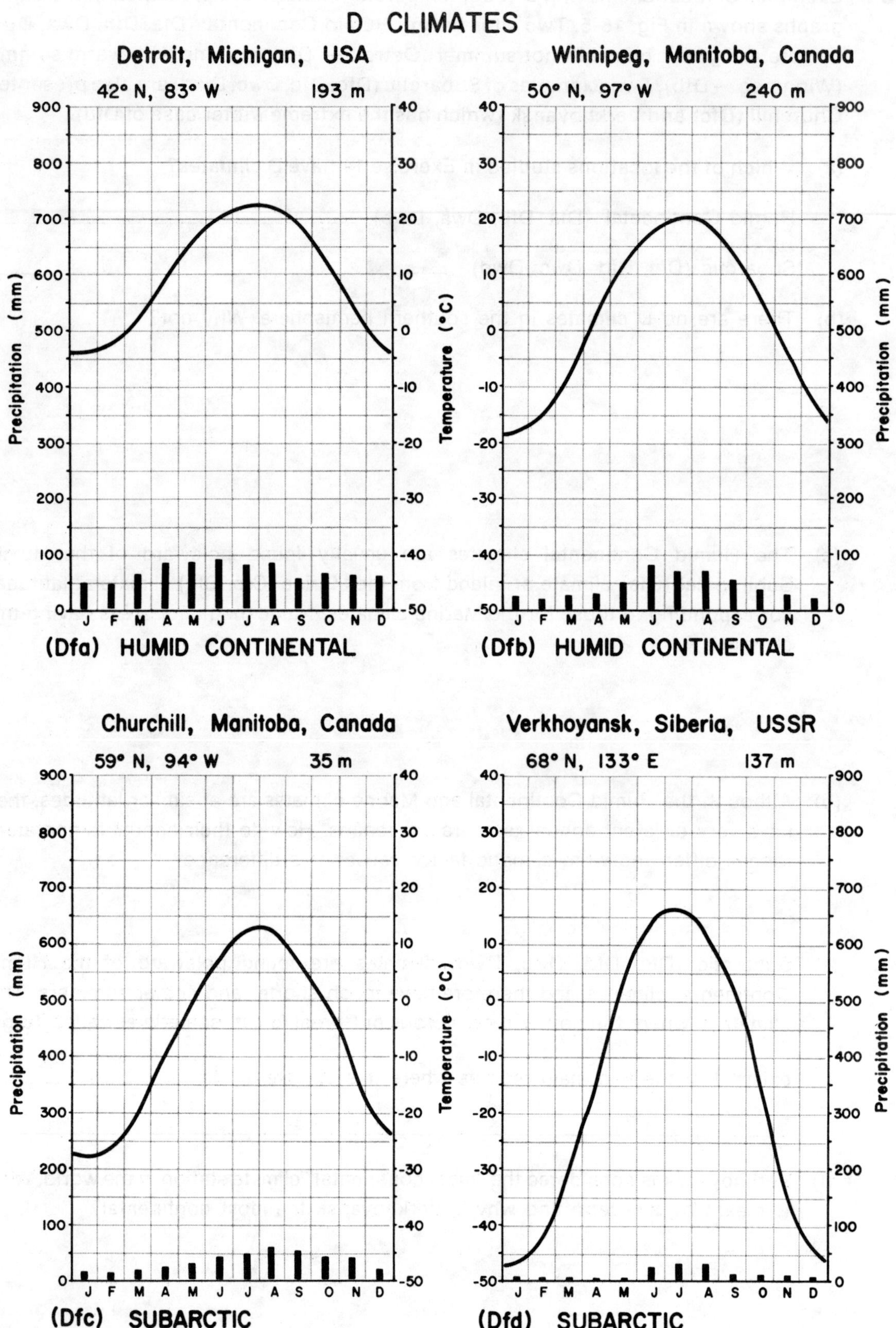

D CLIMATES
Detroit, Michigan, USA
42° N, 83° W
193 m
Winnipeg, Manitoba, Canada
50° N, 97° W
240 m
Precipitation (mm)
Temperature (°C)
J F M A M J J A S O N D
(Dfa) HUMID CONTINENTAL
(Dfb) HUMID CONTINENTAL
Churchill, Manitoba, Canada
59° N, 94° W
35 m
Verkhoyansk, Siberia, USSR
68° N, 133° E
137 m
(Dfc) SUBARCTIC
(Dfd) SUBARCTIC

Name________________________________ Date________________

Instructor______________________________ Section______________

(g) The Dw climates are confined to Asia where the huge Siberian high pressure area develops in winter creating dry conditions. However, even North American locations with the Humid Continental climate often have a summertime peak of precipitation. What genetic type of precipitation causes this extra rainfall in summer? ____

____

15.6 Two examples of E (polar) climates are shown in the climographs presented in Fig. 15-6. One is for a Tundra (ET) climate and the second is for an Ice (EF) climate.

(a) List the locations studied in exercise 14 that classified as E climates.

Tundra (ET): ____

Ice (EF): ____

(b) Where are Tundra (ET) climates found in:

North America: ____

Europe and Asia: ____

the southern hemisphere: ____

(c) Most ET climates have small annual totals of precipitation. What factor causes these low precipitation values?

(d) Even though the world's Tundra climates are further poleward than the Subarctic climatic regions, the annual temperature ranges for Tundra locations are often smaller than in the Subarctic climate zone. What factor accounts for this situation?

(e) Ice (EF) climates are colder than ET climates with no average monthly temperatures above 0°C implying no growing season. Where are the two main regions of EF climate? ____ ,

____ .

Figure 15-6

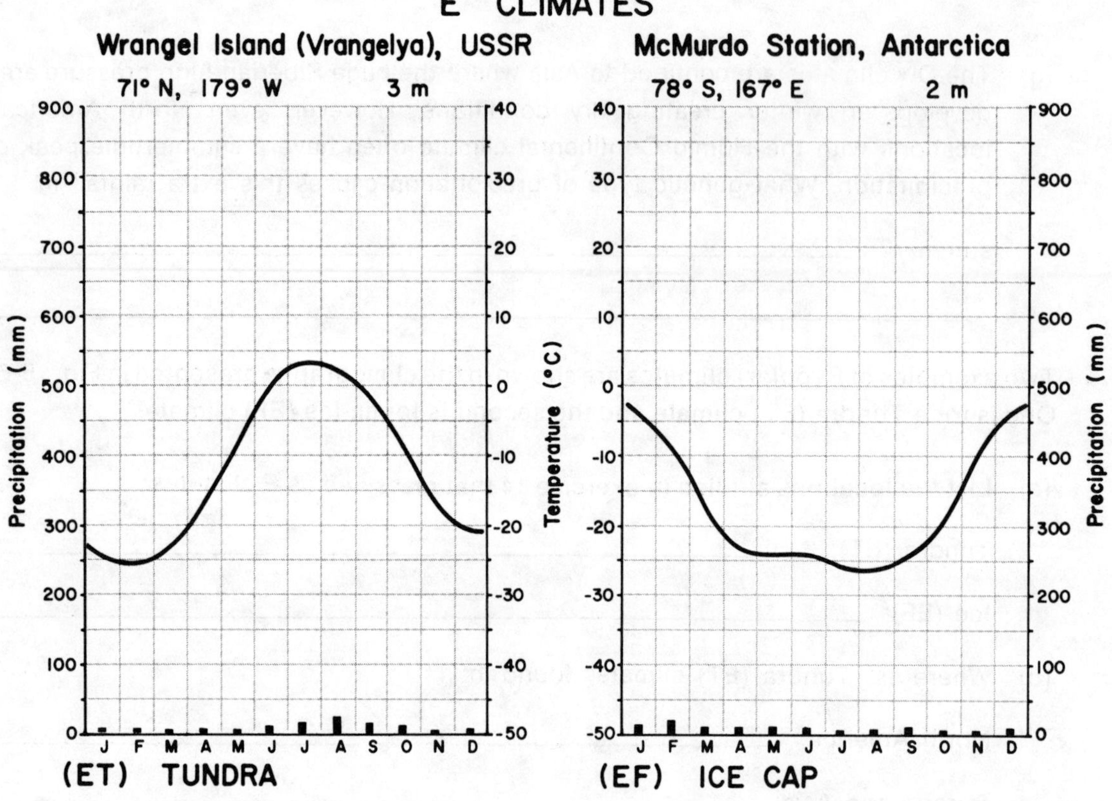

15.7 In locations with very high altitudes such as the Himalayan Mountains in Asia or the Andes Mountains of South America, Köppen's climate classification can result in sites at equatorial latitudes classifying as C or E climates or locations in subtropical latitudes (such as the Himalayas) being E climates. Rather than using these designations, the term H climate (for Highland) is adopted as illustrated in the map of world climates given in Appendix F.

The climate data for Quito, the capital of Ecuador, is given below. Quito is located at a latitude where Af climate is expected. However, its elevation of 2766 m changes its Köppen classification. What would be Quito's climate according to the Köppen classification? ______________________

Quito, Ecuador 0° S, 79° W; 2766 m

13	13	13	13	14	13	13	13	14	13	13	14	13
99	112	142	175	137	43	20	30	69	112	97	79	1115

Many of the other capital cities of South American countries in the Andean region are also at very high altitudes requiring considerable physical adaptation by humans. In addition to Quito, La Paz, Bolivia (17° S) is at 3658 m elevation and Bogota, Colombia (5° N) is at 2654 m. All three of these cities classify as C climates in regions where A climates are expected. Therefore, the designation H climate is used to signify the cooling effects of altitude.

EXERCISE **16**

GLOBAL TECTONICS

The average density of the earth is 5.5 gms/cc. and that of surface rocks 2.8 gms/cc indicating that material within the earth is of high density. Most knowledge of the interior has been gained through the study of *seismic* or shock waves released during *earthquakes* which occur when the crust ruptures along faults. The point of the earth's surface immediately above the *focus* of an earthquake is the *epicenter* and is the area most damaged by shock waves. The point on the earth diametrically opposite the epicenter is the *anticenter.* Three varieties of seismic wave are recognized. Primary or *P-waves* are compressive, can pass through solids and liquids and travel at 6.0-8.5 kms/sec. Secondary *S-waves* are shear, pass only through solids and travel at 3.6-5.1 kms/sec. Two other wave types collectively known as Long or *L-waves,* which travel in the surface rock layers of the earth, are responsible for the majority of damage by earthquakes. By noting the time interval on a *seismogram* between the arrival of P- and S-waves the distance from the *seismograph station* to the earthquake epicenter can be calculated. Records from three stations can therefore be used to determine the exact locations of earthquakes.

The behaviour of shock waves as they travel through the earth has revealed that the planet is encased in a rigid, light *crust* ranging in thickness from 10-35 km. Beneath the crust to a depth of 2,900 kms. is a high-density solid *mantle* which at least in its upper layers can "flow." Below the mantle lies the *core* which to 5,000 km. depth is a liquid, explaining why S-waves never reach the anticenter of an earthquake, and at the very center it is a solid of density 14-16 gms/cc. The major break between the crust and the mantle is called the *Mohorovičić Discontinuity* or 'Moho' and that between the mantle and the core the *Oldham or Gutenberg Discontinuity.*

The *magnitude* of an earthquake as defined by Charles F. Richter in 1935 is "the logarithm to base 10 of the maximum seismic wave amplitude measured in thousandths of a millimeter recorded on a Wood-Anderson seismograph at 100 km. from the earthquake epicenter," and is a measure of the amount of energy released. The definition has been extended so that any calibrated seismograph may record magnitude at any distance from an earthquake epicenter. An earthquake registering 5.0 on the *Richter Scale* releases the same energy as 20,000 tons of T.N.T. An increase of one magnitude step corresponds to an increase of 30 times in the amount of energy released as seismic waves. An earthquake of magnitude 8.6 releases almost one million times as much energy as one of magnitude 4.3. The largest earthquake ever recorded had a Richter magnitude of 8.9. The San Francisco earthquake of 1906 registered 8.3 and 700 people died. The 1953 San Francisco earthquake registered 5.3, there were no deaths. It is estimated that a future earthquake in San Francisco of magnitude 8.0 lasting for one minute would kill or injure 350,000 persons.

With such great potential for damage, achieving some measure of protection from earthquakes assumes high priority. If the time and place of damaging earthquakes can be *predicted,* stiff zoning controls and building codes can minimize damage and evacuation save lives. Some earthquakes have already been successfully predicted to within a few hours by observing physical changes in the crust. The *prevention* of earthquakes involves the lubrication of active faults to prevent their lockup. The intent is to induce a series of small displacements along a fault with the accompanying harmless *microearthquakes* and so prevent sudden major displacements which produce damaging earthquakes. Although this technique has proved successful in laboratory experiments its application to a stressed fault in a populated area carries obvious, and perhaps unacceptable risks.

The concept of *continental drift* is credited largely to Alfred Wegener who wrote at length upon the subject in the early 1900's. His main thesis was that the continents were once joined in a single landmass called *Pangaea* which began to fragment approximately 200 million years ago. (Appendix C) The northern portion of Pangaea consisting of North America and Eurasia is known as *Laurasia,* the southern portion including South America, Africa, India, Australia, and Antarctica as *Gondwanaland.* Although joined these two large supercontinents were indented by a hugh stretch of ocean called the *Tethys Sea.* Wegener's ideas were ridiculed for many years despite a considerable body of supporting geographic, geologic, and paleontologic evidence. Extensive research in the world's ocean basins and a revolutionary theory in geology have now conclusively established the reality of continental drift.

Ocean basins cover 71% of the earth's surface, have a relative relief of more than 40,000 feet (compared to 30,000 feet on the continents), and can be subdivided into five major topographic zones. The submerged margins of the continents are characterized by a gently sloping platform, the *continental shelf,* which is generally less than 600 feet below sea level. Seawards of it is the *continental slope* (average gradient 4.3°) which leads down to a deep ocean floor at an average depth of 12,500 feet. Research has revealed that the continents fit together almost perfectly if their edges are considered to be at a depth of 3,000 feet on the continental slopes. In some places continental slopes are deeply dissected by sinuous and often dendritic *submarine canyon* systems, a few of which appear to be continuations of major continental rivers. Where *turbidity currents* have funnelled large volumes of sediment down these canyons *submarine fans* have been built up on the ocean floor.

The deep ocean floor, once thought to be a featureless plain, is now known to be traversed by great submarine mountain ranges of volcanic origin which extend down the axes of the major oceans and which often branch into less prominent *plateaus* or *rises.* These mountain chains, which vary from 400-1,000 miles wide and from 5,000-10,000 feet high, form the *mid-ocean ridge* systems. Some of the higher peaks rise above sea level to form *volcanic islands.* Sea floor or *abyssal plains* are best developed between continental slopes and mid-ocean ridges, typical depths are from 10,000-20,000 feet. Although some plains are flat and featureless most are dotted with submarine volcanoes which commonly occur in rows or clusters. These are known as *seamounts* or if they have flat tops as *guyots.* *Deep-ocean trenches* are long, narrow gashes in the ocean floor which are from 25,000-30,000 feet deep and like the mid-ocean ridge systems define narrow belts of intense earthquake and volcanic activity. Ocean trenches are particularly common around the margins of the Pacific Ocean.

Figure 16.1

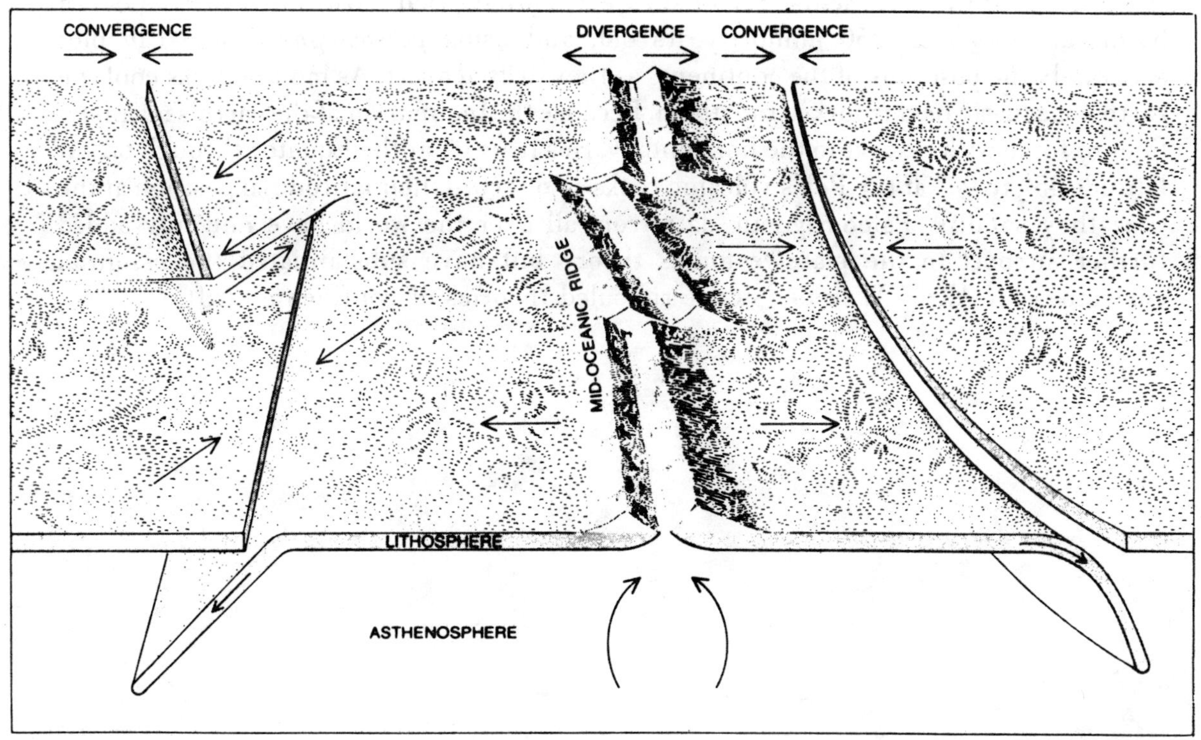

Convergent and divergent plate boundaries according to the plate tectonics theory (from Rona 1973).

It is now known that the outer shell of the earth is composed of 6-8 *lithospheric plates* which can move over the underlying 'plastic' upper mantle material of the *asthenosphere* (Figure 16.1). These plates are 70-100 kms. thick and may contain segments of both continental and oceanic material. Plates are believed to move in sympathy with slow-moving convection currents in the upper mantle (Fig. 16.2). At the mid-oceanic ridges, which mark *divergent plate boundaries,* plates are moving apart. Fissures opened up are filled with molten magma injected into the crust. At the deep-ocean trenches, which mark *convergent plate boundaries,* plates are in collision, one plate passing beneath the other in the *zone of subduction*. Crustal material is therefore created at the mid-oceanic ridges and destroyed in zones of subduction centered on the deep-ocean trenches. Where vast quantities of sediment have accumulated between two plates in collision folded and faulted mountains may be built at the leading edge of the overriding plate. *Parallel plate boundaries* occur where one plate simply slides past another. The San Andreas fault of California delimits just such a boundary between the Pacific and American Plates. Plate boundaries whether convergent, divergent or parallel understandably define narrow zones of intense earthquake and volcanic activity.

The *plate tectonics theory* has revolutionized man's understanding of the earth's surface by explaining in a single concept the existence of mid-oceanic ridges, deep-sea trenches, and the major mountain belts; the extreme localization of tectonic activity; and the drifting of the continents. Scientists are now aware that most of the world's ocean basins are less than 250 million years old, and using *paleomagnetic data* can plot accurately the positions of the continents as they drifted apart. As molten rocks cool the earth's magnetic field is locked into them. Because the inclination of the earth's magnetic lines of force with respect to the horizontal varies from 0° at the Equator through 63° at latitude 45° to 90° at the Poles volcanic rocks also record their latitudinal position at the time they solidify. Paleomagnetic data from all over the world have not only added further support to the plate tectonics theory but have substantiated the essential correctness of Alfred Wegener's once ridiculed theories of continental drift.

Name______________________________________ Date____________________

Instructor__________________________________ Section__________________

QUESTIONS

16.1 The first seismic waves from an earthquake to reach a seismograph are the P-waves and lagging behind are the S-waves which can be recognized on the seismogram (the record made by the seismograph) by their greater amplitude. The difference in arrival time between the slower S- and the faster P- waves can be used to calculate the distance of an earthquake epicenter from the seismograph station. After the distance to the earthquake has been determined from at least three widely-spaced seismograph stations, circles are drawn with these distances as radii. The common intersection of these circles locates the epicenter.

S-waves travel at 2.7 miles/second and P-waves at 4.8 miles/second (that is P-waves are 2.1 miles/second faster than S-waves). The time T_1 (seconds) taken by S-waves to travel a distance D (miles) from the epicenter of an earthquake to a seismograph station is

$$T_1 = D/2.7$$

The time T_2 for P-waves to travel the same distance is

$$T_2 = D/4.8$$

The difference in arrival time (Δ T) between the P- and the S-waves is

$$\Delta T = T_1 - T_2 = D/2.7 - D/4.8$$

$$\text{or } \Delta T = 2.1D/13$$

The distance D from the earthquake epicenter to the seismograph station is, therefore, given by

$$D = 13\Delta T/2.1$$

Using this relationship complete Table 16.1 and locate the epicenter of the earthquake on Figure 16.3 (in pocket).

Table 16.1 Distance of epicenter of an earthquake from three seismograph stations

Seismograph Station	Time Between Arrival of P- and S-waves	Estimated Distance To Epicenter
Cape Town, South Africa	5 mins. 31 secs.	
Rio de Janeiro, Brazil	7 mins. 21 secs.	
Luanda, Angola	7 mins. 45 secs.	

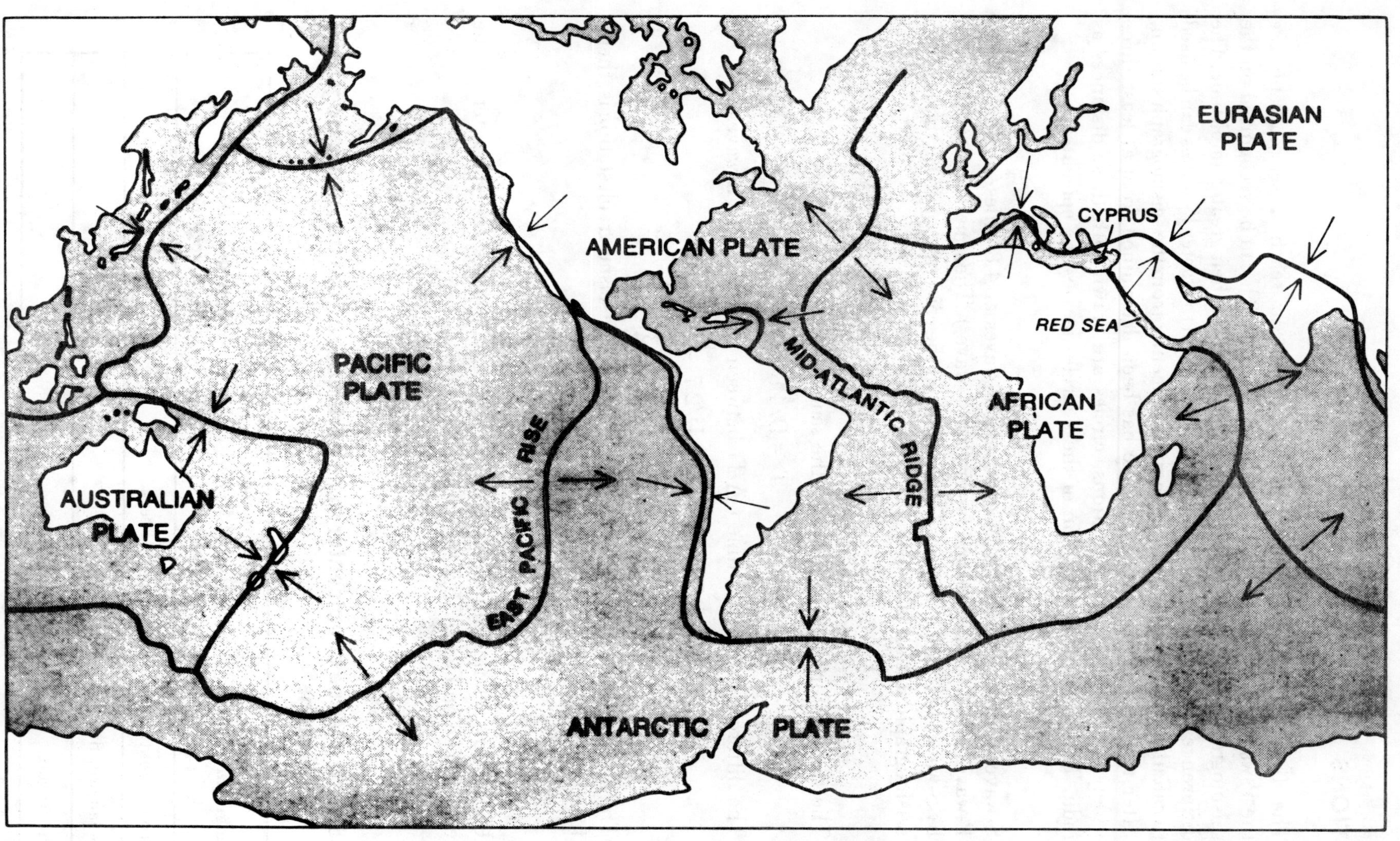

Figure 16.2 **The six principal tectonic plates of the lithosphere (from Rona 1973).**

Name________________________________ Date________________

Instructor____________________________ Section______________

16.2 The relationship between the inclination of the earth's magnetic lines of force (I) measured downward from the horizontal and latitude (L) is given by

$$\text{Tan (I)} = 2 \text{ Tan (L)},$$

where both I and L are in degrees.

In one region paleomagnetic evidence from a series of basalt flows shows that 100 million years ago the inclination of the earth's magnetic lines of force was 5° to the north. Fifty million years ago it was 35°, and 20 million years ago it was 45°. At present the region has a latitude of 32°N. Describe what has happened to this region during the last 100 million years in terms of both its geographic position and climate.

16.3 Examine Figure 16.3.

(a) Identify and mark on the maps two examples of each of the following features.

A Continental shelf
B Continental Rise
C Abyssal Plain
D Fracture Zone
E Deep Ocean Trench
F Guyot
G Seamount (SMT)

(b) In several areas large 'cones' are marked on the Continental Rise. Examples include the Mississippi Cone in the Gulf of Mexico, the Laurentian Cone off southeastern Canada, the Indus Cone off Pakistan and the Ganges Cone off Bangladesh. What are these features and how have they formed?

(c) The island of Iceland in the North Atlantic has formed where the Mid Atlantic Ocean Ridge has built itself above sea level. From Iceland follow the ridge south into the Mid Indian Ocean Ridge and so into the East Pacific Ocean Ridge and name several other islands or island groups that appear to have originated in the same way. What relationship, if any, is there between these islands and seamounts and guyots?

(d) Locate the Hawaiian Islands in the central Pacific and determine the approximate height of the shield volcano which has formed them.

(e) Estimate the relative relief of the Mid Atlantic Ocean Ridge by comparing elevations along its crest with those in the abyssal plains to east and west.

(f) Noting that elevations given in Figure 16.3 are in meters complete Table 16.2.

Table 16.2 Water depths in the world's oceans

Location	Average Depth of Water	
	Meters	Feet
Argentine Abyssal Plain		
Bellinghausen Abyssal Plain		
Angola Abyssal Plain		
Mariana Trench, Western Pacific		
Kuril Trench, Western Pacific		
Java Trench, Indian Ocean		
Mid Atlantic Ocean Ridge		
Mid Indian Ocean Ridge		
East Pacific Ocean Ridge		

Name________________________ Date______________

Instructor____________________ Section____________

16.4 Table 16.3 is a selected list of severe earthquakes that have occurred since 1290 A.D. Mark their approximate locations on Figure 16.2. What correspondence, if any, does there appear to be between earthquake activity and tectonic plates?

Table 16.3 Selected severe earthquakes

Date	Approximate Location	Estimated Death Toll
July-November, 1976	Tanghsan-Tientsin, China	100-200,000
May, 1976	Northeast Italy	900
February, 1976	Guatemala City	23,000
September, 1975	Eastern Turkey	2,000
December, 1974	Bangladesh	5,200
December, 1972	Nicaragua	6,000
April, 1972	Southern Iran	5,057
February, 1971	San Fernando, California	65
June, 1970	Northern Peru	66,794
March, 1970	Western Turkey	1,086
August, 1968	Northeastern Iran	11,588
March, 1964	Alaska	131
July, 1963	Yugoslavia	1,100
September, 1962	Northwest Iran	10,000
May, 1960	Southern Chile	5,700
June, 1956	Northern Afghanistan	2,000
August, 1950	Assam, India	1,500
June, 1948	Fukui, Japan	5,131
December, 1932	Kansu, China	180,000
May, 1875	Venezuelo, Colombia	16,000
November, 1755	Lisbon, Portugal	60,000
October, 1737	Calcutta, India	300,000
January, 1556	Shensi, China	830,000
April, 1293	Kamakura, Japan (Tokyo)	30,000
September, 1290	Chihli, China	100,000

16.5 Compare Figures 16.2 and 16.3.

(a) How has the Red Sea formed?

(b) What is happening to the Pacific Plate along its northern border with the American Plate and what feature has formed at this junction?

(c) In Figure 16.3 the Mid Atlantic Ocean Ridge seems to have a distinctive central rift, why?

(d) Vast amounts of sediment have already accumulated in the Mediterranean Basin which marks the boundary between the Eurasian and African Plates. The European Alps to the north are an indication of what might happen to the Mediterranean in the future. Explain.

16.6 Read the 1974 *Science News* article "Final Piece in the Gondwana Game" reproduced below. Locate and mark the area discussed on Figures 16.2 and 16.3 and sketch in its original position on Figure 16.3.

Final Piece in the Gondwana Game

Article reproduced from *Science News*, V. 106, p. 54, 1974.

The Gondwanaland Game is a little like that geometrical brain-teaser in which a T-shaped figure can be taken apart and reassembled into a square. Nature has played it for the past 200 million years, breaking up the ancient supercontinent of Gondwanaland and shifting around the pieces atop plates of the earth's crust until they reached their present arrangement.

Ever since they realized the game was going on, earth scientists have been trying to play it backwards, reuniting the pieces in their original configuration to see how the game began. A major sub-plot in the Reverse Gondwanaland Game has been to rejoin Africa and South America, by hunting up the missing continental bits that would make their facing coastlines fit exactly instead of just approximately. (Figure 16.4)

Last week, researchers from Columbia University and the University of Birmingham, England, announced that they had dropped in the subplot's final piece—a 750-mile tongue of submerged land thrusting eastward from the Falkland Islands off the Argentine coast that used to border what is now the southeastern coast of South Africa.

"It completes the puzzle," says geologist Ian W. D. Dalziel of Columbia. "All the other points along the Atlantic contours of the two continents have been proved by various scientific means to have been joined long ago. Now, with deep sea drilling, we've identified a large area of foundered continent, and the last piece is in place."

The last piece was discovered during Leg 36 of the remarkable journey of the research ship Glomar Challenger as part of the Deep Sea Drilling Project. With Dalziel and Peter Barker of the University of Birmingham as co-chief scientists for the leg, the drill team had to send their drill string down through more than a mile and a half of water and 1,835 feet of bottom sediment before they struck the hard base of continental granite that would later prove to be the missing fragment.

For the first 50 million years or so of the Gondwanaland Game, starting about 200 million years ago, the African-South American rift-to-be remained locked together and dry, in fact with a rather balmy, Mediterranean-type climate. When it finally began to separate, the land along the facing edges no longer supported at the junction, started sinking, reaching its present depth about 80 million years ago.

The missing piece was discovered when Dalziel and Barker realized that they had struck continental granite in an extension of the Falkland Plateau that reached so far to the east that it must formerly have filled in a vast, canyon-shaped space northwest of the tongue of the Mozambique Plateau.

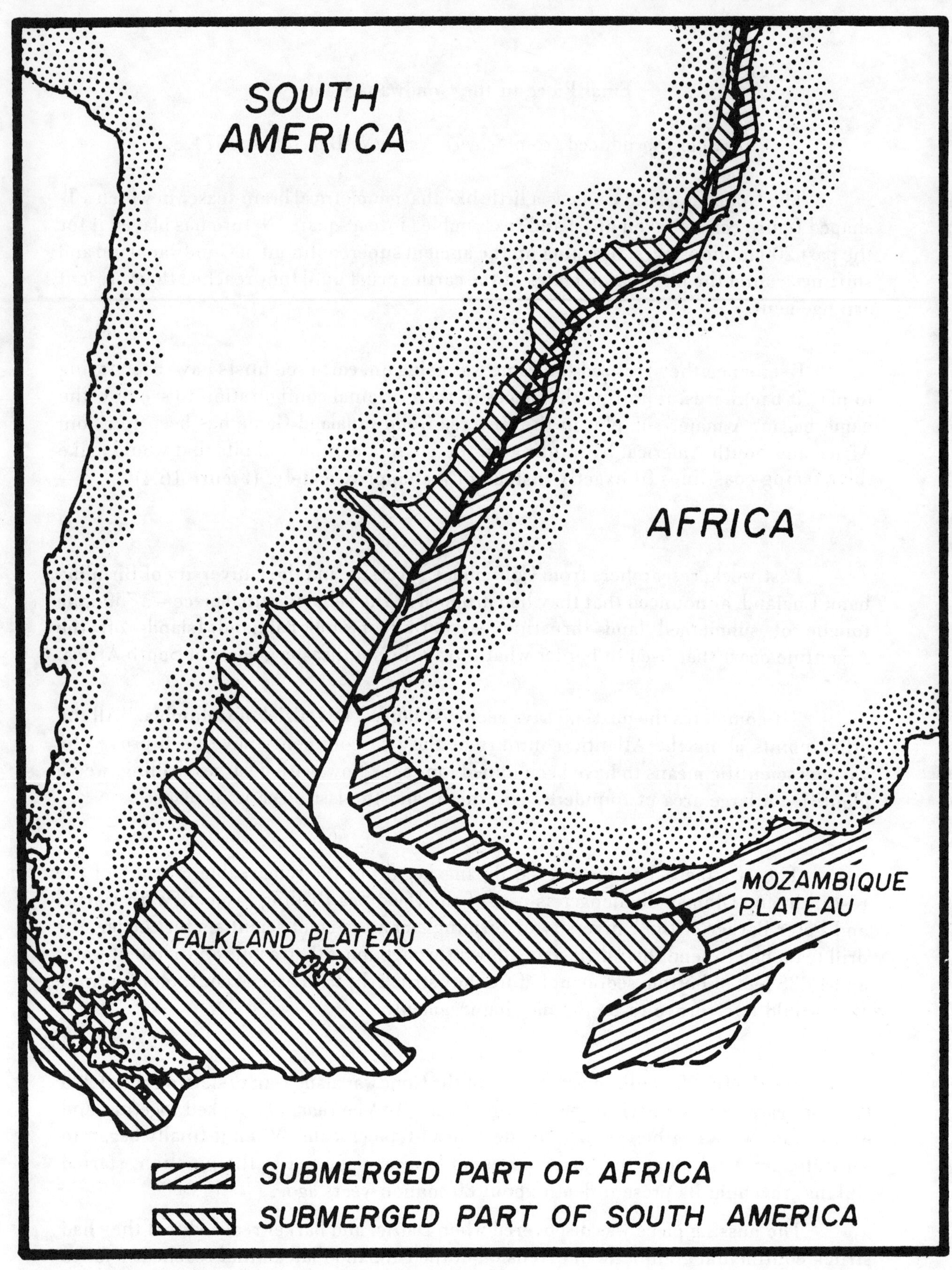

Figure 16.4 Africa and South America: A comfy fit

EXERCISE **17**

LANDFORMS OF INTRUSIVE AND EXTRUSIVE VULCANISM

The combination of molten rock, dissolved gas, gas bubbles and suspended crystals is known as *magma.* Magmas that are injected into the earth's crust but which fail to reach the surface before cooling form *intrusive volcanic* structures such as dikes, sills, laccoliths, lopoliths, and volcanic necks. *Extrusive volcanic structures,* the principal amongst them being the volcano, are formed where magma reaches the surface and becomes *lava* (Figure 17.1). The temperature of molten lava lies between 900-1200° C. The solid products of successive eruptions pile up around the *vent* to form a *volcanic cone* with a central *crater.* The form of the crater may change during each eruption as parts of it are either destroyed or added to. Magma frequently breaks through weaknesses in the flanks of the main cone in *flank eruptions* with the formation of flank or *parasitic cones* which may be of different character to the parent structure. The vast majority of volcanoes are located at plate boundaries where magma is created at depth. They are ubiquitous on the ocean floor and form volcanic islands along the oceanic ridge systems. Examples of such volcanic islands include Iceland and Hawaii.

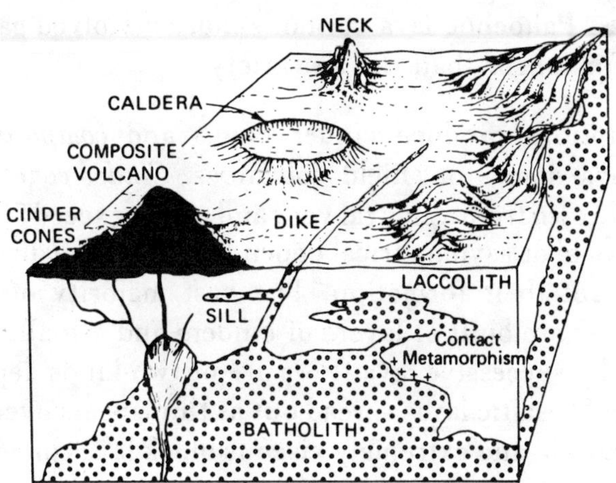

Figure 17.1 Landforms associated with intrusive and extrusive vulcanism (from Butzer, 1976)

Volcanic eruptions are either *explosive* or *effusive* depending upon the viscosity of the magma when it reaches the surface (a substance of low viscosity is extremely runny). In general viscosity increases with increase in the proportion of silica in the magma. As magma approaches the surface the gas pressure may build up so that an explosive eruption occurs and fragments of lava and older solid rock are thrown into the air. These fragments, whether liquid or solid, are called *pyroclasts* or *tephra* and eventually fall back to the ground. Small pyroclasts form *volcanic dust*; particles of lead shot size, *lapilli*; and large blocks *volcanic bombs.* Lapilli and bombs may be full of gas bubbles in which case they are called *scoria* or *cinder.* Lava extruded from explosive volcanoes is usually siliceous and extremely viscous and is called *block lava.* Flows may be one or two thousand feet thick and seldom extend more than 5-6 miles from the vent.

One of the most dangerous of explosive volcanic eruptions is that involving the formation of a *nuées ardentes* or *glowing avalanche.* These often develop in association with highly viscous block lava which plugs the vent and allows gases to build up pressure within. The principal part of a nuées ardentes is an avalanche of incandescent lava blocks, sand and dust beneath a rapidly expanding cloud of dust. Speeds may exceed 75-100 m.p.h. because the fragments in the avalanche are suspended in a cushion of expanding gas, often extremely hot and noxious, which eliminates frictional effects.

Effusive eruptions produce mostly basaltic lavas, poor in silica and relatively rich in iron and magnesium, which flow out of the main vent, a flank fissure, or a parasitic cone. Gases escape so easily from these lavas that explosive activity becomes subordinate. Fluid basaltic lavas are of two main types. *Pahoehoe lava* has a relatively smooth but wrinkled surface in contrast to *aa lava* which has a surface of rough, jagged fragments called clinker, resting on a solid mass of lava. Block lava resembles aa lava except that the surface fragments are more rounded. Pahoehoe and aa flows are usually 5-20 feet thick and may extend tens of miles from the fissure or vent. Aa lava is partly crystallized when it leaves the vent and gas escapes in sudden bursts, breaking the hardened crust into irregular jagged pieces. Pahoehoe lava contains more dissolved gas, cools more slowly, and remains mobile for longer than the aa variety.

Explosive eruptions produce cinder cones and *composite volcanoes,* effusive eruptions produce lava domes or shield volcanoes. *Cinder cones* are built entirely of pyroclastic material thrown from a central vent and rarely exceed 500-1000 feet in height. They may be symmetrical or asymmetrical depending upon the direction and strength of the wind at the time of their formation. The vast majority of volcanoes are of the composite variety. They are built of layers of cinders and ash alternating with layers of lava. Volcanoes built by successive lava flows are of two kinds depending upon the viscosity of the lava. Fluid basaltic lavas spread out as thin sheets over great areas and build *shield volcanoes.* Silica-rich, and extremely viscous block lavas pile up around the vent to form steep-sided *volcanic domes.* Domes may also develop in the summit crater, or on the flanks of composite or other volcanoes. Volcanoes surrounding the Pacific Ocean have moderately to highly silicic lavas and tend to erupt explosively, in contrast to the gentle eruptions of Hawaii and other predominantly basaltic areas.

Some central vent volcanoes have gigantic depressions within the walls of their truncated summits. Such a depression is called a *caldera* and its diameter (10 miles or more) is many times greater than that of an eruptive vent. Nested calderas are also commonly encountered in single complex volcanoes. Most calderas are formed by the collapse and engulfment of the original superstructure during great eruptions. This process is called *cauldron subsidence. Explosion calderas* formed by the violent removal of the upper part of the cone are uncommon and generally of relatively small size.

Name_______________________________ Date_______________

Instructor _______________________________ Section_______________

QUESTIONS

17.1 Examine the map and aerial photographs of the Menan Butte area of Idaho.

(a) The Menan Buttes are examples of what kind of volcano? (Figures 17.2 and 17.3)

(b) What relationship is there between the Menan volcanoes and the extremely irregular plain to the northwest?

(c) Examine the topographic profile of one of the Menan volcanoes you drew in answer to question 2.8. Explain why the cone and crater in your diagram are asymmetrical in the southwest to northeast direction but not in the northwest to southeast direction.

(d) What effect does the formation of the Menan Buttes appear to have had on the course of the Snake River?

17.2 The aerial photographs in Figure 17.4 show Asama-yama one of Japan's most active and dangerous volcanoes which is located on Honshu Island.

(a) What kind of volcano is Asama-yama?

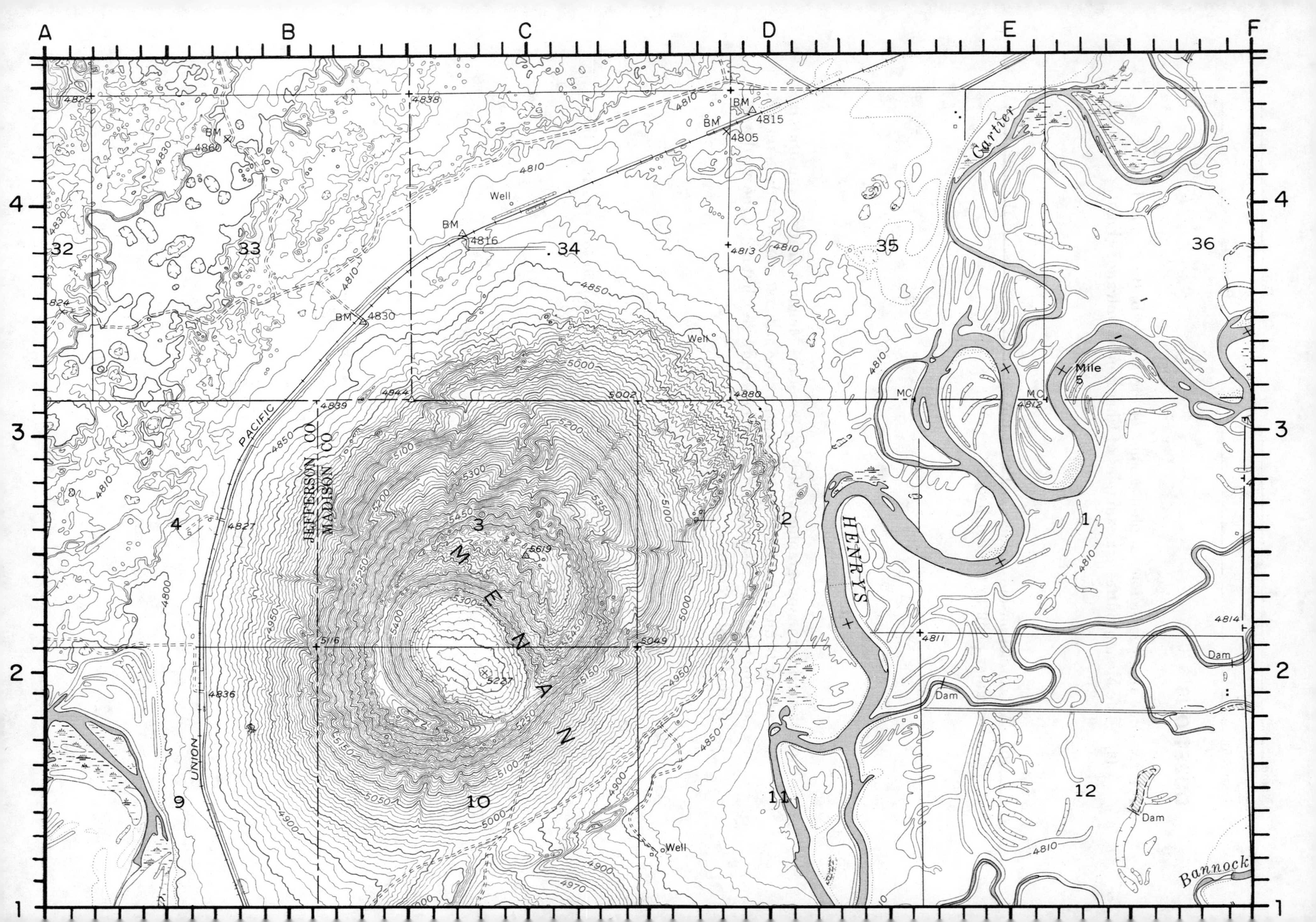

MENAN
HENRYS
Cartier
Bannock
JEFFERSON CO
MADISON CO
UNION
PACIFIC
Mile 5
Well
Dam
BM
4815
4805
4816
4830
4860
5619
5227

Name______________________________ Date______________

Instructor______________________________ Section______________

(b) The photographs show evidence of three main stages in the history of the main volcanic cone. Outline these stages.

(c) On the photographs mark the boundary of the most recent lava flow, the Onioshidashi-iwa, which occurred in 1783. This flow partly fills the head of a flat-floored, steep-sided trench cut in pyroclastic material (the Kambara ditch) which was carved by a nuée ardente (sometimes called a glowing avalanche) that immediately preceded the lava.

(d) Mark in a tongue of lava that clearly pre-dates the 1783 flow.

(e) Did the newest cone of Asama-yama form prior to or later than 1783?

(f) Why is the pyroclastic material west of the 1783 lava flow highly dissected and that east of it not?

(g) The small dome east of the main volcanic vent is Ko-asama-yama. What relationship does that feature have to the main cone, and given that it is built entirely of lava, why does it have such steep sides?

17.3 The intersecting narrow, linear ridges of Huerfano County, south-central Colorado (Figure 17.5) control the drainage pattern. What has produced such an unusual topography?

Figure 17.2 Menan Buttes, Idaho (1:24,000, A.0-1.0 in southwest corner).

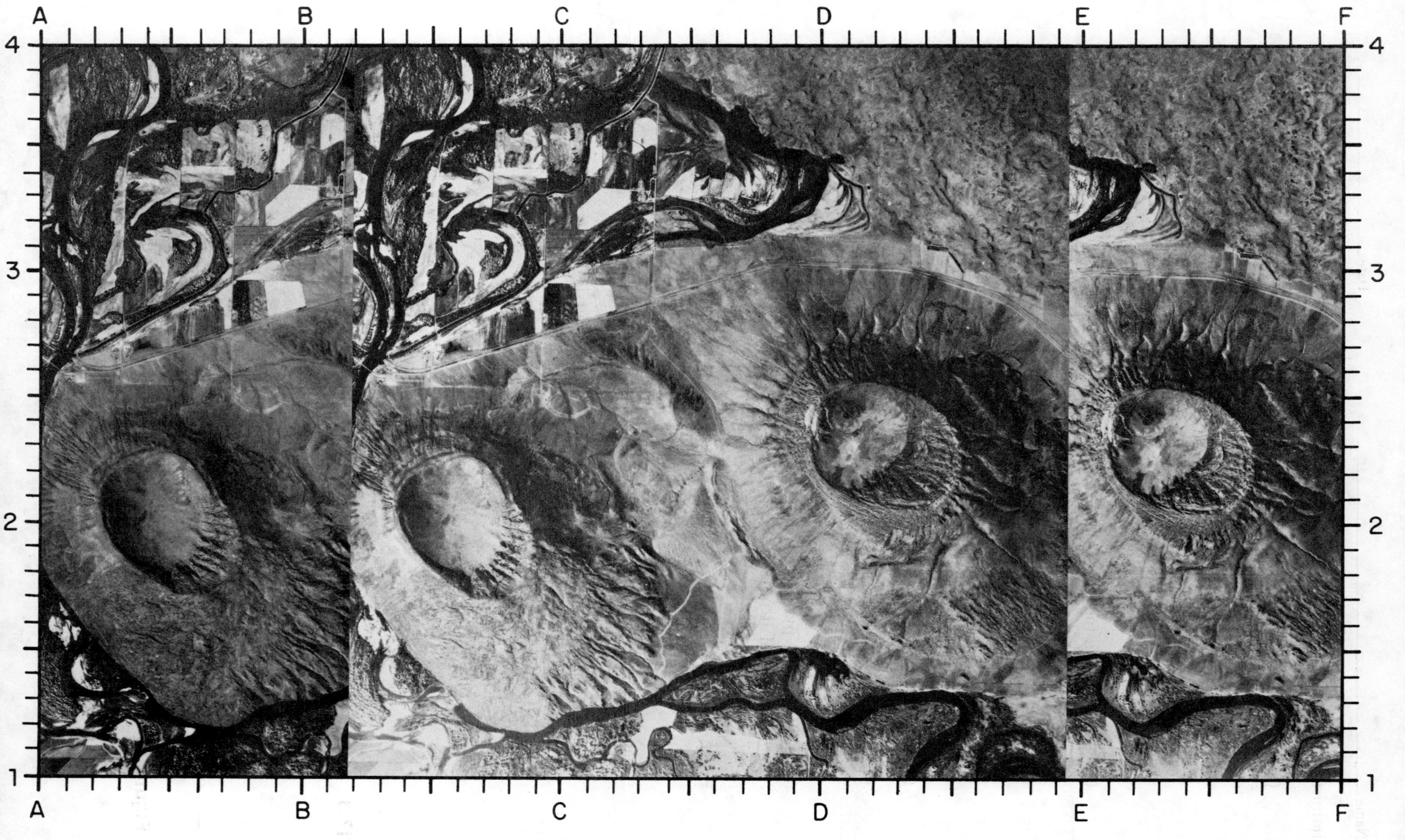

Figure 17.3 **Menan Buttes, Idaho (A.0-1.0 in southeast corner).**

Figure 17.4 **Asama Volcano, Japan (1:52,800, A.0-1.0 in southwest corner).**

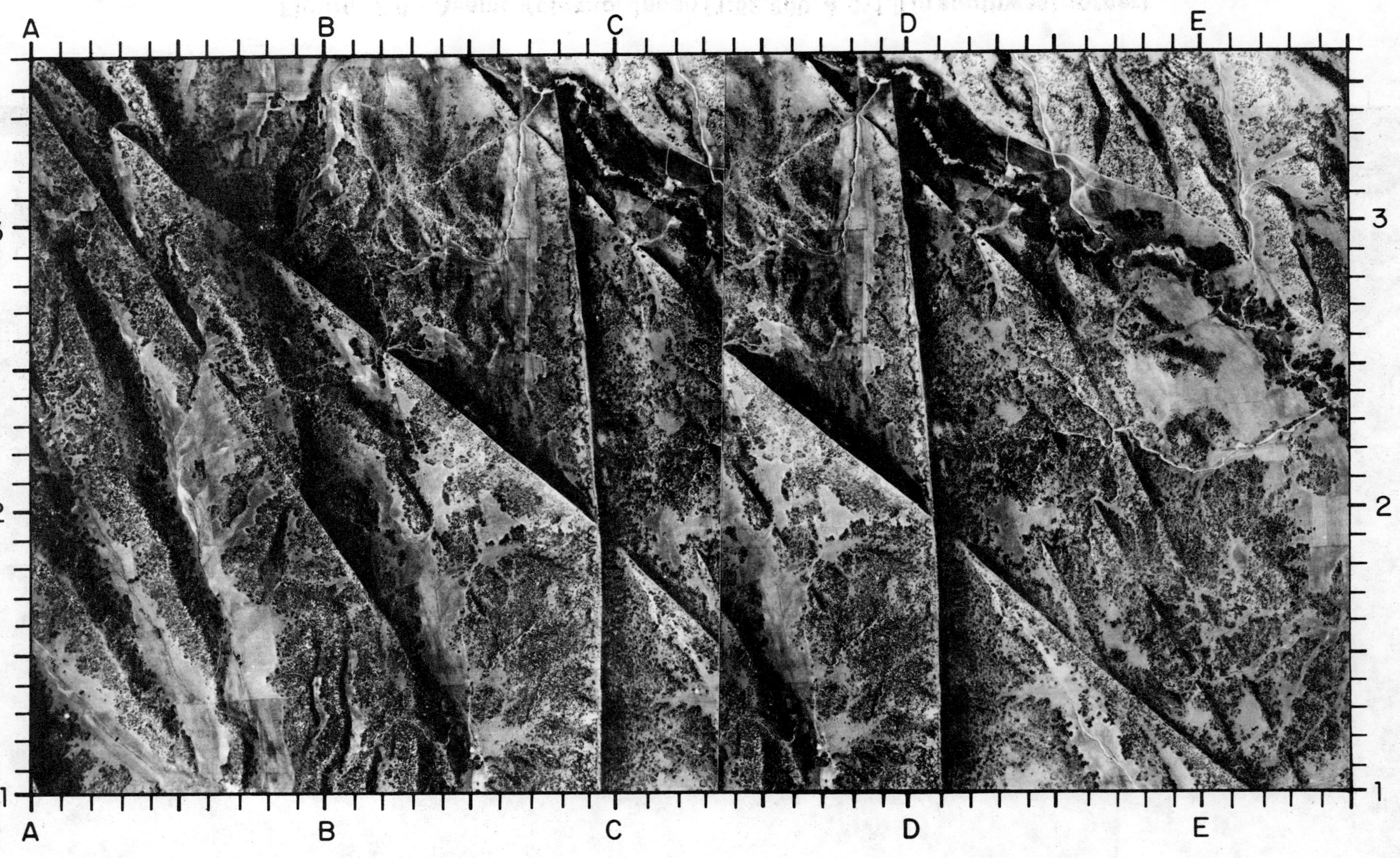

Figure 17.5 Spanish Peaks, Colorado (1:26,600, A.0-1.0 in southeast corner).

EXERCISE **18**

STRUCTURAL LANDFORMS

Variations in rock composition and structure exert a powerful control upon landform and landscape morphology. Before they are modified by erosion fault and fold structures impart *primary relief* to an area. As erosion progresses areas underlain by weaker rocks (eg. shale), are lowered more rapidly than areas underlain by resistant rocks (eg. sandstone and conglomerate) so that the resulting relief is determined by both structural and lithological characteristics. Over time fold and fault structures are gradually reduced to surfaces of faint relief but at all stages of their evolution the landscape reflects the underlying structure.

Uplifted areas underlain by horizontally-bedded sedimentary rocks with alternate weak and resistant layers develop a unique set of landforms. Slope retreat from stream valleys or canyons is controlled by rock hardness. The resistant layers form cliffs above debris slopes developed in the weaker layers, in come cases cliffs are completely undercut. The softer rocks are rapidly removed so that a series of broad *stripped structural surfaces* develop on the resistant beds and a stepped topography results. As an uplifted plateau is dissected extensive table-top residual hills called *mesas* are isolated and as slope retreat reduces their size they become steep-sided rock pinnacles or *buttes* (Figure 18.1).

Sedimentary rock layers may be gently tilted or warped with domes, basins, monoclines, and anticlines and synclines. Many mountain belts are little more than a series of parallel to subparallel folds with the anticlines forming ridges or *anticlinal mountains* and the synclines lowlands or *synclinal valleys.* As *fold mountains* of this kind are dissected streams cut into the flanks of the anticlines, which are frequently weakened by networks of open joints, and expose the underlying weaker rock layers. In some cases a narrow valley may be incised parallel to the heavily fractured crest of the anticline and an *anticlinal valley* formed. A synclinal valley on resistant rocks between two rapidly expanding anticlinal valleys cut in softer rocks may eventually become a ridge called a *synclinal mountain* and a *reversal of topography* will have occurred.

Where folded rock structures are eroded and the strata exposed, denudation of the weaker layers is rapid with the formation of lowlands separated by low ridges developed in the more resistant beds. Where the strata in a ridge dip is one direction only, for example on the flank of an anticline or a syncline, the ridge is a *homoclinal ridge.* Valleys cut in softer rocks which also dip in only one direction are *homoclinal valleys.* If the beds dip at low angles (less than about 10°) broad low ridges with one steep face and a gently sloping backslope develop, these are referred to as *cuestas.* Where the strata dip at greater angles they form sharp-crested *hogback ridges* which alternate with narrow valleys. A series of parallel ridges and valleys form where horizontal anticlines and synclines are eroded, while sinuous to zigzag ridges dominate landscapes cut across plunging folds. The hard rock ridges at the noses of eroded anticlines and synclines are called *anticlinal or synclinal coves.* A concentric pattern of cuestas or hogbacks is typical of dissected domes and basins. In domes the steeper faces of the upstanding ridges face towards the center of the structure and retreat outwards from it. In basins the reverse is the case.

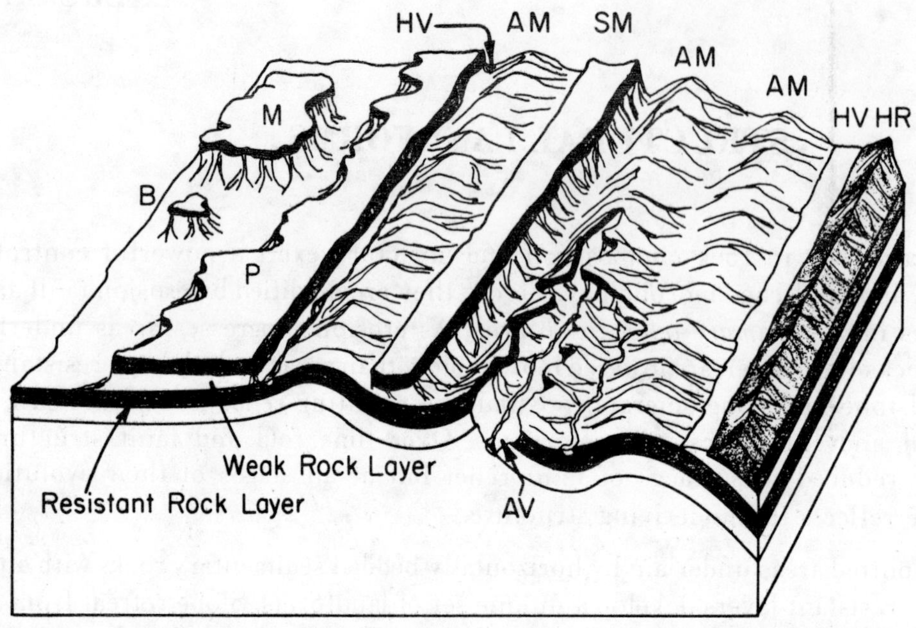

Figure 18.1 Structural landforms of horizontal and folded sedimentary rocks (M = mesa, B = butte, P = stripped structural plateau, HV = homoclinal valley, HR = homoclinal ridge, AM = anticlinal mountain, AV = anticlinal valley, SM = synclinal mountain).

The relative uplift of huge blocks of rock as a result of vertical faulting creates linear to slightly sinuous *fault scarps* in the landscape similar to those produced by monoclinal flexing. As streams dissect an upthrown fault block the fault scarp is fretted by stream valleys and canyons. The portions of the fault scarp that remain intact have a triangular shape and are called *triangular facets.* Ultimately even these are destroyed and the eroded fault scarp becomes a *fault line scarp* located at some distance from the original fault but approximately parallel to it. Faults frequently occur in large numbers and it is common for huge blocks of rock to be raised or lowered between two faults. A narrow block dropped down between two normal faults is called a *graben,* a block raised between two faults a *horst.* When formed, grabens form topographic depressions and where outlet streams do not develop, such as may be the case in an arid area where water is evaporated or in a karst area where it drains underground, they may remain depressions. Uplifted horsts form *fault block mountains* which like grabens have linear sides. The Basin-and-Range Province of the western United States is an erosionally modified horst and graben terrain. Intersecting networks of faults and joints (fractures along which there has been no vertical or lateral displacement) represent weaknesses in the rock along which groundwater can penetrate. When these are etched out by chemical and mechanical weathering, or by stream erosion they impart a distinctive grain to the landscape and are normally responsible for *lineations* visible in aerial photographs.

Name ________________________ Date ____________

Instructor ________________________ Section ____________

QUESTIONS

18.1 (a) What geologic and topographic features are labelled A through M in Figure 18.2?

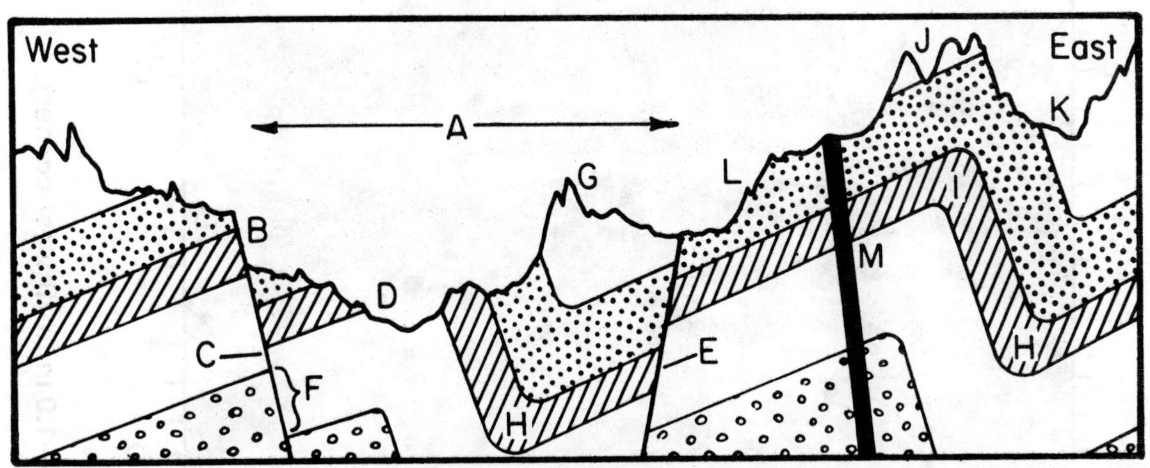

Figure 18.2 The structural landforms of a hypothetical terrain.

LABEL	DESCRIPTION
A	
B	
C	
D	
E	
F	
G	
H	
I	
J	
K	
L	
M	

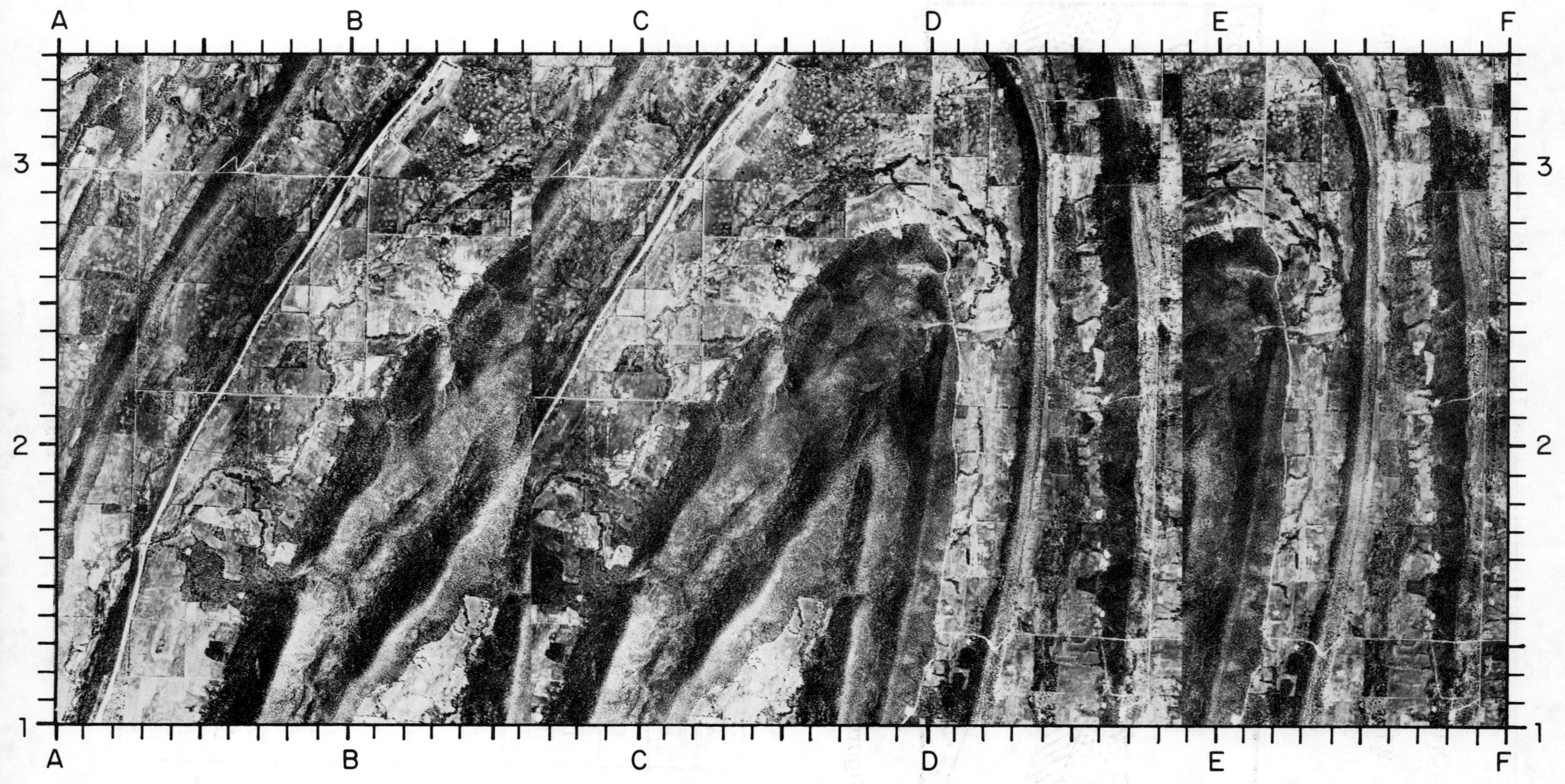

Figure 18.3 Pine Mountain, Oklahoma (1:48,300, A.0-1.0 in northwest corner).

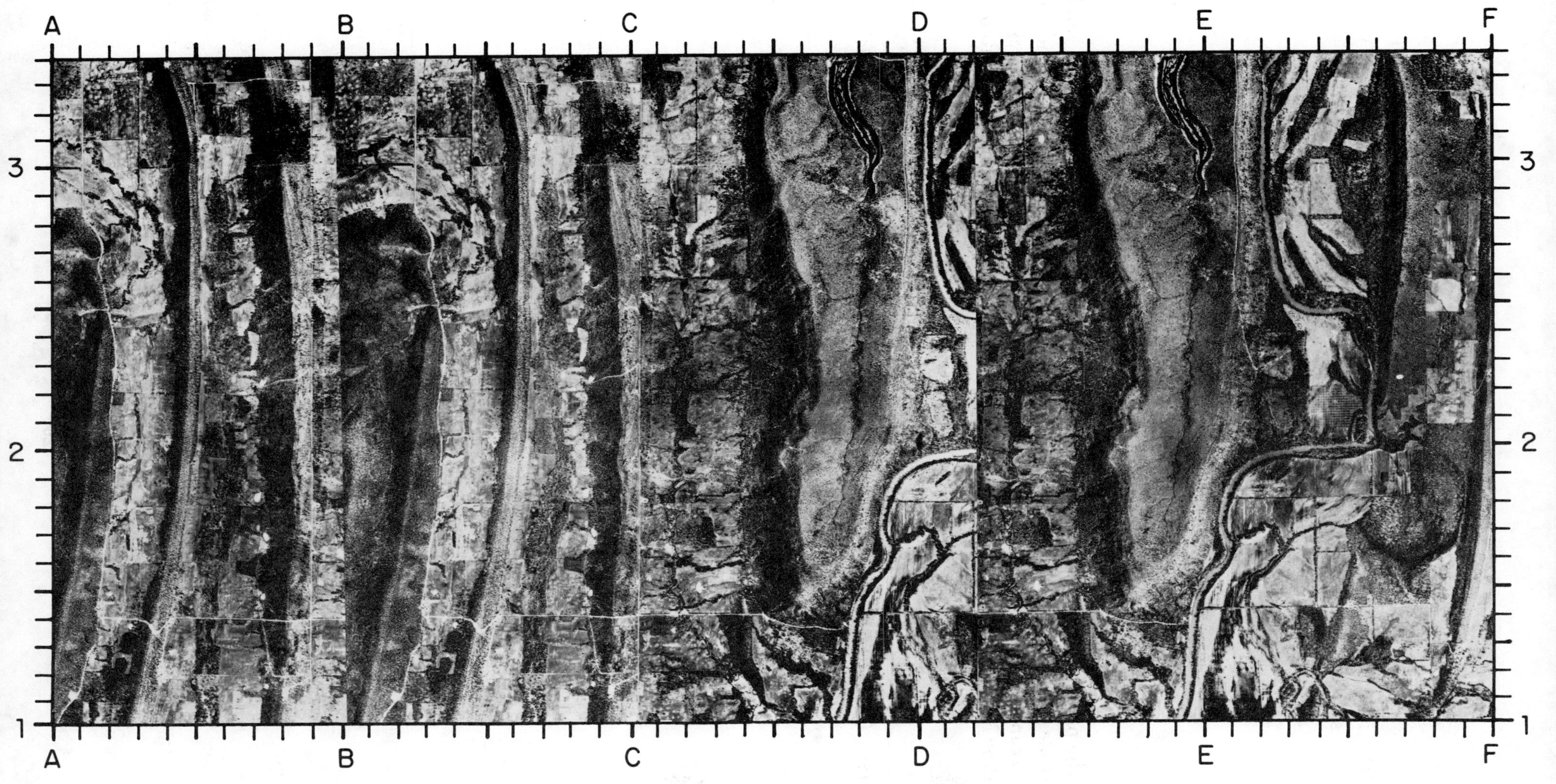

Figure 18.3 (continued). Pine Mountain, Oklahoma (1:48,300, A.0-1.0 in northwest corner).

(b) Measure the magnitude and direction of dip of the faults at (C) and (E) and show the relative motion of the fault blocks.

18.2 The photographs in Figure 18.3 show a folded sequence of Pennsylvanian sandstones and shales on the boundary between the Ouachita Mountains and the Arkansas Valley in Oklahoma. Strip mines have been opened up in the rocks of Pine Mountain which is bordered in the south by the meandering Poteau River.

(a) Give a detailed account of the geological structure of the area noting anticlinal and synclinal structures and their direction of plunge.

(b) On the photographs label the following features:

A	Synclinal Mountain	D	Hogback Ridge
B	Homoclinal Valley	E	Synclinal Nose
C	Homoclinal Ridge	F	Anticlinal Nose

Name______________________ Date______________

Instructor______________________ Section______________

(c) Describe and explain the ridge-and-valley topography.

(d) Use a schematic topographic and geological section to explain the presence of Pine Mountain (C.7-2.0).

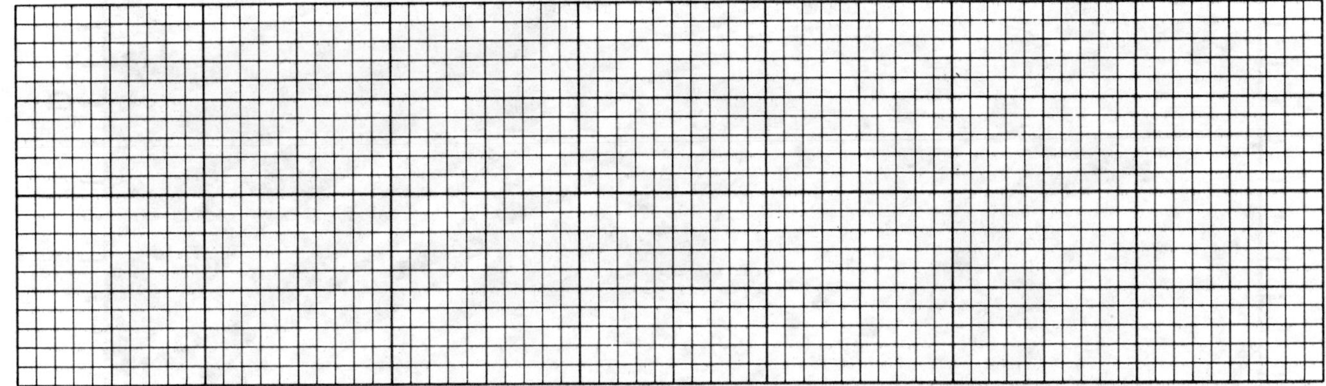

(e) What rock type underlies the floodplain of the Poteau River (D.9-1.0) and why does the river follow the outcrop of this particular rock?

18.3 The central lowland of the geologic structure shown in the photos of Fremont County, Wyoming (Figure 18.4) is underlain by Triassic shale and sandstone and is surrounded by alternating sandstones and shales of Jurassic age.

(a) On the photographs label:

A Hogback Ridge	C Hard Rock Layer
B Homoclinal Valley	D Soft Rock Layer

(b) Describe the geologic structure and in doing so take particular note of variations in the dip of the various strata.

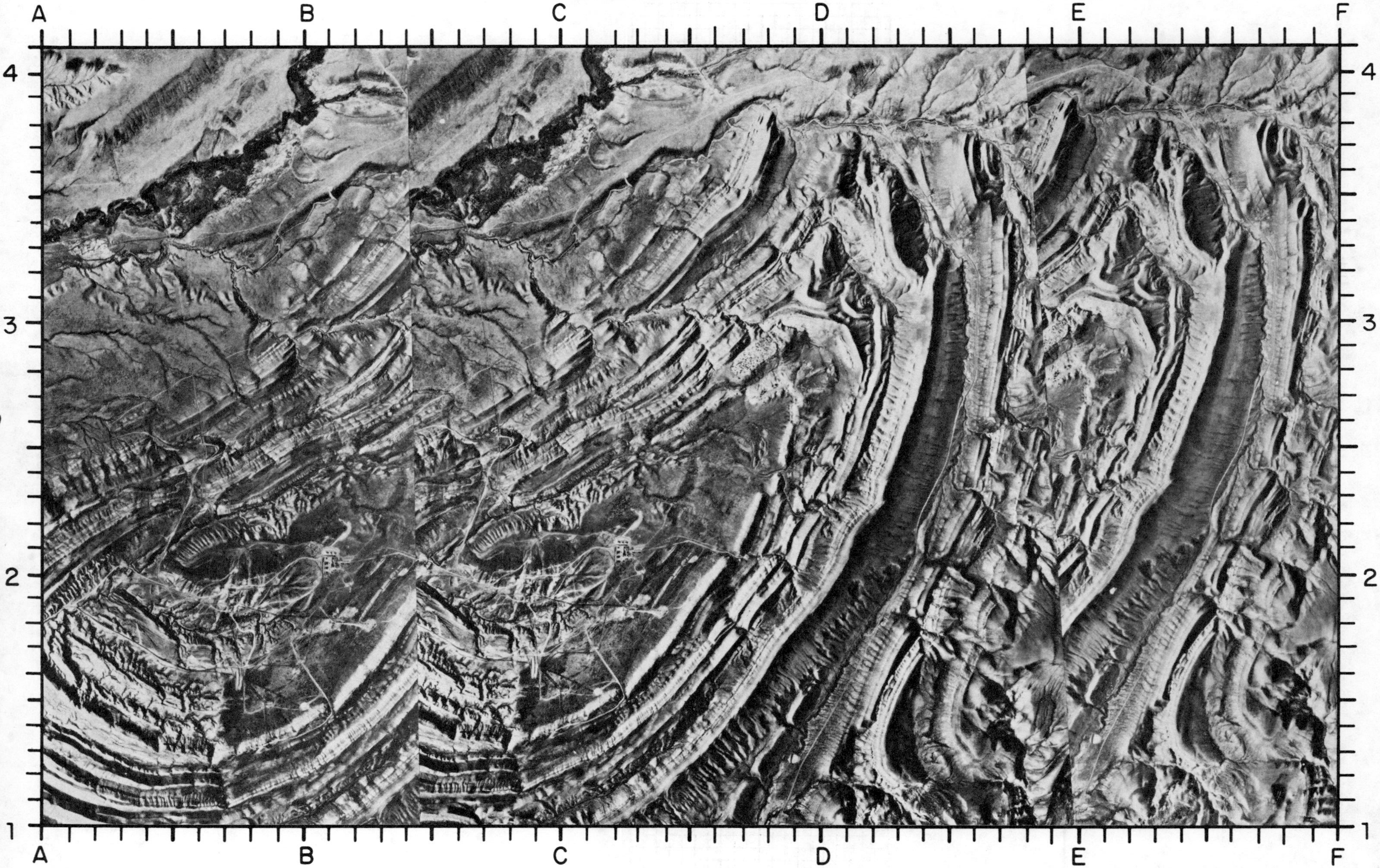

Figure 18.4 Circle Ridge, Wyoming (1:31,400, A.0-1.0 in southeast corner).

Name ______________________ Date ______________

Instructor ______________________ Section ______________

(c) How has the geological structure influenced the topography and drainage pattern?

18.4 The portion of the 1:250,000 scale Harrisburg, Pa. map shown in Figure 18.5 shows excellent examples of eroded plunging fold mountains in a variety of Pennsylvanian, Mississippian, Devonian, Silurian and Ordovician rocks. The contour interval on the map is 100 feet and you are given that the town of Shamokin (D.3-4.1) occupies a synclinal structure.

(a) Identify a further synclinal and two anticlinal structures and indicate their direction of plunge.

(b) Using letters indicate on your map at least one of each of the following.

A	Anticlinal nose	F	Synclinal mountain
B	Synclinal nose	G	Anticlinal valley
C	Homoclinal ridge	H	Synclinal valley
D	Homoclinal valley	I	Resistant layers of rock
E	Anticlinal mountain	J	Weak layers of rock

(c) How do you know that two of the rock types in the area are limestone and coal?

Selinsgrove
Middleburg
Shamokin
Trevorton
Uniontown
Marion Heights
Kulpmont
Mount Carmel
Atlas
LINE MOUNTAIN
MAHANTANGO MOUNTAIN
HOOFLANDER MOUNTAIN
FISHER RIDGE
NORTHUMBERLAND COUNTY
DAUPHIN COUNTY
SCHUYLKILL COUNTY
GOOD SPRING MTN
BIG LICK MOUNTAIN
BEAR MTN
Valley View
Tremont
Pine Grove
Wiconisco
Williamstown
Tower City
Lykens
Elizabethville
Millersburg
Grafts Island
Byers Island
HALDEMAN STATE FOREST
BERRY MOUNTAIN
BROAD MOUNTAIN
PETERS MOUNTAIN
STONY MOUNTAIN
HALF FALLS MOUNTAIN
BUFFALO MOUNTAIN
DIVIDING RIDGE
LITTLE MOUNTAIN
SWATARA GAP
Halifax
Berrysburg
Gratz

Name________________________________ Date______________

Instructor ____________________________ Section______________

18.5 The photographs in Figure 18.6 show the eastern half of the Santa Ana Mesa near the confluence of the Jemez River and the Rio Grande in New Mexico. The mesa is built of a series of basalt flows which issued from a north to south trending string of vents to the west. Numerous faults displace the basalts from a few to 150 feet.

(a) Given that the faults dip towards the west identify the upthrown and downthrown blocks and determine what kinds of faults they are.

(b) Have the fault blocks been uplifted or depressed evenly or are they tilted and if so, in which way?

(c) Identify on the photographs a fault scarp and a fault line scarp. Label these (A) and (B) respectively.

(d) In what ways has the faulting in this area influenced the drainage pattern?

Figure 18.5 Harrisburg, Pennsylvania (1:250,000 A.0-1.0 in southwest corner).

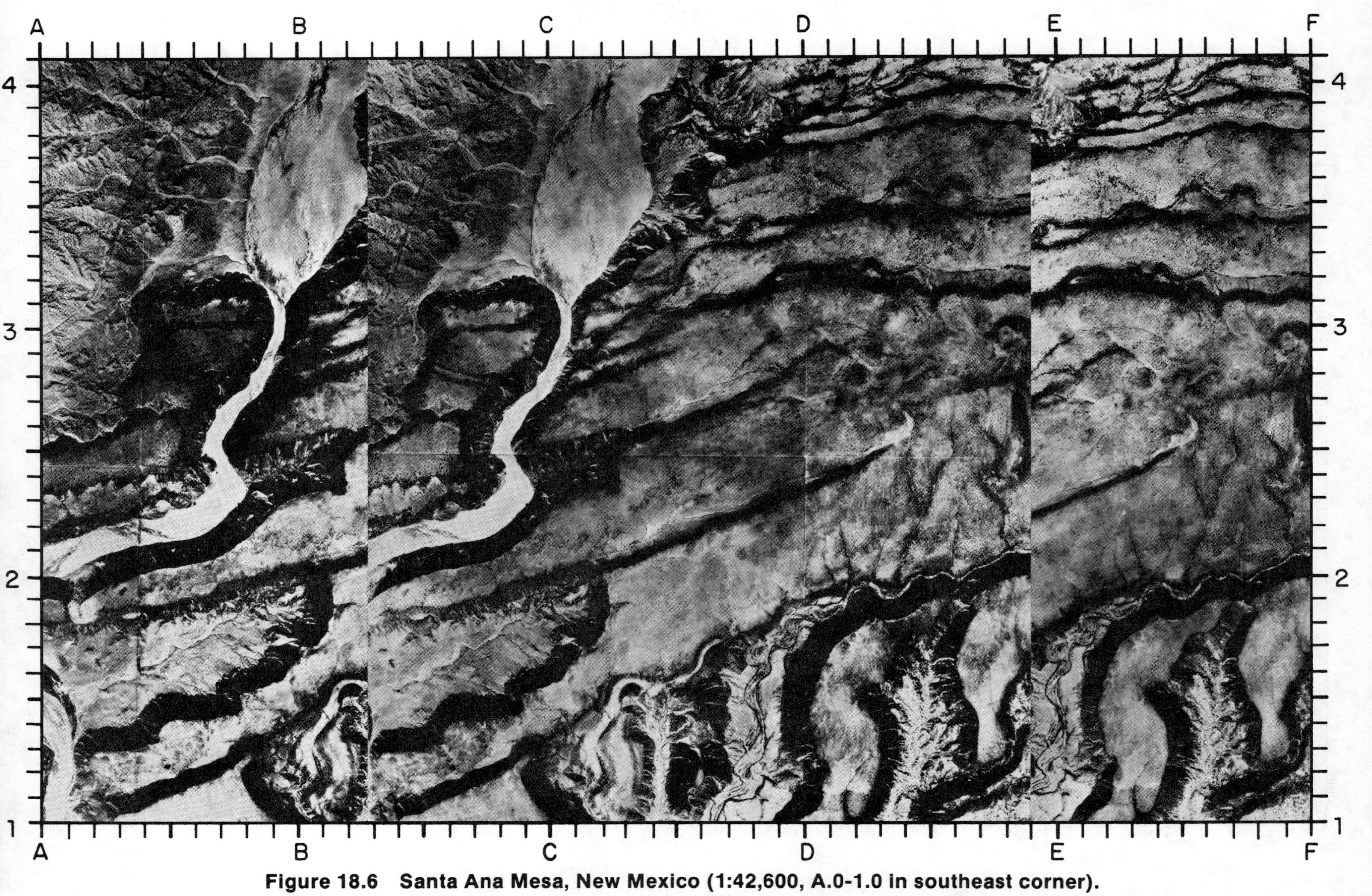

Figure 18.6 Santa Ana Mesa, New Mexico (1:42,600, A.0-1.0 in southeast corner).

EXERCISE 19

HILLSLOPE FORM AND MASS WASTING

Hillslopes are that part of the landscape between the crests of hills and nearby drainage lines and therefore include the majority of the earth's surface. The form of any slope is ultimately determined by the relationship between the rate of weathering of the underlying rock and the rate of removal of rock debris from the surface. Movement of debris under the influence of gravity is called *mass wasting* and occurs when the downslope gravitational stress on the material exceeds its frictional resistance to movement. Hillslopes have four main elements which are perhaps best called the *crest*, the *scarp*, the *debris-slope*, and the *foot-slope* (Figure 19.1). One or more of these elements may be absent from any given hillslope.

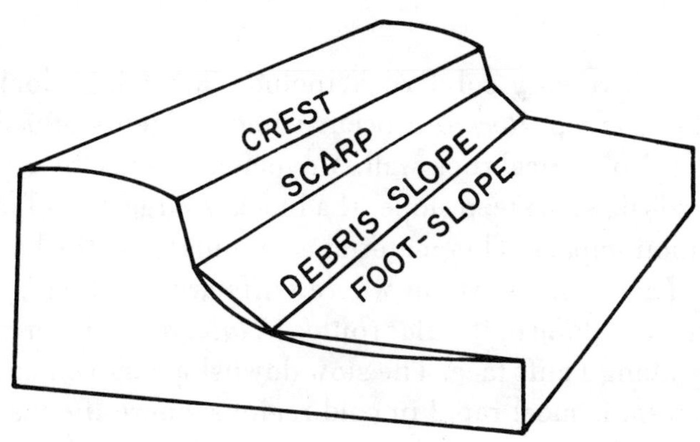

Figure 19.1 The four main hillslope elements.

Although forming a continuous series from the standpoint of slope formation, mass movements are generally classified according to whether they involve compact rock or unconsolidated materials; the rapidity of movement; and whether material slides or falls in discrete units or whether it flows and is thus internally deformed. The various types of mass movement are listed in Table 19.1, and the most important illustrated in Figure 19.2. Some mass movements occur suddenly and involve vast amounts of material, while others are barely or not at all perceptible. Catastrophic movements have blocked valleys with the formation of lakes, diverted drainage, and taken lives; and like other mass movements have occurred more frequently where natural slopes have been undercut by groundwater seepage, rivers or waves.

Table 19.1 Classification of mass movements (after Butzer (1976))

Type of Movement	Type of Material: Bedrock	Type of Material: Soil or Regolith
Falls	Rockfall	Soilfall
	Block slump (rotational)	
Slides	Block glide (planar)	Soil slump
	Rockslide	Debris slide
Slow flows	Talus creep	Soil creep
	Block streams	Solifluction
Rapid flows	Dirty snow avalanche	Rapid earthflow Mudflow
Complex	Combinations of materials or type of movement	

Mass movements involving solid rock include rockfalls, block slumps, block glides, rockslides and talus creep. *Rockfalls* occur when segments of bedrock of any size are detached from a steep to vertical rock wall. Movement may be by vertical fall or by a series of leaps and bounds down a steep slope. If a block is large or its fall great it may be completely shattered upon impact. This material accumulates at the base of the slope in talus cones or aprons. *Talus cones* develop where cliffs are notched by narrow ravines which funnel rock fragments along particular routes. A *talus apron* is formed when several cones coalesce laterally along a cliff face. The slow downslope movement of fragments in talus, known as *talus creep,* is most rapid in cold regions where the major trigger is the alternate freezing and thawing of ice in the interstices of the rock waste. The angle at which falling debris comes to rest is called the *angle of repose* and usually varies from 25°-40° depending upon the size, shape, and density of debris; the roughness of particle surfaces; and how well sorted the fragments are. In general an increase in fragment size or in the height of fall reduces the angle of repose.

On moderately steep slopes masses of bedrock may become dislodged and slide downslope. *Block gliding* is said to occur when blocks move along a flat, inclined plane and *block slumping* when blocks rotate backward on a concave-up slip plane. Block gliding and slumping normally occur intermittently over a long period although some movements are rapid and final. Movements are usually caused by the removal of support from beneath the toe of the slip block. The most destructive mass movements involving solid rock are rockslides which occur on very steep bedrock slopes and so are most common in mountainous areas. *Rockslides* are downward and usually rapid movements of detached segments of bedrock which slide on bedding, joint, or fault surfaces. Movements may involve thousands of tons of rock and minor amounts of soil and other debris. Rockslides occur most frequently in soluble rocks such as limestones and dolomites which are weakened by the action of solution which gradually widens joint and bedding plane structures.

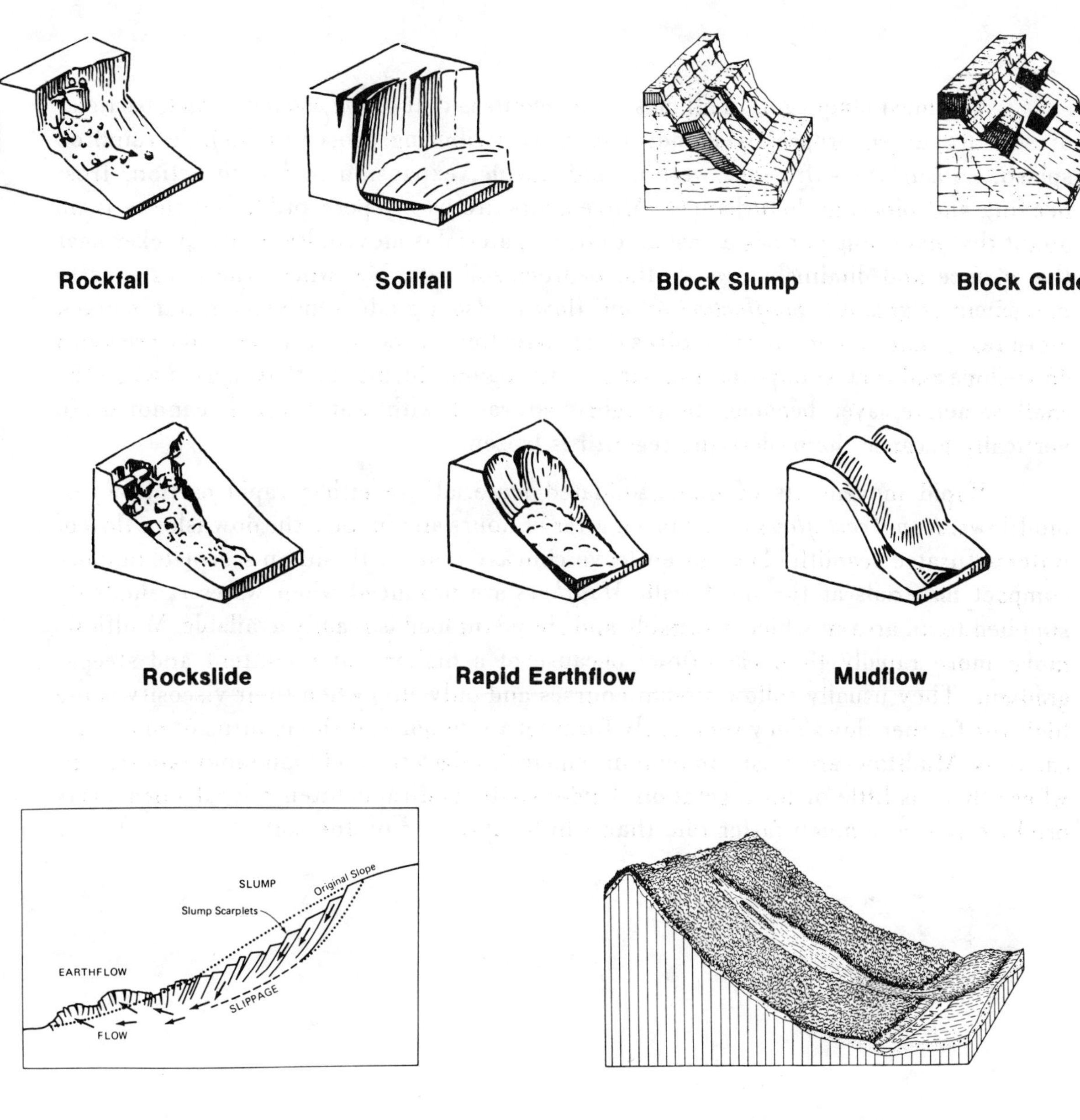

Figure 19.2 Types of mass movement (after Leopold et al. 1964, Butzer, 1976 and Sharpe 1960).

Mass movements in soil or regolith differ from those in bedrock because most involve some internal deformation or flow. However, when soil moves as one, or a series of rigid units, movements are similar to those that occur in solid rock. *Soilfalls*, *soil slumps* and *debris slides* in regolith are equivalent to rockfalls, rock slumps, and rockslides in solid rock. Debris slides are often called *debris avalanches* or *debris flows* depending upon the relative importance of the part played by water in the movement.

The most ubiquitous of all mass movements is that of *soil creep* in which particles of soil are urged progressively downslope. Contributing processes include raindrop splash, subsurface suffosion, wetting and drying, expansion and contraction, frost heaving and biogenic disturbances. Movements are barely perceptible but there is no doubt that over long periods a vast amount of material is moved. Rates are quicker near the surface and diminish towards the bedrock-soil interface where the resistance to movement is greatest. *Solifluction* or soil flow is also a gradual movement but is much more rapid than soil creep. It involves the plastic flow of water-saturated soil or regolith downslope and is most important in permafrost regions during the thaw period when the shallow active layer becomes thoroughly saturated with water which cannot drain vertically because the underlying regolith is frozen.

Rapid movements of unconsolidated material are either rapid earthflows or mudflows. *Rapid earthflows* occur in a matter of hours and involve the downslope flow of water-saturated regolith. They often leave concave scars with slump scarplets in more compact materials at the head wall. *Mudflows* are produced when water is suddenly supplied to an area in which a suitable and abundant load is readily available. Mudflows move more rapidly than earthflows because of a higher water content and steeper gradient. They usually follow stream courses and only stop when their viscosity is too high for further flow. They frequently form lobate tongues at the mouths of mountain canyons. Mudflows are most common in semiarid, subarctic, and high mountain regions where there is little or no vegetation. Under such conditions intense local storms may produce rain at a much faster rate than can be absorbed by the soil.

Name ______________________ Date ______________

Instructor ______________________ Section ______________

QUESTIONS

19.1 As can be seen in Figure 19.3, instability of the southern slope of the basalt-capped mesa north of the settlement of King Hill on the Snake River (just off the photos to the south) and east of King Hill Creek (A.5-3.5) caused a major mass movement. The mesa has a relative relief of approximately 1,000 feet.

(a) Identify the type of mass movement and explain what happened.

(b) On the photographs outline the region covered by collapse debris and calculate its approximate area in square miles.

(c) Three well-defined slump blocks are clearly visible in the photographs, label these.

(d) Did debris reach King Hill Creek and if so where did this occur?

(e) Follow King Hill Creek north and east and mark areas of valley slope collapse. Is there any evidence to suggest that collapse debris may once have blocked the valley.

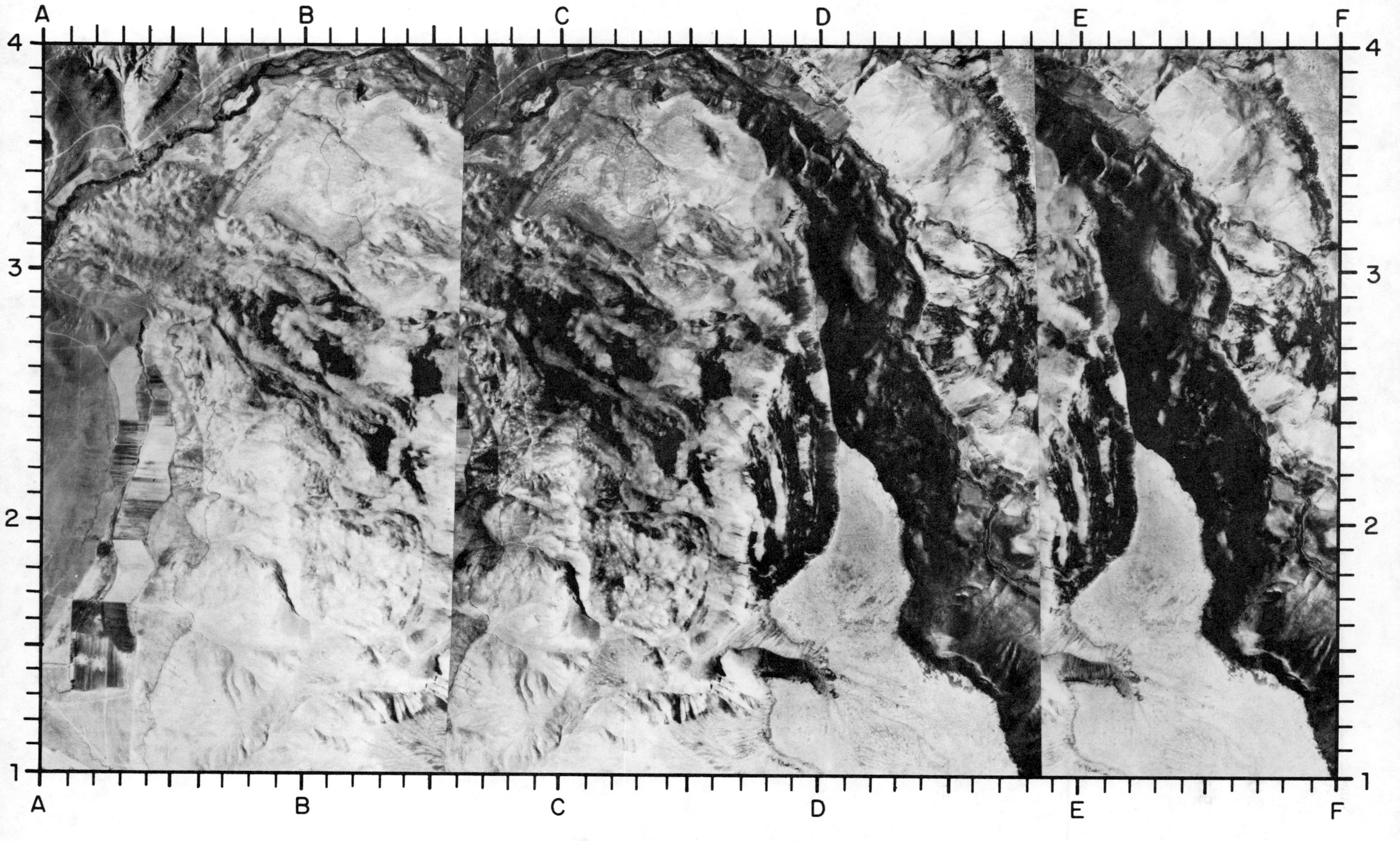

Figure 19.3 King Hill, Idaho (1:26,300, A.0-1.0 in southeast corner).

Name________________________________ Date________________

Instructor_____________________________ Section______________

19.2 The aerial photographs in Figure 19.4 show a prehistoric mass movement (D.5-2.7) that affected the east slope of Lake Fork Gunnison River Valley. The lake in the photograph is Lake San Cristobal.

(a) What kind of mass movement occurred?

(b) What relationship is there between the moved debris and Lake San Cristobal?

(c) The lower one third of the debris is stable and contains wood that has a radiocarbon age of about 700 years. The oldest trees in the upper two-thirds of the debris are approximately 350 years old. What do these dates tell us about the age of the original mass movement and subsequent activity?

19.3 Among the geomorphic features that are visible in the aerial photographs of the southern Mackenzie Mountains region of Canada (Figures 19.5 and 19.6) are two major mass movements.

(a) What kind of mass movement has occurred in the unconsolidated sands (D.1-2.0) in Figure 19.5 and what triggered it?

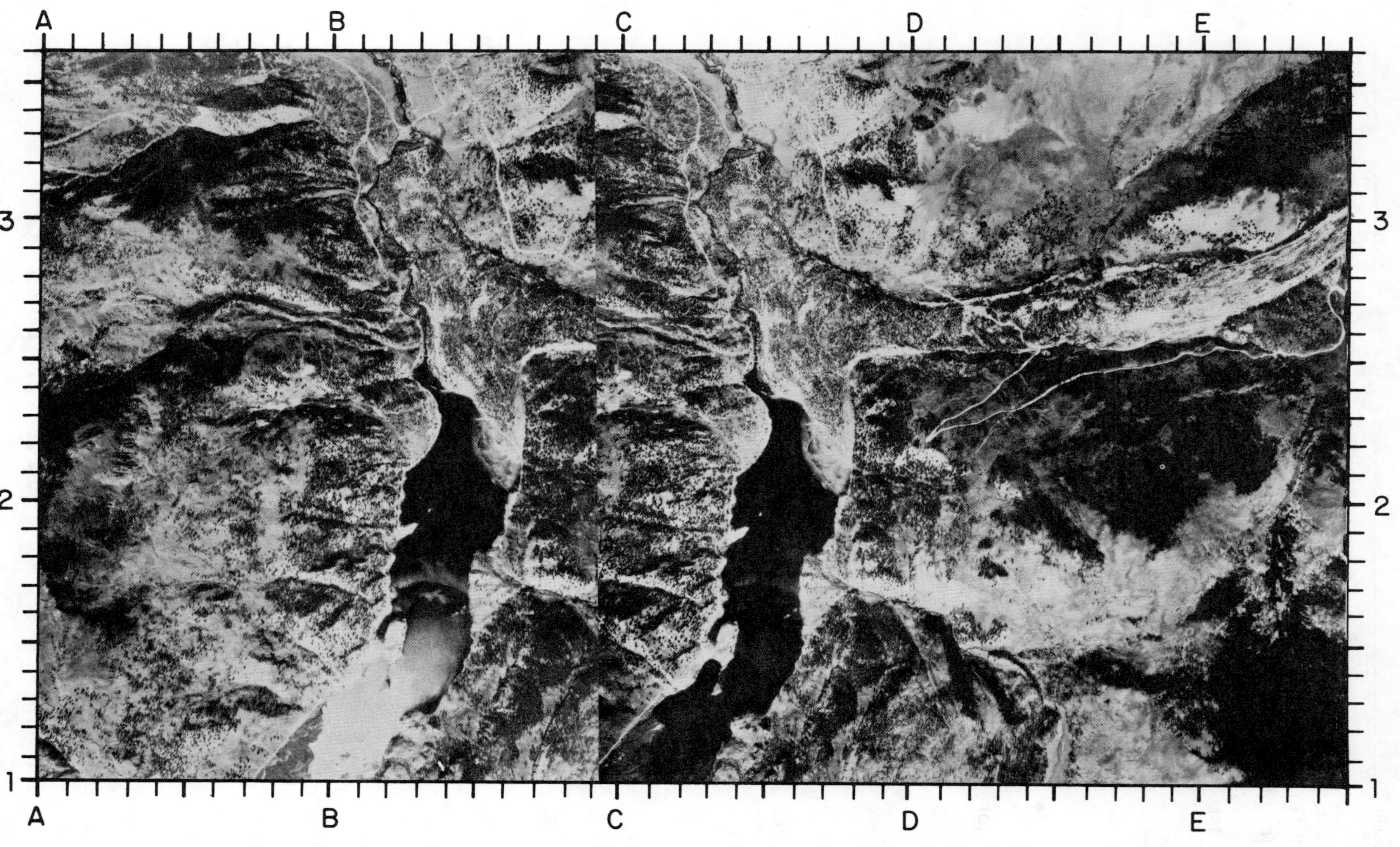

Figure 19.4 Slumgullion, Colorado (1:62,700, A.0-1.0 in southwest corner).

Name________________________________ Date__________________

Instructor ___________________________ Section________________

(b) Given that the bedrock of the area shown in Figure 19.6 is largely limestone and dolomite explain in detail what has happened at C.0-1.9. If the approximate scale of the photographs is 1:55,000 what is the area covered by collapse debris, and is there evidence that the mass movement was structurally controlled?

19.4 Figure 19.7 shows a 600 feet high granite inselberg in the Transvaal Province of South Africa. The domal form of this inselberg is due to curved sheeting structures which develop in the rock. These are clearly visible in the photograph. On the photograph delimit and label the following hillslope elements.

A Crest
B Scarp
C Debris Slope
D Foot-Slope or Pediment

19.5 Figure 19.8 is a view of Rio de Janeiro, Brazil as seen from Sugar Loaf Mountain. The local bedrock is granite and sheeting structures frequently develop parallel to the hillslopes. In some cases rock sheets may exceed 20 feet in thickness. In terms of slope stability explain the presence of the white stone wall and white concrete pillars on the steep granite slope at the left of the photograph.

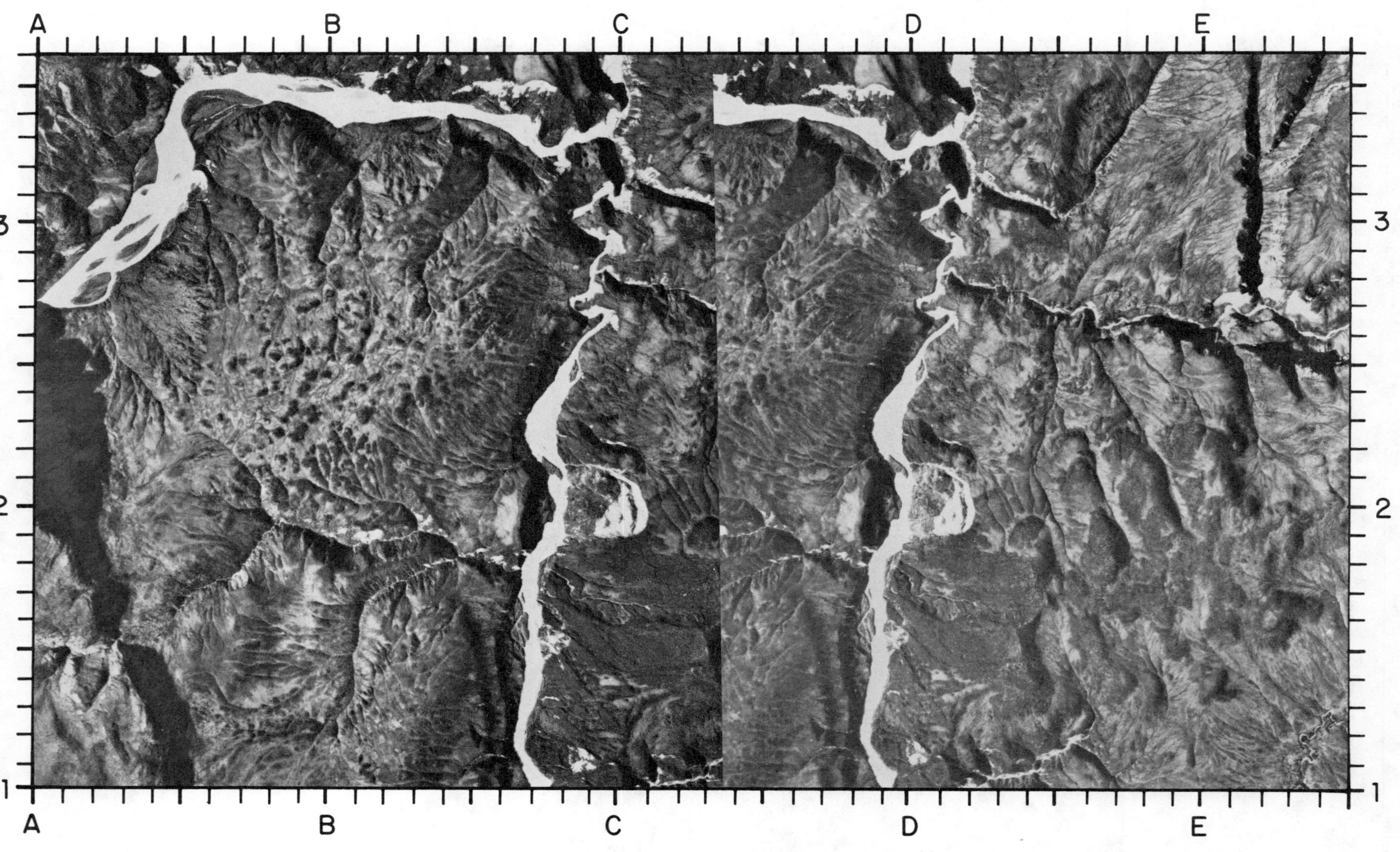

Figure 19.5 Sundog Basin, Canada (1:55,000 A.0-1.0 in southwest corner).

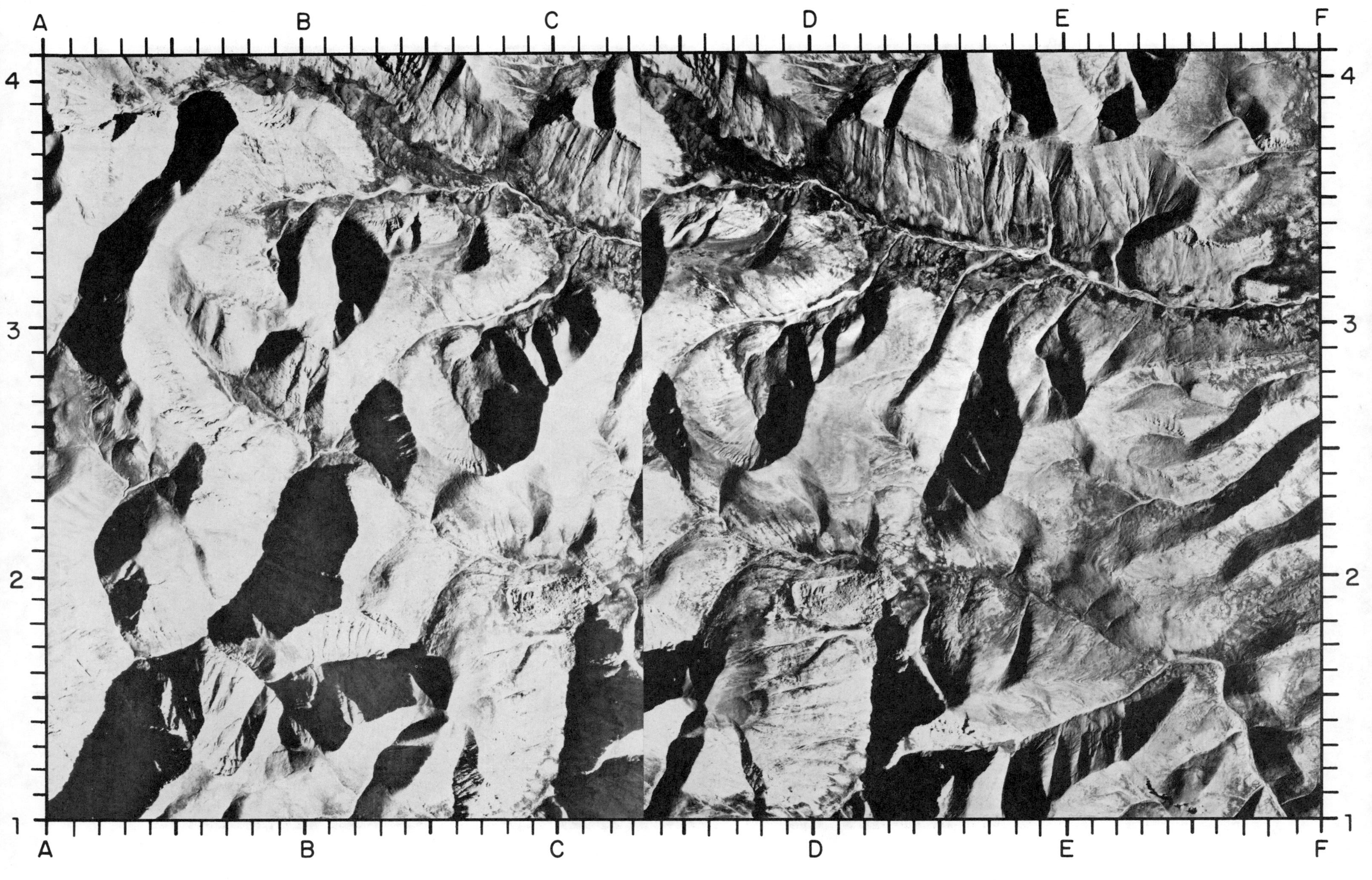

Figure 19.6 Nahanni Plateau, Canada (1:55,000, A.0-1.0 in southwest corner).

Figure 19.7 Granite inselberg, Transvaal, South Africa.

Figure 19.8 Rio de Janeiro, Brazil as seen from Sugar Loaf Mountain.

Name ____________________ Date ____________________

Instructor ____________________ Section ____________________

19.6 What evidence is there in Figure 19.9 to indicate that there has been a relatively recent rockslide in Surprise Valley, near Jasper in the southern Canadian Rockies? Given that the local bedrock is limestone briefly outline the events that probably preceded this catastrophic mass movement.

Figure 19.9 Rockslide in Surprise Valley, Southern Canadian Rockies.

19.7 Figures 19.10 and 19.11 show the results of two mass movements that occurred fairly recently in the MacKenzie Mountains region of northwest Canada. With reference to Table 19.1 and Figure 19.2 classify these mass movements.

Figure 19.10 Recent mass movement in Sundog Basin, Mackenzie Mountains, Canada.

Figure 19.11 Recent mass movement in Sundog Basin, Mackenzie Mountains, Canada.

EXERCISE **20**

FLUVIAL LANDSCAPES

Of all the processes which sculpture the earth's surface the single most important is that of running water. *Streams* drain water from the land via the valleys which they occupy. A stream whose course developed when water first flowed down a newly created landsurface is a *consequent stream* and normally parallels rock dip. Streams which later adjust their courses to weak rock belts or fracture lines by differential erosion are *subsequent streams* and frequently these parallel the direction of strike. Rill, gully, creek and river are terms applied to streams of various sizes which together make up *drainage systems*. The arrangement of streams in a drainage system constitutes the *drainage pattern* and usually reflects the structural and lithological characteristics of the underlying rocks. Branching or tree-like *dendritic* drainage systems are the most common. *Trellis* and rectangular patterns develop where rocks are either well jointed or in belts of alternate weak and resistant beds. *Radial* consequent stream patterns develop on volcanic cones and structural domes and sometimes becomes *annular* as subsequent streams develop in exposed softer rock layers. *Parallel* drainage develops where main and tributary streams nearly parallel one another in the direction of the regional slope (Figure 20.1).

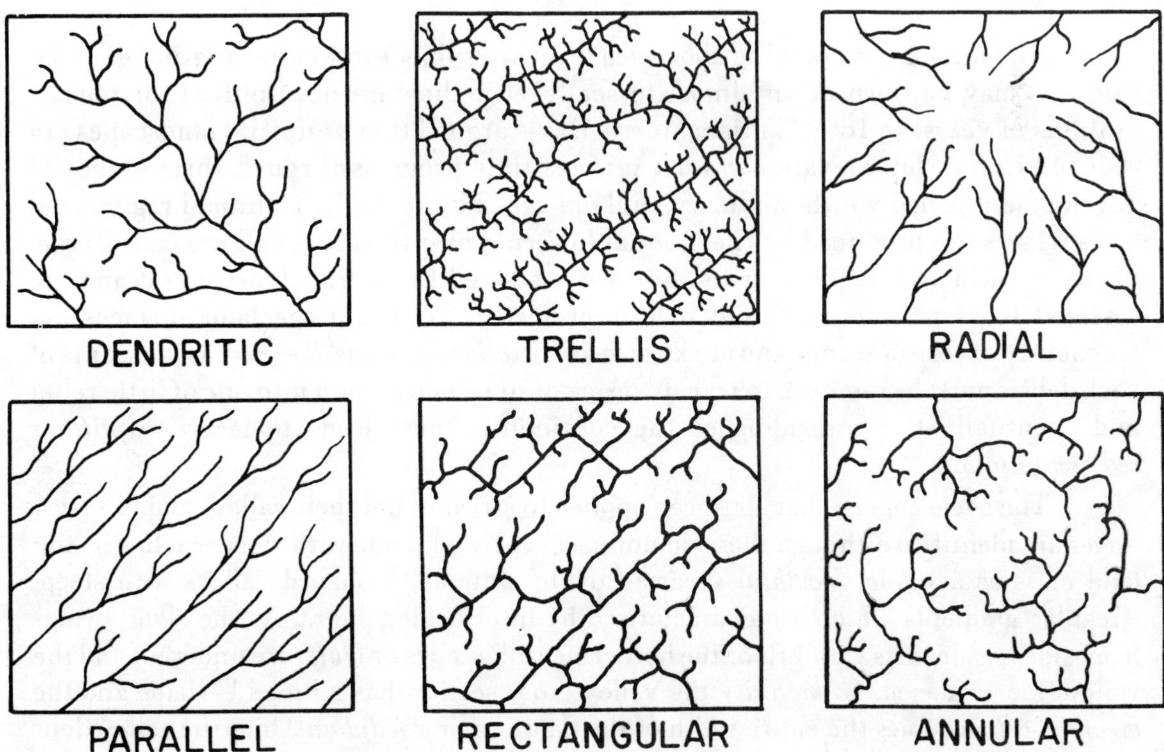

Figure 20.1 Common stream drainage patterns.

If rates of erosion in the *headwaters* of different drainage systems are not the same *stream capture* may take place by which a stream in one system captures or diverts part of the drainage in another system. In areas of alternate hard and soft rock layers it is common for a subsequent stream eroding its course in less resistant rock to capture part of the drainage of a consequent stream flowing across resistant rock ridges through *water gaps.* At the *elbow of capture* the consequent stream meets the subsequent stream forming a sharp bend in the course of the new trunk river. The consequent stream down-valley of the point of capture is said to be *beheaded* and, deprived of much of its drainage, it appears too small for the valley it occupies and is called a *misfit* stream. The deep gorge cut by the consequent is left virtually dry when drainage is diverted and becomes a *wind gap* (Figure 20.2).

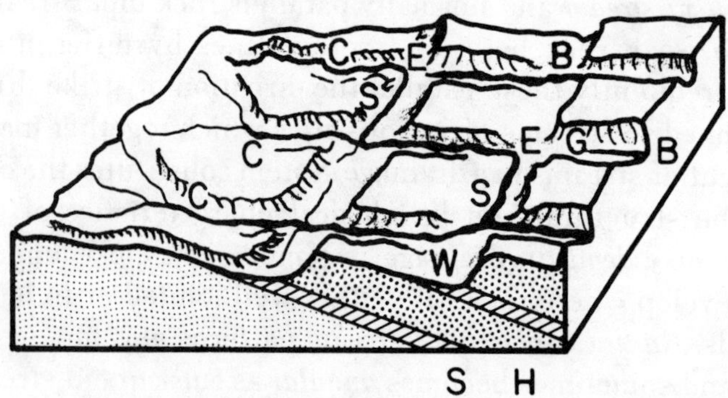

Figure 20.2 Stream Capture in an area of gently dipping hard and soft rocks (C = consequent stream, S = subsequent stream, E = elbow of capture, W = water gap, G = wind gap, B = misfit stream, S = soft rock layer, H = hard rock layer).

All land masses standing above sea level are subject to weathering and denudation and may be worn down almost to sea level if they are not uplifted for tens of millions of years. In 1899 William Morris Davis argued that as upland landscapes are reduced to flat, low-elevation *erosion surfaces* they progress through three stages of development called youth, maturity, and old age (Figure 20.3). In humid regions old age surfaces are produced by the gradual lowering of interfluves and are called *peneplains.* In arid and semiarid areas they are produced by lateral slope retreat and are referred to as *pediplains.* Residual hills projecting from old age land surfaces are frequently of resistant rock and are known as *monadnocks* or *inselbergs.* Vast amounts of rock debris must be removed to transform a mountainous terrain into one of little relief and eventually this unloading of the continental mass leads to *tectonic uplift* or *sedimentoisostasy.*

The cycle concept has also been applied to streams and their valleys and the same stages are identified although these do not necessarily coincide with stage reached by the land mass as a whole. *Youthful streams* flow in sinuous, V-shaped valleys with steep, irregular gradients which are characterized by interlocking spurs. As the river swings from side to side in its valley floor the bordering valley sides or *bluffs* are undercut and the valley floor widened. In *maturity* the valley cross section has a broad U-shape and the *meander belt* occupies the entire width of the flat floor or *floodplain.* Once the valley floor has been widened the river is free to flow in any direction and takes a highly sinuous or *meandering course.* The floodplain is widened so that in *old age* it may be several times as wide as the meander belt.

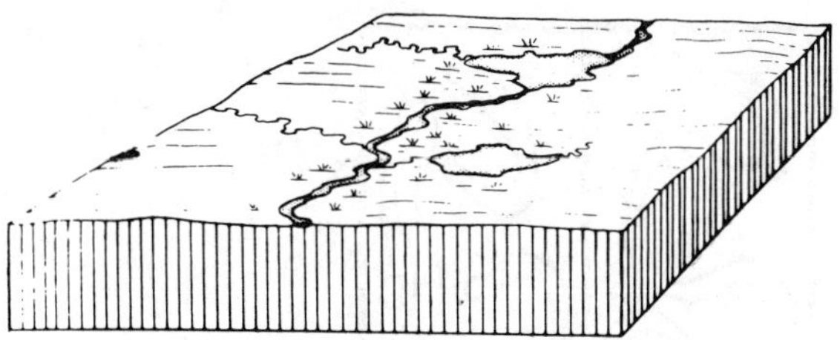

A. In the initial stage, relief is slight, drainage poor.

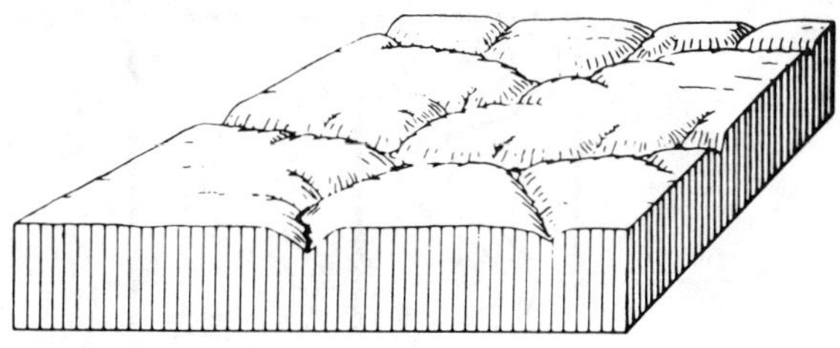

B. In early youth, stream valleys are narrow, uplands broad and flat.

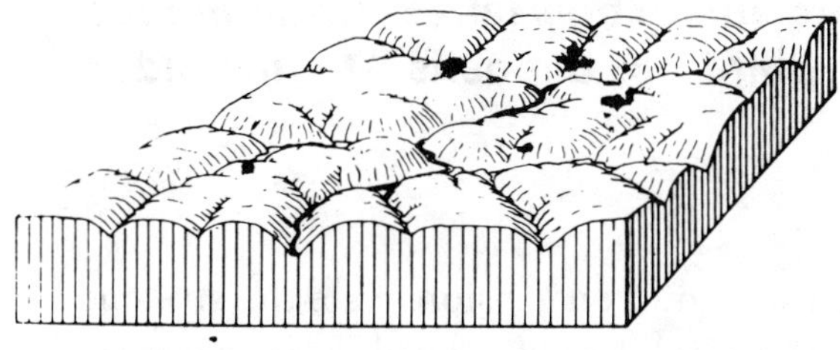

C. In late youth, valley slopes predominate but some interstream uplands remain.

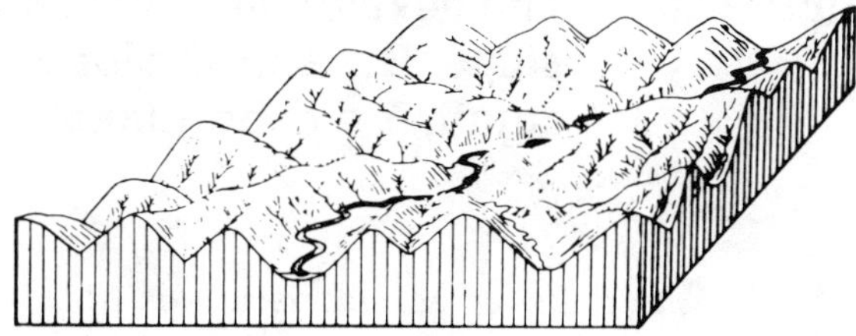

D. In maturity, the region consists of valley slopes and narrow divides.

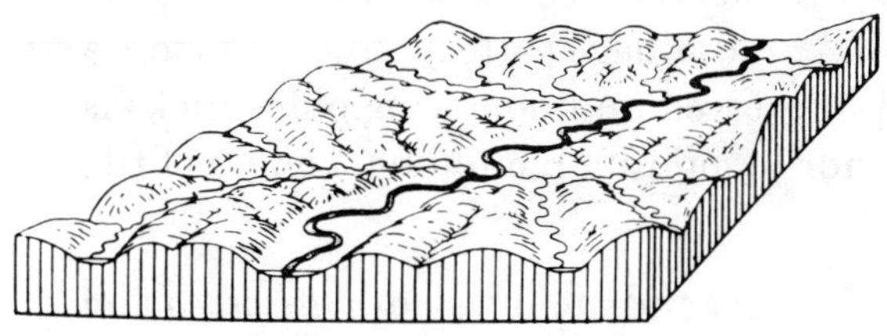

E. In late maturity, relief is subdued, valley floors broad.

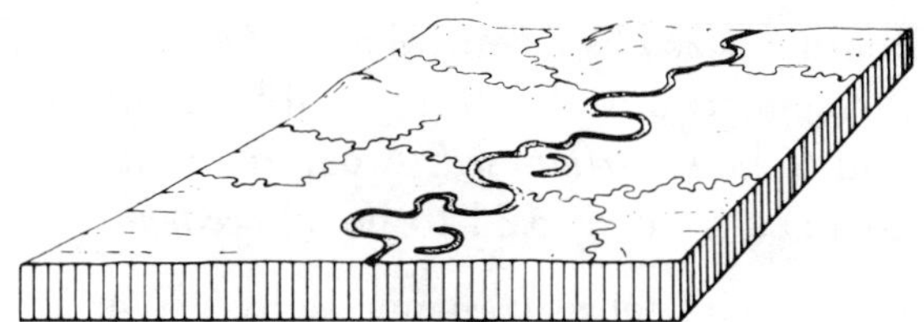

F. In old age, a peneplain with monadnocks is formed.

Figure 20.3 The cycle of land mass denudation in a humid climate (from Strahler 1975).

Meanders develop as a sinuous stream erodes its bank or *undercut slope* on the outside of a bend and accumulates alluvium or *point bar deposits* on the *slipoff slope* on the inside of the bend. As meandering continues the undercut slopes of two bends may intersect causing the intervening *meander loop* to be abandoned. The abandoned channel left by the meander *cutoff* becomes an *oxbow* lake or marsh (Figure 20.4).

Sediments laid down on the floodplains of mature and old age rivers during periods of flood are known as *overbank deposits.* The coarsest part of the load is dropped near the banks of the river channel eventually building a low embankment or *levee* on either side. The finer particles settle out more evenly across the flooded area. During floods levees may block the junctions of small tributaries which then may have to flow for miles parallel to the main channel before they can find a new entrance. Such streams are called *yazoo streams.*

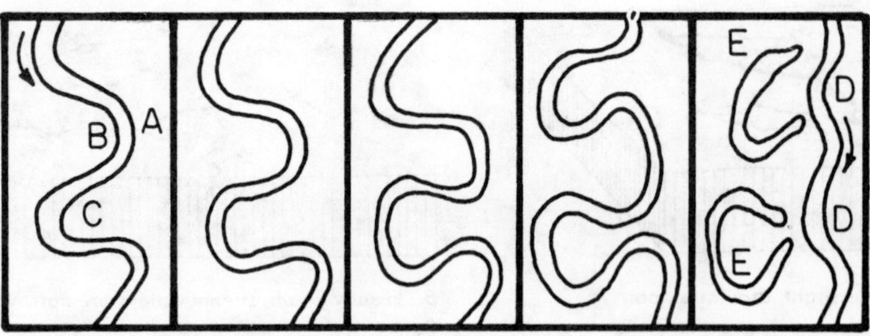

Figure 20.4 The development of meanders and oxbow lakes (A = undercut slope, B = slipoff slope, C = point bar deposits, D = meander cutoff, E = oxbow lake).

When an old age landscape is uplifted at the end of an erosion cycle the streams begin rapid downcutting to grade their profiles to the new lower base level. This process is called *rejuvenation.* If an old age stream is uplifted but not tilted it may cut a deep gorge into its former floodplain and carry the meanders into bedrock as *entrenched meanders.* Entrenched meanders also produce cutoffs, the abandoned meander loop being characterized by a central *meander core* in solid rock. A drainage pattern inherited from a former structural or erosional land surface and impressed on underlying older rocks is said to be *superimposed.* A pattern that existed prior to mountain building, and was little changed by it, is said to be *antecedent.*

The debris transported by a river is called its *load* and includes both *dissolved* and *solid* material. The proportion of solid to dissolved load is extremely variable both in space and time. On average, the major rivers of the world carry about 30% of their total annual load in solution. The major proportion of the total load carried by streams consists of silt and clay particles carried in *suspension* by the turbulence of the water. Particles too large to be carried in suspension constitute the *bedload* of a stream. Sand particles may be whirled up into the stream and carried along with it before settling on the bed once more. These particles move downstream in a series of jumps, a process known as *saltation.* Pebbles and cobbles too large to be lifted from the bed may be rolled or pushed along the channel floor by *traction, abrading,* or *corrading* the stream bed as they go. Rivers also *corrode* their channels. When a river contains debris of various shapes and sizes the maximum load that can be carried is proportional to the third or fourth powers of the velocity. For fragments of a given shape the largest size that can be moved is proportional to the *sixth power of the velocity,* which explains how enormous boulders weighing many tons can be moved during periods of flood.

Deposition of particles of a given size occurs when the current velocity falls below the minimum required for transport. When a river flows into a lake or into the sea its velocity is checked and sediment deposited with the formation of a *delta.* Deposition is more rapid in salt than in fresh water because the finer particles of clay in the suspended load *flocculate* into *aggregates* that are too large to remain in suspension. The velocities of streams flowing from mountains into nearby lowland areas are also checked with the formation of steep *alluvial fans* characterized by *braided streams.*

Name________________________ Date____________

Instructor ________________________ Section____________

QUESTIONS

20.1 Examine the maps and aerial photographs in this manual and select from them good examples of dendritic, trellis, radial, parallel, rectangular and annular drainage patterns. (Figure 20.1).

20.2 Look at the maps or aerial photographs of the following areas and determine which stage of the geographical cycle (youthful, mature, old age) each one has reached.

(a) Harrisburg, Pennsylvania (Figure 18.5).
(b) Philipp, Mississippi (Figure 2.4).
(c) Renova, Pennsylvania (Figure 20.5).
(d) Sacaton Mountains, Arizona (Figure 27.4).
(e) Pine Mountain, Oklahoma (Figure 18.3).
(f) Circle Ridge, Wyoming (Figure 18.4).
(g) Mackenzie Mountains, Canada (Figure 19.6).

20.3 Examine the Harrisburg, Pennsylvania area shown in Figure 18.5.

(a) Is the Susquehanna River an antecedent, a consequent or a superimposed stream?

(b) On the map identify good examples of a water gap and a subsequent stream.

(c) What has controlled the course of Mahantango Creek (A.3-3.2) in the west of the map-area?

Roundtop
Summerson Cem
Hoover Hollow
Bearfield
Butler Hol
Crawford Hol
Honey
Run
KETTLE
CREEK
Summerson
Owl
Hollow
Fivemile Hol
Middle
Branch
Huling
Shintown
Jordan Hol
EAST
EATING
Rock
Wildcat Hol
Camp
Run
Cooks
Crowley Hollow
Short Bend Run
Twomile
Branch
Westport
Robbins Cem
Noyes Cem
Butler Hol
Bitumen
Round Top Mountain
Milligan
Run
SINNEMAHONING
Grass Flats
Wistar
Run
CREEK
Little Round Top
Keating
BUCKTAIL
TRAIL
PENNSYLVANIA
WEST
BRANCH
Smith
Fish
Dam
Bear Trap Hol
Run

Name________________ Date________

Instructor________________ Section________

20.4 The Philipp area shown in Figure 2.4 is a part of the Mississippi floodplain although the Mississippi River is more than 100 miles distant to the west. The region is dominated by an indeterminate drainage with yazoo streams.

(a) On the map draw the boundary between the alluvial floodplain of the Mississippi River and the ancient river bluff and calculate an average east to west gradient for the floodplain in this region.

(b) Identify and explain the existence of alluvial fans at the base of the ancient Mississippi bluff.

(c) Delimit the meander belt of Tippo Bayou (B.1-3.9) and measure its average width in miles.

(d) Why do elevations on the east side of the Tallahatchie River (A.2-1.2) increase towards the stream channel?

(e) Identify and label good examples of the following:

A	Large meander	G	Slipoff slope
B	Small meander	H	Oxbow lake
C	Meander cutoff	I	Oxbow swamp
D	Meander scars	J	Abandoned channel
E	Point bar deposits	K	Floodplain swamp
F	Undercut slope	L	Yazoo stream

20.5 The map in Figure 20.5 shows the West Branch of the Susquehanna River in Pennsylvania and two of its tributaries Kettle Creek and Sinnemahoning Creek. The stream valleys are cut largely in bedrock and the stream pattern traverses the geologic structure.

(a) On the map identify two abandoned entrenched meander loops, their respective meander cores, and their points of cutoff.

Figure 20.5 Renova West, Pennsylvania (1:62,500, A.0-1.0 in southeast corner).

(b) Mark on the map the former courses of Sinnemahoning Creek and the Susquehanna River and measure how much each stream was shortened by the cutoffs.

(c) Draw a topographic section across the meander loop immediately northeast of Little Round Top (A.5-3.2) from A.3-2.9 to B.4-2.5. Label your diagram and comment on the asymmetry of this stretch of the Susquehanna Valley.

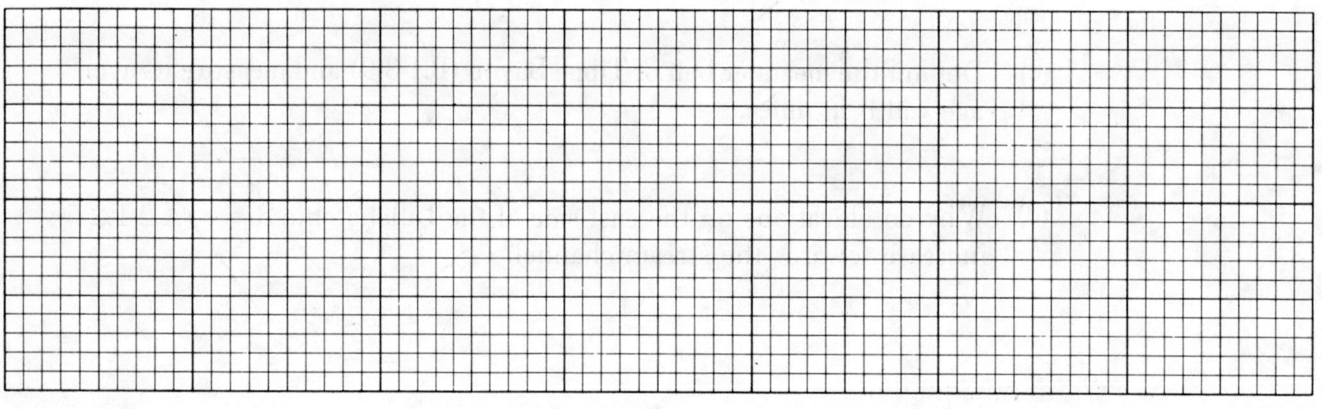

(d) Account for the entrenched creeks, rivers and river meanders in this area. Is it possible, for example, that the stream pattern is inherited from an earlier time when the river flowed at a higher level, perhaps on an erosional or depositional surface no longer visible in the present topography? If so, identify two locations where ancient floodplain deposits may have been preserved.

Name________________________ Date____________

Instructor________________________ Section____________

20.6 Meandering streams are highly sinuous channels in which the curves are nearly symmetrical. Measures which are commonly used to describe river meanders are illustrated in Figure 20.6 and Table 20.1. Sinuosity or tightness of bend is the ratio of the length of the channel in a given curve to the wavelength of the curve (L/λ), the higher the value the more sinuous the channel. It has been found that in many meandering streams the ratio of radius of curvature to width (r_c/w) lies in the narrow range 2 to 3 with a median value of 2.7. This explains why meandering streams closely resemble one another even though they may vary considerably in size.

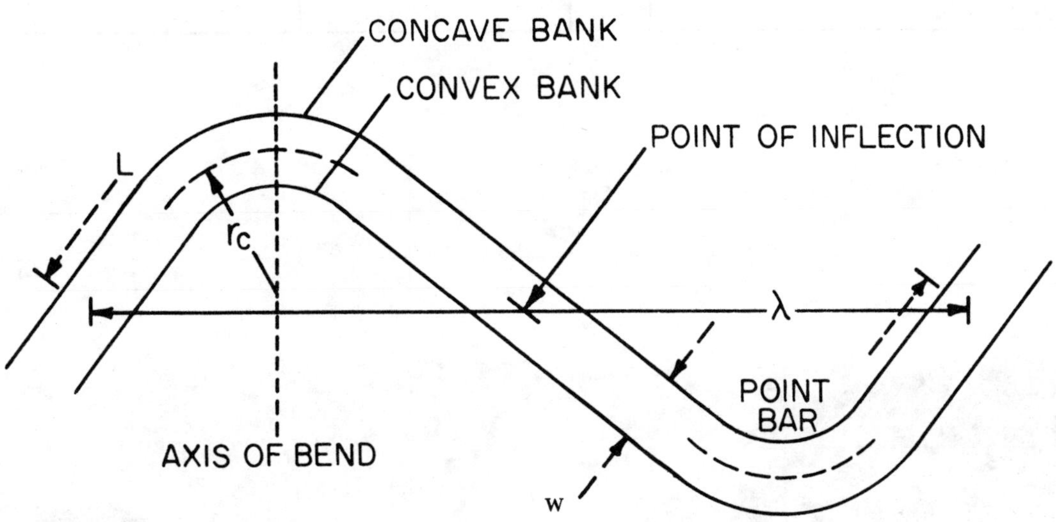

w = Width of Channel
λ = Wavelength
L = Length of Channel
r_c = Radius of Curvature

Figure 20.6 Common measures used to describe river meanders.

COMMON MEASURES	RATIO	TYPICAL FIGURES
Length of channel to wavelength (sinuosity)	$\frac{L}{\lambda}$	1.3 to 4.0
Wavelength to width	$\frac{\lambda}{w}$	7 to 15
Radius of curvature to width	$\frac{r_c}{w}$	2 to 3
Wavelength to radius of curvature	$\frac{\lambda}{r_c}$	Average 4.7

Table 20.1 Properties used to describe river meanders (after Leopold and Langbein 1966).

(a) Measure the meander properties and stream gradients of Henry's Fork (Figure 17.2) and the Rio Grande de Arecibo (Figure 23.5), and complete Table 20.2. Comment on your results.

Table 20.2 Meander Characteristics of Selected Streams

RIVER	λ	L	r_c	w	L/λ	L/w	r_c/w	λ/r_c	Stream gradient
Henry's Fork									
Rio Grande de Arecibo									

Figure 20.7 Dry Valley in the Sierra de El Abra range, Mexico.

Name________________________________ Date________________

Instructor ____________________________ Section______________

20.7 Figure 20.7 shows a high-level dry valley in the Sierra de El Abra of eastern Mexico. This range is composed of resistant limestones while the plain in the foreground is cut across shales. What geomorphic feature is the dry valley and how did it form?

Table 20.3 Dissolved and suspended load in selected rivers in different climatic regions of the United States (modified after Leopold et al. 1964).

River and Location	Drainage Area (sq mi)	Average Discharge, Q (cfs)	Discharge ÷ Drainage Area (cfs/sq mi)	Average Suspended Load	Average Dissolved Load	Total Average Suspended and Dissolved Load	Total Average Load ÷ Drainage Area (tons/sq mi/yr)	Dissolved Load as Percent of Total Load (%)
				(millions of tons/yr)				
Little Colorado at Woodruff, Arizona	8,100	63.3	.0078	1.6	.02	1.62	199	1.2
Canadian River near Amarillo, Texas	19,445	621	.032	6.41	.124	6.53	336	1.9
Colorado River near San Saba, Texas	30,600	1,449	.047	3.02	.208	3.23	105	6.4
Bighorn River at Kane, Wyoming	15,900	2,391	.150	1.60	.217	1.82	114	12
Green River at Green River, Utah	40,600	6,737	.166	19	2.5	21.5	530	12
Colorado River near Cisco, Utah	24,100	8,457	.351	15	4.4	19.4	808	23
Iowa River at Iowa City, Iowa	3,271	1,517	.464	1.184	.485	1.67	510	29
Mississippi River at Red River Landing, Louisiana	1,144,500[b]	569,500[b]	.497	284	101.8	385.8	337	26
Sacramento River at Sacramento, California	27,000[c]	25,000[c]	.926	2.85	2.29	5.14	190	44
Flint River near Montezuma, Georgia	2,900	3,528	1.22	.400	.132	.53	183	25
Juniata River near New Port, Pennsylvania	3,354	4,329	1.29	.322	.566	.89	265	64
Delaware River at Trenton, New Jersey	6,780	11,730	1.73	1.003	.830	1.83	270	45

20.8 A large percentage of the sediment transported by streams is carried in suspension or in solution.

(a) Examine Table 20.3 and determine the relative importance of the suspended and dissolved loads of streams. Give some typical figures to support your conclusion.

(b) In Table 20.4 the discharge per square mile within a drainage basin is listed as a climatic indicator. As can be seen in Table 20.3 low discharges occur in arid and semiarid areas and higher discharges in humid temperate areas. Why is the dissolved load in streams less important in arid and semiarid terrains than it is in humid temperate regions?

Table 20.4 Variation with climate of ratio of dissolved load to total load. (modified after Leopold et al. 1964).

Climatic Indicator, Discharge per Square Mile (cfs/sq mi)	Dissolved Load as a Percentage of Total Load
0–0.1	9
0.1–0.3	16
0.3–0.7	26
>–0.7	37

EXERCISE **21**

THE DRAINAGE BASIN

The work of Horton and Strahler has shown that a stream network is a hierarchy of *channel segments* of different importance or *order*. Using Strahler's method of subdivision each fingertip tributary is a *first order stream*. When two streams of similar order combine the resulting stream is an order higher so that two first order streams become a second order channel and two fourth order streams become a fifth order channel. When a lower order stream joins one of higher order there is no increase in order. Drainage basins are also ordered so that the *trunk stream* in a sixth order basin is of order six.

Horton compared the characteristics of stream segments of one order with those of the next highest order in terms of a number of simple ratios which he found remained almost constant from one stream order to the next within the same drainage basin (Table 21.1). The next stage was the realization that stream order is related to number of streams, channel length, basin area, and stream gradient by simple geometric relationships. In fact number, length, area and gradient, the dependent variables (Y), are related to stream order, the independent variable (X), by either a positive or a negative exponential function. Stream order (X), therefore, plots against these variables (Y) as a series of straight lines on semilogarithmic graph paper (Figure 21.1). Horton's exponential drainage basin laws are important because they demonstrate that far from being chaotic, the fluvial landscape is very highly ordered (Table 21.2).

Table 21.1 Important ratios between streams of different order in a drainage basin

Bifurcation Ratio	$R_b = \frac{N_u}{N_{u+1}}$
Length Ratio	$R_L = \frac{\bar{L}_u}{\bar{L}_{u-1}}$
Slope Ratio	$R_s = \frac{\bar{S}_u}{\bar{S}_{u-1}}$
Area Ratio	$R_a = \frac{\bar{A}_u}{\bar{A}_{u-1}}$

Channel density within a drainage basin depends upon the interaction between the eroding force of flowing water and the resistance of the soil to erosive channeling. The eroding force increases with an increase in the slope of the land and in the amount and intensity of precipitation. Variations in the permeability of rock and soil, vegetative cover, and soil cohesion are factors that determine resistance to erosion.

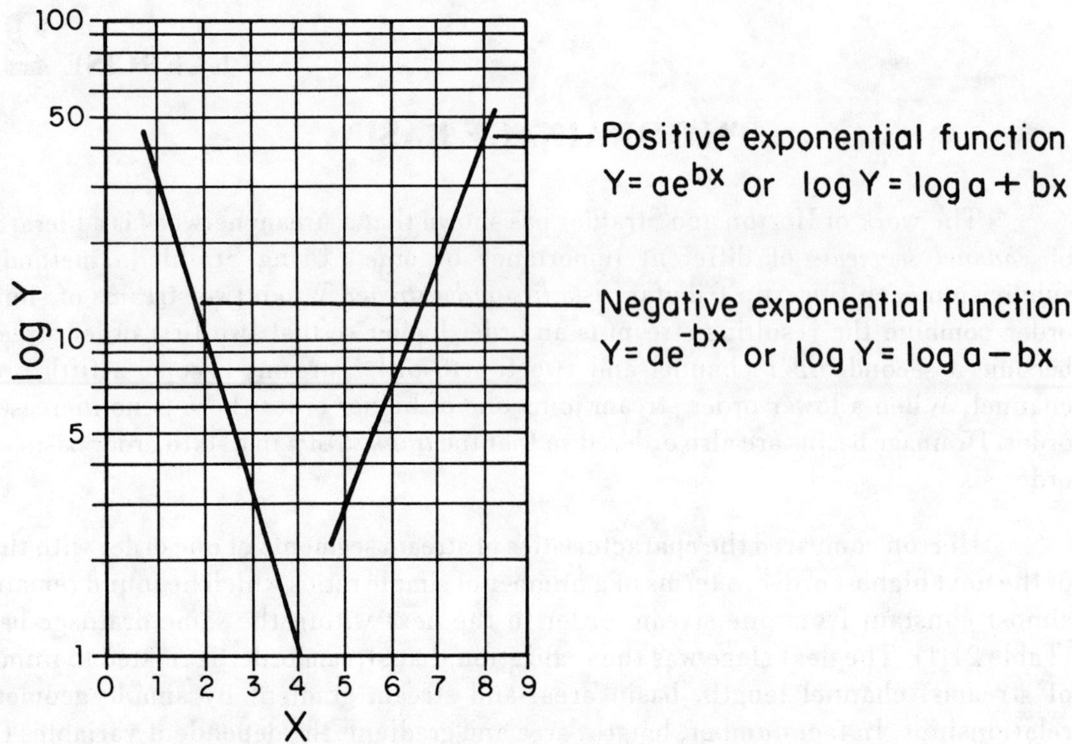

Figure 21.1 Positive and negative exponential functions plotted on semi-logarithmic graph paper.

Table 21.2 Drainage basin network laws.

Law of Stream Numbers	$N_u = R_b^{(k-u)}$
Law of Mean Stream Lengths	$\bar{L}_{u^*} = \bar{L}_1 R_L^{(u-1)}$
Law of Stream Gradients	$\bar{S}_u = \bar{S}_1 R_s^{(u-1)}$
Law of Basin Areas	$\bar{A}_u = \bar{A}_1 R_a^{(u-1)}$

u	order
N_u	number of streams of order u
$\bar{L}_u$	mean stream length of order u
$\bar{S}_u$	mean gradient of stream of order u
$\bar{A}_u$	mean area of basin of order u
k	is highest order of the basin
$\bar{L}_1$	mean length of first order segments
$\bar{S}_1$	mean slope of first order segments
$\bar{A}_1$	mean areas of first order basins
$\bar{L}_{u^*}$	cumulative mean length of streams up to order u

Name ____________________ Date ____________

Instructor ____________________ Section ____________

QUESTIONS

21.1 Order the stream network shown in Figure 21.2 by color coding stream segments of the same order and providing a key. What order drainage basin is depicted?

21.2 Count the number of streams of each order. Add these to Table 21.3 and calculate bifurcation ratios. In regions of uniform climate, rock type and relative relief the value of R_b tends to remain constant from one order to the next. Values of R_b between 3 and 5 are characteristic of natural stream systems. In light of these comments, discuss your results.

Table 21.3 Bifurcation ratios of streams in a hypothetical drainage basin

Stream Order	Number of Streams	Bifurcation Ratio
1		
2		
3		
4		

21.3 Plot number of streams (logarithmic axis) against stream order (arithmetic axis) on semilogarithmic graph paper. Does Horton's law of stream numbers hold for this drainage basin and if so is the relationship a positive or a negative exponential one?

21.4 Measure the lengths of all streams by order and calculate the mean stream length for each order. Add these data to Table 21.4 and calculate length ratios.

Table 21.4 Length ratios of streams in a hypothetical drainage basin

Stream Order	Total Length of Streams	Mean Stream Length	Length Ratio	Cumulative Mean Stream Length
1				
2				
3				
4				

21.5 Plot cumulative mean stream length (logarithmic axis) against stream order (arithmetic axis) on semilogarithmic graph paper. The cumulative mean stream length of second order streams is the mean stream length of second order channels added to the mean stream length of first order channels. Does Horton's law of stream lengths hold for this drainage basin and if so, is the relationship a positive or negative exponential one?

Name____ Date____________________

Instructor ___________________________________ Section__________________

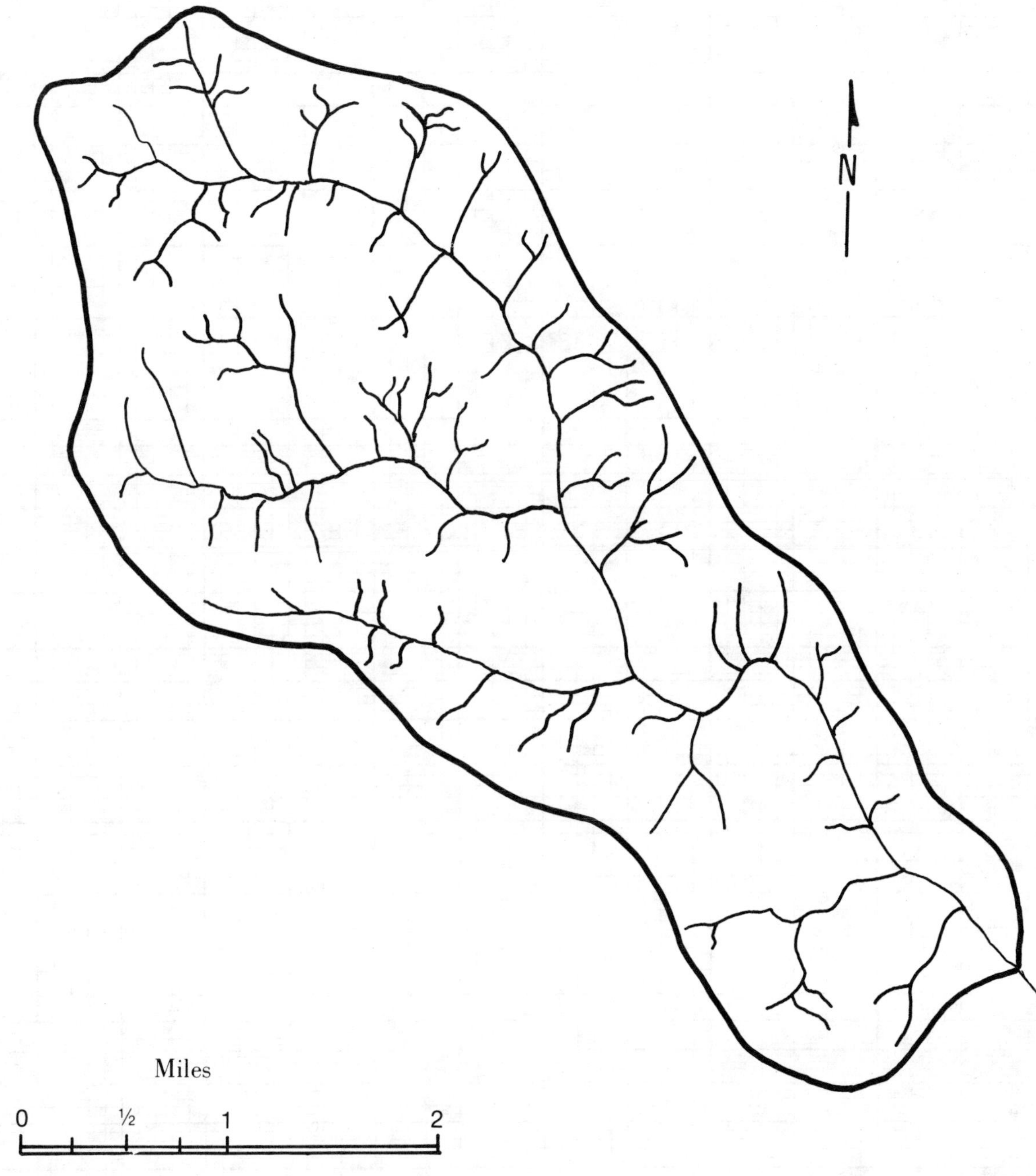

Figure 21.2 Parks Creek, Georgia

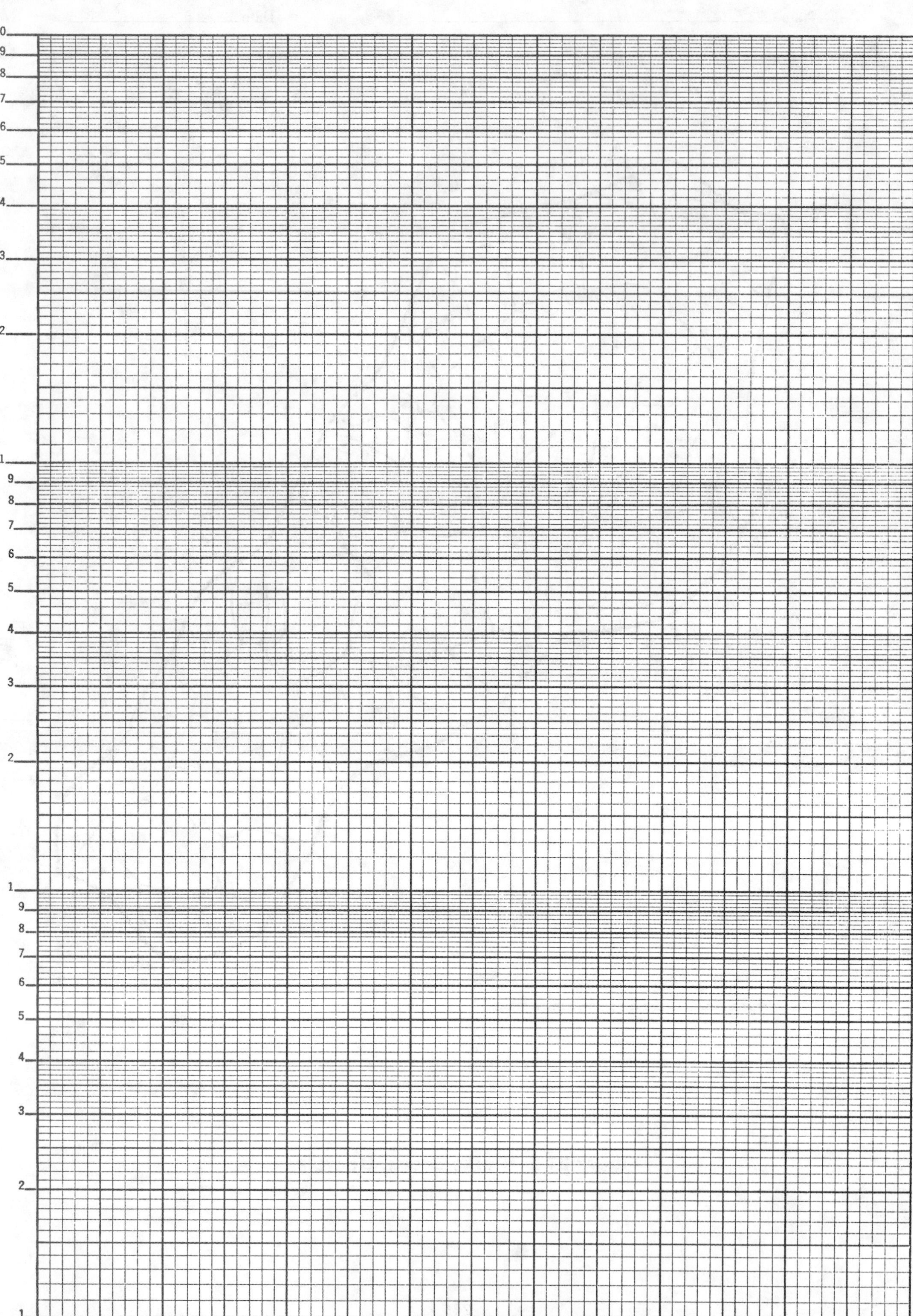

Name ______________________ Date ______________

Instructor ______________________ Section ______________

21.6 Drainage density is given by $D_d = \frac{\Sigma L}{A}$ where D_d is the drainage density in miles/square mile, ΣL represents the total length of all channels of all orders, and A is the total basin area. Table 21.5 summarises information contained in Strahler (1957) concerning ranges in drainage density and the texture of topography. The more highly dissected a terrain the finer the topographic texture. Calculate the area of the drainage basin (use the transparent graph paper in the pocket of the manual) and the total length of stream channels in it and estimate the drainage density in miles/square mile. Examine Table 21.5 and determine the density and texture classes into which this basin falls.

Table 21.5 Drainage density and texture classes of topography

Drainage Density Class	Topographic Texture Class	Drainage Density (miles/sq. mile)
Low	Coarse	< 8
Medium	Medium	8 - 24
High	Fine	24 - 200?
Badlands	Ultrafine	> 200

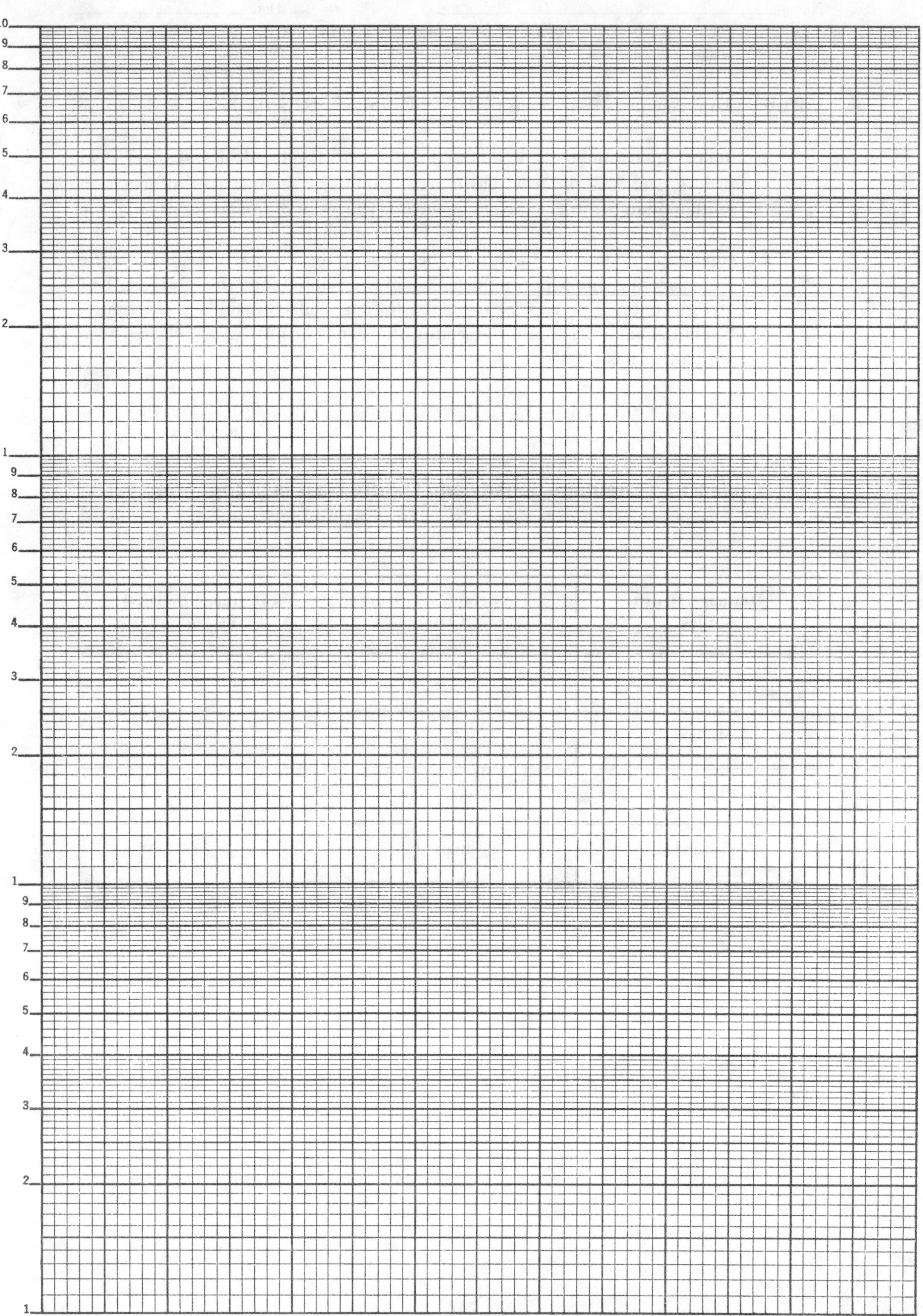
10
9
8
7
6
5
4
3
2
1
9
8
7
6
5
4
3
2
1
9
8
7
6
5
4
3
2
1

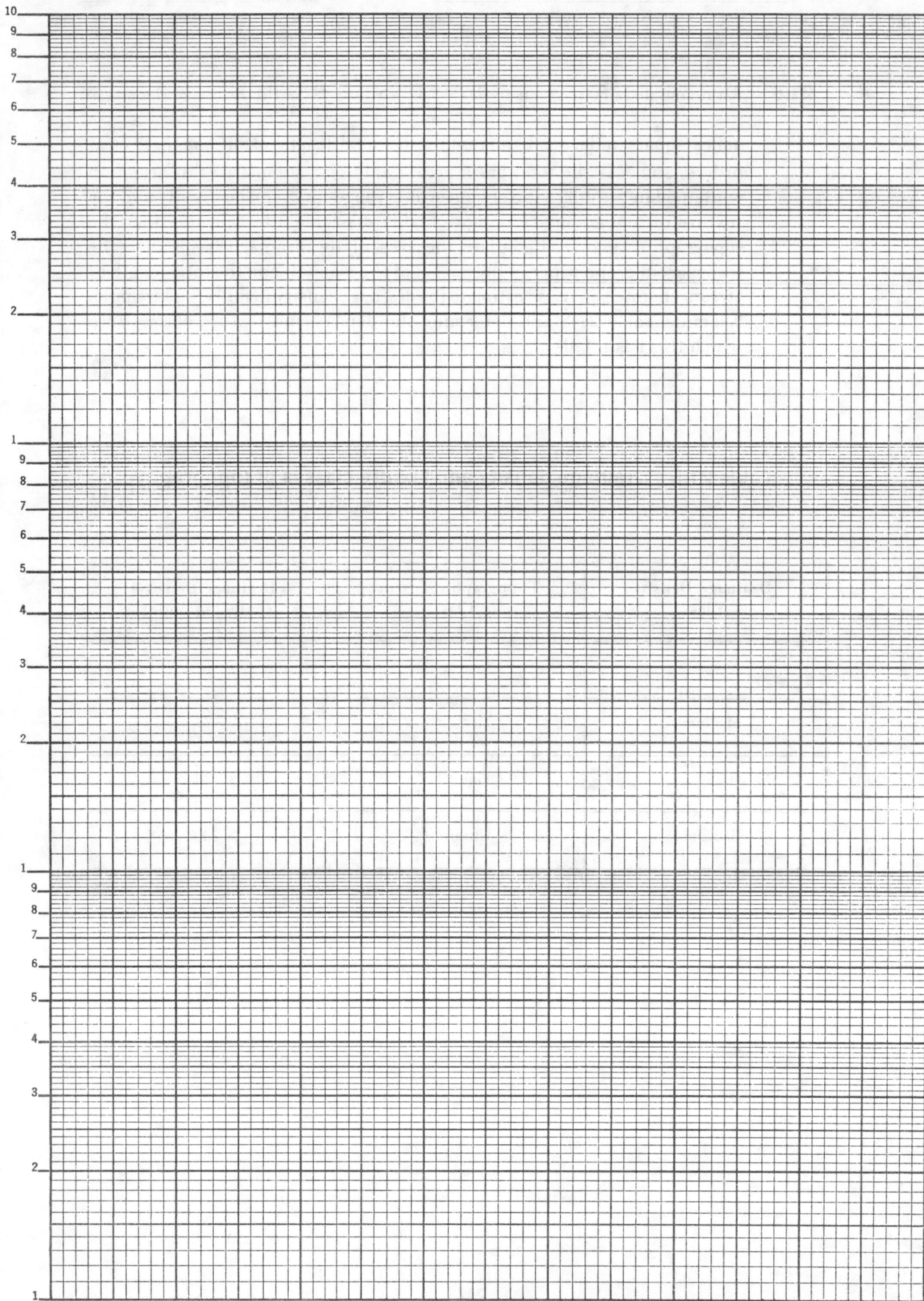

EXERCISE **22**

STREAM DISCHARGE AND FLOOD FREQUENCY

The *discharge* of a stream is its volume of flow per unit of time. Normally this flow is expressed in cubic feet per second (c.f.s.). At any point along the stream channel the discharge (Q) in cubic feet per second is given by

$$Q = A\ V,$$

where A is the cross sectional area of the stream in square feet and V is the mean velocity in feet per second. When a stream is at the brink of overflowing its banks, discharge is *bankfull*; when the flood spills over, there is *overbank flow*. On average, the bankfull stage occurs 1.5 times a year in humid mid-latitudes, with the typical range of variation lying between 1 and 4 times. Generally ungrassed channel banks delimit the bankfull level.

Water flowing in a stream channel is impelled forward by the force of gravity. The water is retarded in its movement by friction between the moving water and the floor and sides of the channel and by internal friction between water particles. As the discharge of a stream increases the cross sectional area increases at a more rapid rate than the *wetted perimeter* resulting in a relative decrease in frictional retardation by the floor and banks of the channel. For this reason the velocity of flow increases with discharge. At any point the *mean velocity* occurs at approximately 0.6 of the distance from the surface of the stream to the bed. Velocities vary considerably with the *maximum velocity* usually near the middle of the channel just below the surface. It is common for this value to be 0.5 to 0.25 greater than the average velocity of a cross section.

Most rivers show a seasonal variation in flow which is largely a reflection of climatic conditions. This pattern which tends to be repeated year after year is the *regime* of the river or stream. *Floods* are usually high rates of discharge which can lead to the inundation of land adjacent to streams. They are almost always caused by intense rainfall, snowmelt, or a combination of the two. In analyzing the occurrence of floods the underlying assumption is that the available stream discharge record is a sample of an infinitely large population in time. The *recurrence interval* of a flood of given magnitude is the average time within which that flood will be equalled or exceeded once. For example, if the largest flood recorded at a gauging station during the last 50 years was 200,000 c.f.s., it is to be expected that a flood of equal or greater magnitude will occur at least once in the next 50 years. *Flood frequency* can be calculated by tabulating and ranking the highest discharge in each year of record (Table 22.3). The recurrence interval of a particular flow is then given by the formula.

$$\text{Recurrence interval} = \frac{n+1}{m},$$

where n is the number of years of record and m the rank of the particular discharge. The mean of the sample of maximum discharges is the *mean annual flood.*

Name________________________ Date____________

Instructor ________________________ Section____________

QUESTIONS

22.1 Data were collected on June 22 that would allow the discharge of a stream in a remote location to be calculated later in the laboratory.

(a) The width of the stream at bankfull stage measured with a tape was found to be 120 feet. The depth of water in the stream was measured with a wooden rod at intervals across the channel. Using the data shown in Table 22.1 draw a cross section of the stream channel marking in the bankfull and June 22 levels. Use a horizontal and vertical scale of 1 inch represents 20 feet.

Table 22.1 Cross sectional data of a hypothetical Stream Channel

Distance from left bank at bankfull stage (feet)	Height above water level (*) or depth of water (feet and inches)	Comments
0	3 ft. 1 inch (*)	Left bank at bankfull
6	0	Left bank of present stream
10	2 ft. 2 inches	
20	3 ft. 3 inches	
30	6 ft. 4 inches	
34	6 ft. 8 inches	
40	10 ft.	
50	13 ft. 1 inch	
60	13 ft. 4 inches	
70	12 ft. 1 inch	
80	8 ft. 7 inches	
84	3 ft. 9 inches	
90	2 ft. 5 inches	
100	1 ft.	
105	0	Right bank of present stream
110	1 ft. 1 inch (*)	
120	3 ft. 1 inch (*)	Right bank at bankfull

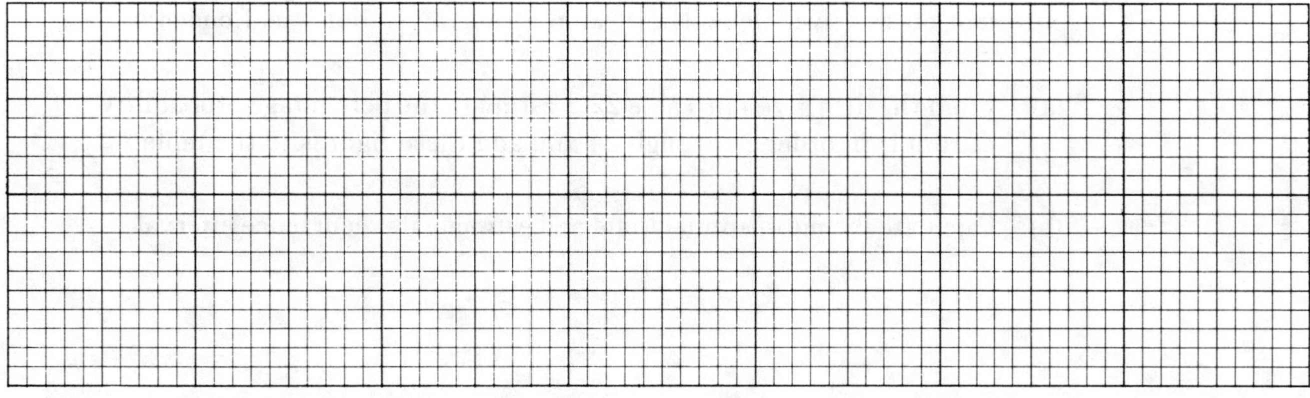

(b) A number of velocity readings were taken across the channel width. Sites were chosen carefully and in particular were located where flow velocities appeared to change substantially. In any depth of water the mean velocity is determined by measuring the speed of flow with a current meter at 0.6 of the depth. On the basis of these results the channel cross section can be divided into four sub-areas or *velocity domains* in which the rate of water flow is approximately constant. Using the information given in Table 22.2 mark the four velocity domains on the stream cross section and using the graph paper divisions measure the area for each. Add these values to the table and calculate the discharge through each sub-area. What is the discharge of the stream and what is its average velocity?

Table 22.2 Velocity and discharge characteristics of a hypothetical stream

Area	Location of velocity domains in terms of distance from the left bankfull position	Average Velocity (ft/sec)	Cross Sectional Area (square ft)	Discharge (A_xV_x) (c.f.s.)
1	6 ft. to 34 ft.	V_1 0.93	A_1	A_1V_1
2	34 ft. to 60 ft.	V_2 3.07	A_2	A_2V_2
3	60 ft. to 84 ft.	V_3 3.14	A_3	A_3V_3
4	84 ft. to 105 ft.	V_4 0.51	A_4	A_4V_4

Total Discharge $Q = A_1V_1 + A_2V_2 + A_3V_3 + A_4V_4 =$ ________ c.f.s.

$$\text{Average Velocity} = \frac{\text{Total Discharge (Q)}}{\text{Cross sectional area } (A_1+A_2+A_3+A_4)} = \_\_\_ \text{ feet/sec.}$$

22.2 The discharge of a stream has been monitored over a period of 24 years and every time the flow exceeded 115,000 cubic feet per second (c.f.s.) a small settlement on the river floodplain near the gauging station was flooded.

(a) From the data given in Table 22.3 estimate the percentage probability of runoff in order 2 through 24 and add these figures to the table.

(b) Calculate the mean annual flood and estimate its recurrence interval.

Name _______________ Date _______________

Instructor _______________ Section _______________

Table 22.3 Annual runoff peaks for a hypothetical stream arranged in descending order of magnitude

Runoff Peak (thousands of cubic ft. per second)	Order of Magnitude (m)	% Probability of runoff being equalled or exceeded ($\frac{m}{n+1}$ x 100)*
149.0	1	4
134.0	2	
127.0	3	
118.0	4	
113.0	5	
109.0	6	
104.0	7	
102.0	8	
100.0	9	
96.5	10	
94.5	11	
91.5	12	
88.0	13	
85.5	14	
82.0	15	
79.5	16	
78.5	17	
76.0	18	
73.0	19	
71.5	20	
67.0	21	
64.0	22	
57.5	23	
55.0	24	

* n is the number of years of record (24)

(c) What is the percentage probability of flooding in the settlement and how many times can it expect to be flooded in the next 20 years if measures are not taken?

(d) Plot the discharge-probability data on log-probability graph paper. Connect the points with a straight line and extend it to cover the range of probabilities from 0 to 100%. This line can be used to estimate the probability of annual runoff peaks outside those recorded in the 24 years that records have been kept. From this graph estimate the probability of an annual runoff peak of 200,000 c.f.s.

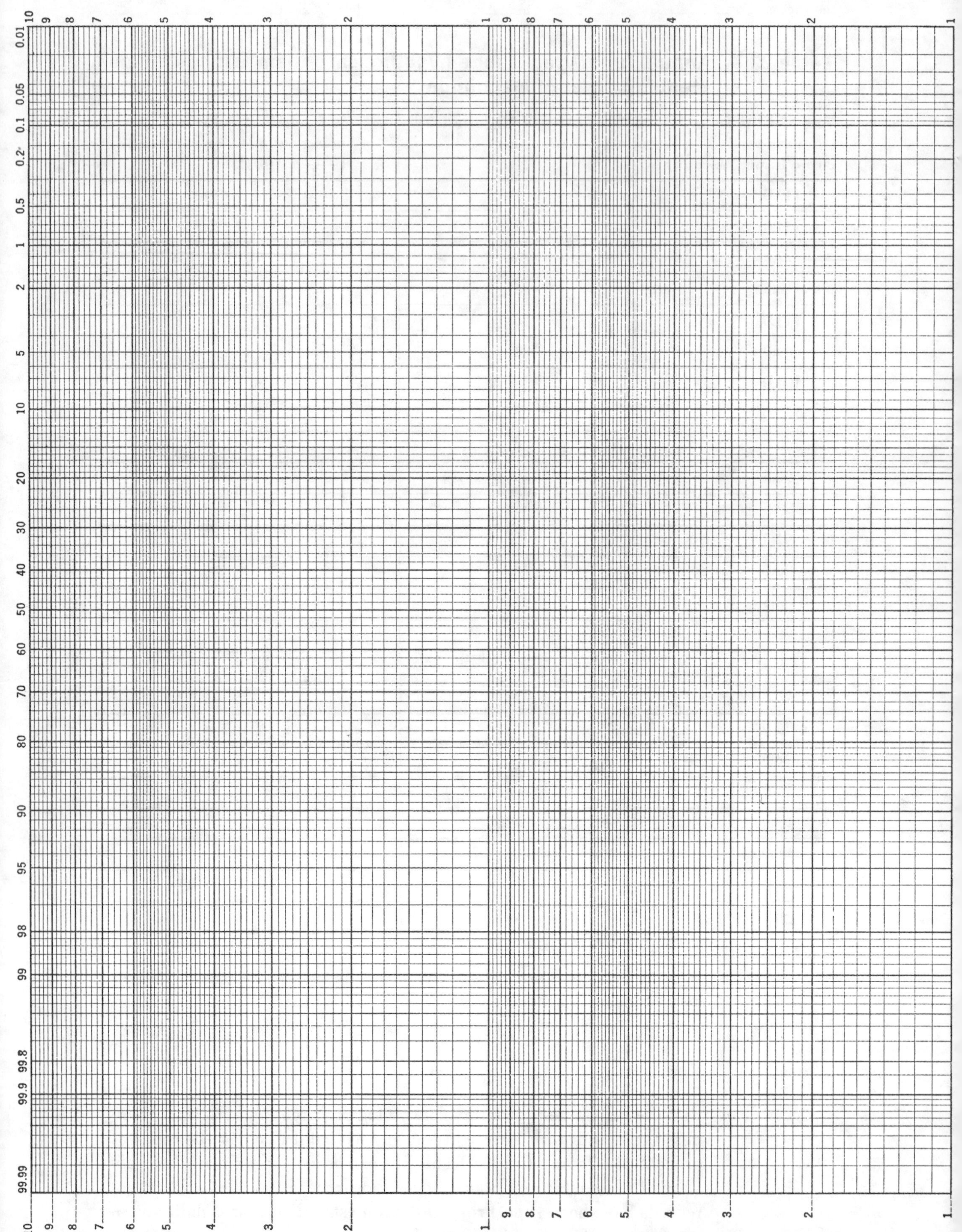

EXERCISE **23**

KARST LANDSCAPES

The German term *karst* derived from the Slav word krš meaning crag refers to terrain that has largely been molded by the action of solution. Solution is never the only process that has fashioned a karst landscape but it is always a most important one. The most common karst rocks are limestones ($CaCO_3$) and dolomites ($CaMg(CO_3)_2$) although evaporites such as halite (NaCl), anhydrite ($CaSO_4$) and gypsum ($CaSO_4 2H_2O$) also form karst landscapes but these are less common.

Pure water containing no carbon dioxide will dissolve approximately 15 milligrams/liter (or parts per million (p.p.m.)) of limestone. Water in contact with atmospheric carbon dioxide (CO_2) which makes up 0.03% by volume of the atmosphere is capable of dissolving 65 mg./liter but waters in contact with the soil, where CO_2 levels are much higher (commonly 0.1-1.0%), frequently dissolve 100-300 mg/liter. Enrichment of the soil atmosphere in CO_2 is due to biogenic and microbial action. The importance of CO_2 in the solution of limestone stems from the fact that when mixed with water it becomes a weak solution of *carbonic acid.* This acid attacks the *calcium carbonate* ($CaCO_3$) with the formation of *calcium bicarbonate* Ca $(HCO_3)_2$ which is carried away in solution. The basic equation for this process is:

$$CaCO_3 + H_2O + CO_2 \leftrightharpoons Ca(HCO_3)_2.$$

The most important effect of solution lies in the enlargement of subsurface voids with the steady increase in *secondary permeability*. This enables underground channels, which are called *caves* when large enough for a man to enter, to transmit water out of the karst area. Rainwater falling on limestone will rarely flow over the surface for any great distance before sinking underground. As secondary permeability increases and the limestone has the capacity to transmit large amounts of water underground, surface drainage systems are disrupted and rivers entering karst are liable to lose all or part of their water. Underground water follows caves which commonly form in the *phreatic zone*, but which upon a lowering of base level come into the *vadose zone* and are often occupied by free-flowing streams. The two longest caves in the world are the Mammoth-Flint Ridge system in Kentucky and Holloch in Switzerland. These are 120 miles and 49 miles long respectively. Drainage water eventually reappears elsewhere in the form of *springs*.

Persistent sinking at a point on a river's course leads to the differential lowering of the river bed by solution. Upstream of the sinking point the flow is greater than it is downstream of it so that a threshold develops in the river profile immediately below the sink. As subsurface voids continue to enlarge the sink will absorb more and more of the river's flow until eventually it may capture all of its flow even during periods of flood. When this happens there is no further surface flow in the valley downstream of the sink so that this becomes a *dry valley* and the valley system upstream of the sink becomes a *blind valley*. The stream itself is usually called a *sinking* or a *disappearing stream*.

In rocks of low solubility the characteristic surface landform is the stream valley. In karst, because of rapid underground drainage and localized surface solution, it is the *closed depression*, the smallest variety of which is called the *doline* or *sinkhole*. Several different types of doline are recognized and the most important of these are shown in Figure 23.1. *Solution dolines* are produced in bedrock by the action of solution from the surface down. *Collapse* and *subjacent karst collapse dolines* are formed by the collapse of the roof of a cave; the latter are produced when this roof is composed wholly or partly of nonsoluble rocks at the surface. *Suffosion* and *subsidence depressions* are developed in surficial unconsolidated, non-soluble material and not in bedrock. Suffosion dolines, which are usually funnel shaped, form as particles of soil are gradually washed into an underground cave. Subsidence depressions develop by the gradual subsidence of regolith, due to collapse or compaction, into a subsurface cavity or depression.

When two or more dolines of any type enlarge and coalesce at the surface, larger complex depressions called *uvalas* are formed. The largest closed depressions characteristic of karst terrains are poljes, which are either of structural origin (that is they were originally synclinal or fault depressions); or they formed by the gradual coalescence of dolines and uvalas. The largest polje in the world is the Livno Polje in Yugoslavia which is 70 kms long and, on average, 10 kms wide. Water entering a polje via surface streams and springs flows across the flat and often sediment-covered floor before sinking in cavities called *ponors*. In many depressions the ponors can not carry away the runoff fast enough after heavy rains and so these depressions are susceptible to flood. It is frequent inundation, which may be annual or once every few years, which leaves polje floors treeless but entirely grass covered when dry. The Slav word polje, meaning field, was used because of the meadow covering polje floors. The boundary walls of poljes are steep and residual hills called *hums* frequently project from their flat floors.

In humid tropical areas with high rainfall, and dense vegetation producing great volumes of biogenic CO_2, karst landscapes are extremely accentuated and varied in nature. Solution is frequently so intense that the landscape is one closed depression after the other in all directions. These depressions, called *cockpits,* are separated by narrow ridges. They differ in form from dolines. In some areas cockpits have enlarged to such an extent that the hills between, known as *cones* or *towers* (sometimes called *mogotes*) depending upon their shape, dominate the landscape. In southern China and Malaysia vertical-walled karst towers rise more than 2,000 feet above the surrounding terrain.

Figure 23.1 Four common varieties of doline to be found in karst areas (after Jennings 1971).

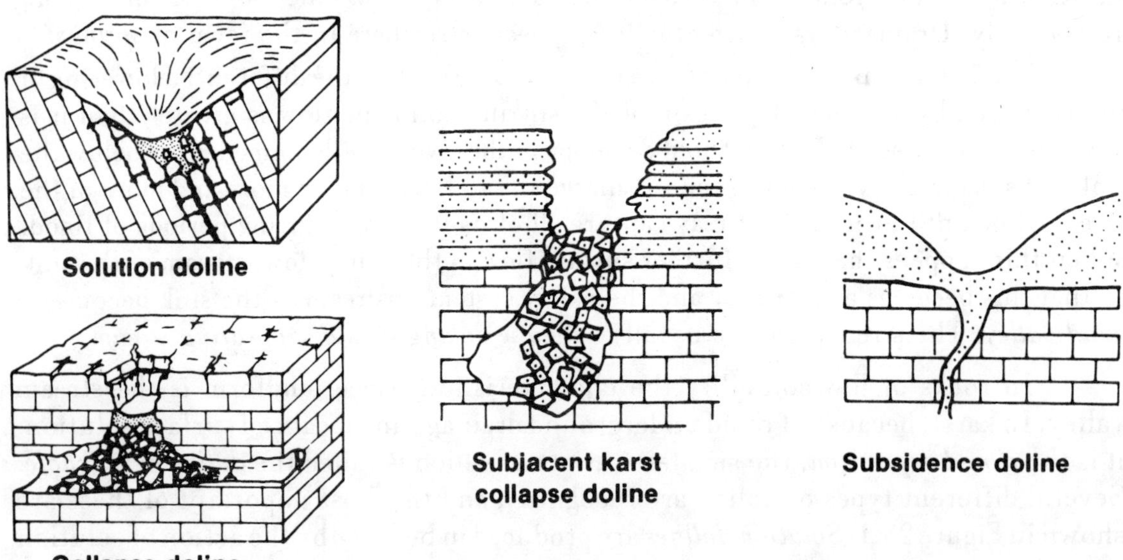

Name______________________ Date______________

Instructor______________________ Section______________

QUESTIONS

23.1 The catchment of a spring is 100 square kilometers in area and there is no surface flow out of it. On average the spring has a discharge of 50 cubic meters/second and carries 200 milligrams/liter (200 parts per million (p.p.m.)) of limestone in solution. If the density of the limestone is 2.5 grams/cubic centimeter, calculate the weight (in kilograms) and the volume (in cubic meters) of rock removed in solution from the catchment each year and if all of this material were removed from the surface of the catchment the amount of surface lowering (in centimeters). Show your calculations in your answer.

23.2 In the southern half of the Mammoth Cave map and in the aerial photographs of the same area (Figures 23.2 and 23.3), the St. Louis Limestone underlies a well developed karst area, the Pennyroyal Plain, with abundant, small closed depressions. Larger depressions to the north in the Mammoth Cave Plateau region have formed in the overlying Ste. Genevieve Limestone also of Mississippian (Carboniferous) age. The wooded hills have sandstone caps of Big Clifty Sandstone. Separating the Pennyroyal Plain and the Mammoth Cave Plateau is the Dripping Springs Escarpment. In the extreme northwest of the map area is a bend of the Green River.

(a) On the photographs draw in the boundary between the St. Louis and Ste. Genevieve Limestones and identify (with an S) a hill capped with sandstone. Also identify on both the photographs and the map the Pennyroyal Plain, the Dripping Springs Escarpment and the Mammoth Cave Plateau.

(b) Where possible, identify on the map a good example of the following in each of the limestone types and label your examples.

A	Doline	E	Disappearing Stream
B	Uvala	F	Dry Valley
C	Pond Doline	G	Cave Entrance
D	Blind Valley	H	Spring location

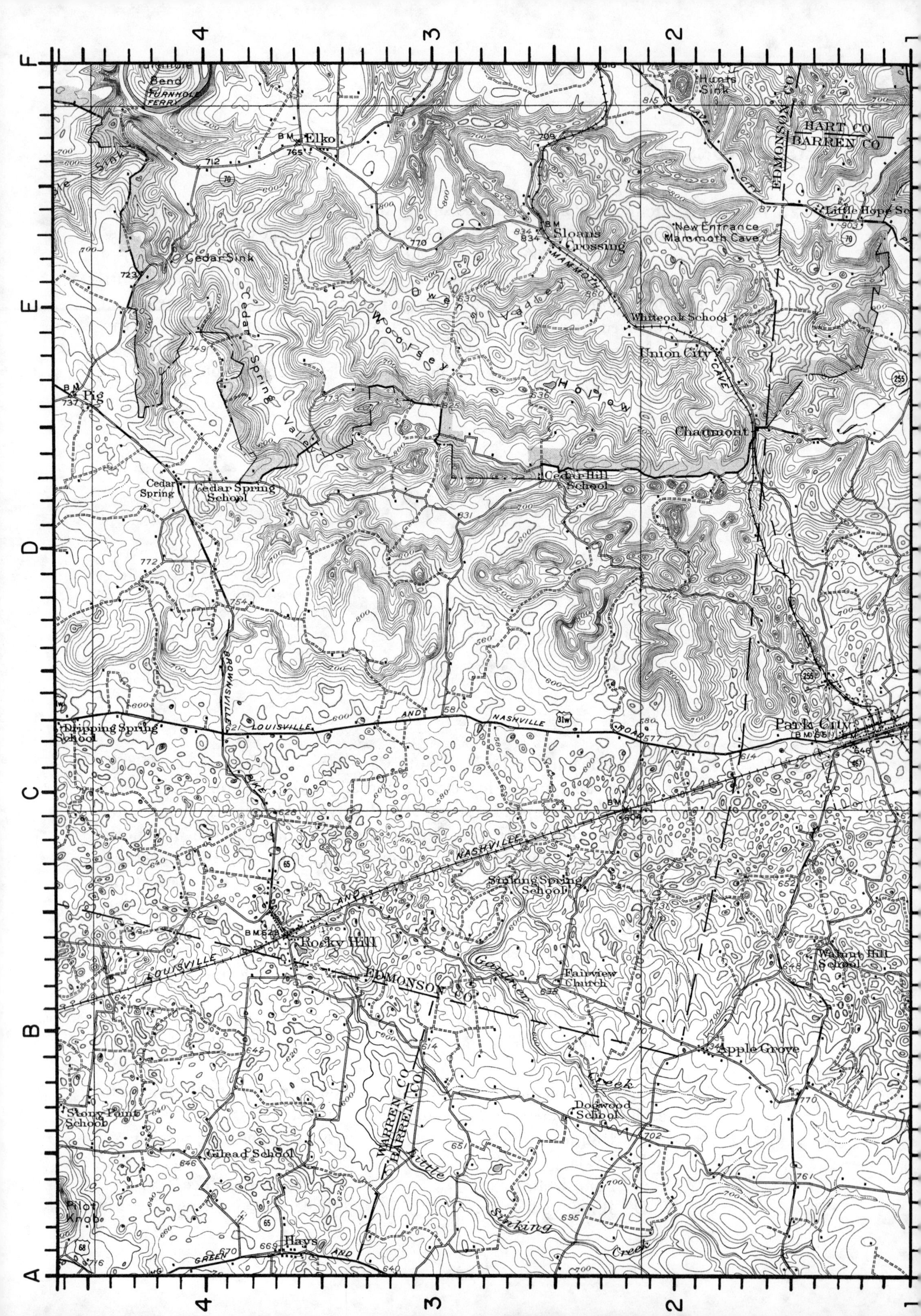

Turnhole Bend
Turnhole Ferry
Elko
Hunts Sink
Edmonson Co
Hart Co
Barren Co
Little Hope Sc
Sloans Crossing
New Entrance Mammoth Cave
Cedar Sink
Mammoth Cave
Owens Valley
Woolsey Hollow
Cedar Spring Valley
Whiteoak School
Union City
Cave City
Pig
Chaumont
Cedar Hill School
Cedar Spring
Cedar Spring School
Brownsville Pike
Louisville and Nashville Road
Dripping Spring School
Park City
Sinking Spring School
Rocky Hill
Louisville and Nashville
Edmonson Co
Barren Co
Gardner Creek
Fairview Church
Walnut Hill School
Apple Grove
Dogwood School
Stony Point School
Gilead School
Warren Co
Barren Co
Little Sinking Creek
Pilot Knob
Hays
Green and

Name ______________________ Date ______________

Instructor ______________________ Section ______________

(c) Carefully examine Cedar Sink (E.2-4.1), Cedar Spring Valley(D.8-3.8), Owens Valley (E.0-2.8), and Woolsey Hollow (D.7-3.0) in the northern portion of the map. How have these features developed and what relationship is there between them and the Green river (F.0-4.0)?

(d) Given that the "karst water table" in this area is not close to the surface, why do some depressions contain ponds?

23.3 Figure 23.4 is a portion of the 1:24,000 scale Lake Wales Quadrangle of central Florida. This area is underlain by up to 1,000 feet of Tertiary limestones which are covered by 100-200 feet of unconsolidated Miocene and Pleistocene sands and clays. The contour interval on the map is 5 feet.

(a) What variety of karstic closed depressions do the following lakes occupy:

Lake Alta ______________	Lake Serena ______________
Crystal Lake ______________	Lake Belle ______________
Twin Lakes ______________	Lake Cooper ______________

(b) From an analysis of lake levels what is a reasonable estimate of the elevation of the "karst groundwater table" in this area?

Figure 23.2 Mammoth Cave, Kentucky (1:62,500, A.0-1.0 in southeast corner).

Figure 23.3 Park City, Kentucky (A.0-1.0 in southwest corner).

Name________________________ Date____________

Instructor________________________ Section____________

(c) What is the most likely reason why some closed depressions contain water and others do not?

(d) What evidence is there to indicate that the Lake Serena depression and several other steep-sided, funnel-shaped depressions are younger than the more numerous broader basin- and saucer-shaped hollows?

(e) Why are there no surface streams marked on the map?

23.4 Both the Utuado (Figure 23.5) and Manati (Figure 23.6) areas of north central Puerto Rico contain highly developed karst terrains. In Figure 23.5 the surface topography is largely developed in Montebello Limestones of Oligocene to Miocene age. The underlying Lares Limestones of Oligocene age are only exposed in the canyon of the Rio Grande de Arecibo where they project from Quaternary sands and gravels. In the southern portion of the Manati photographs is an area of Aguada Limestone of Miocene age, to the north this is overlain by Aymamon Limestone which in this region is heavily mantled by Pleistocene "Blanket Sands".

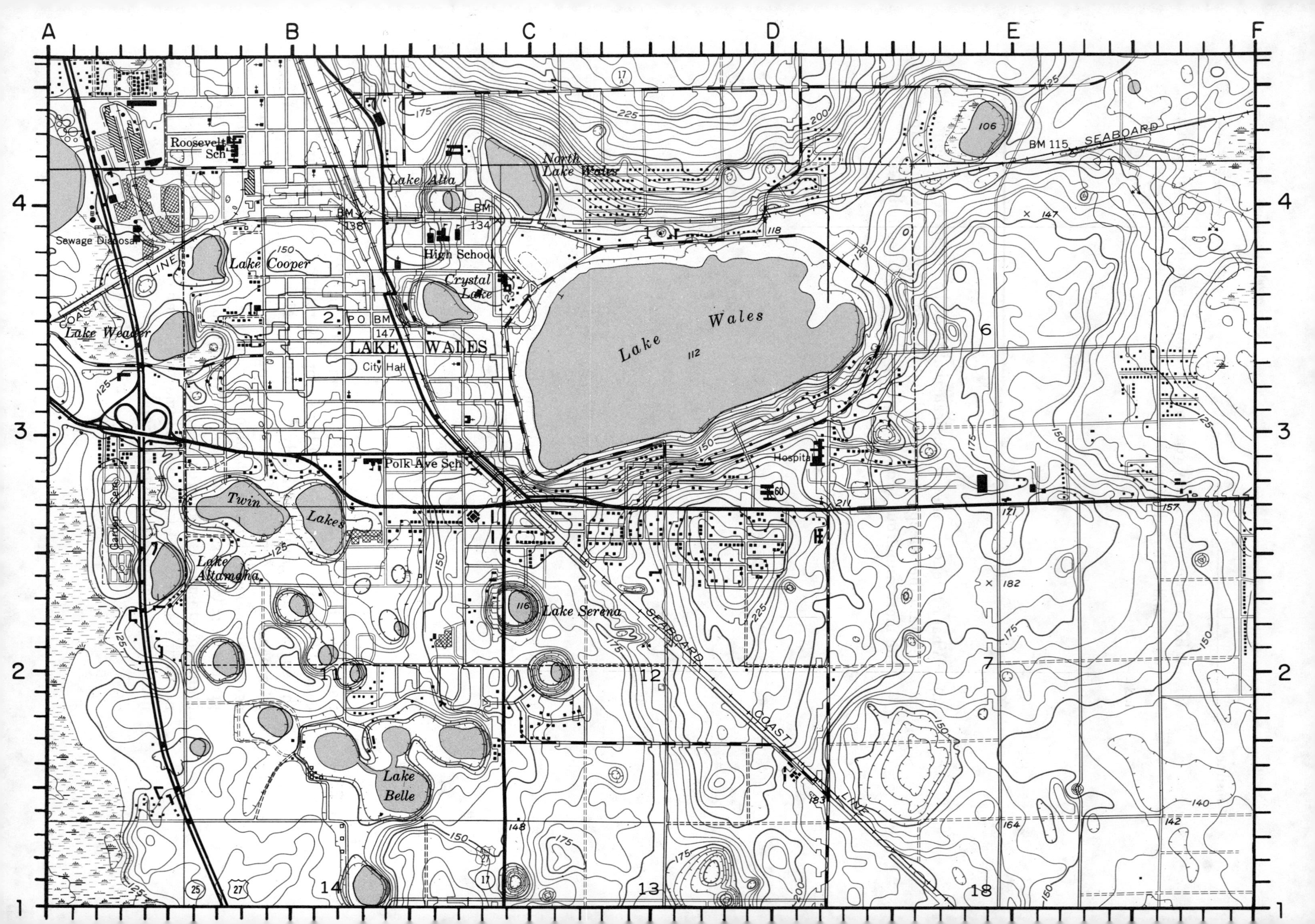

LAKE WALES
Lake Wales
North Lake Wales
Lake Alta
Crystal Lake
Lake Cooper
Lake Weaver
Twin Lakes
Lake Altamaha
Lake Serena
Lake Belle
Roosevelt Sch
High School
Polk Ave Sch
City Hall
PO
Hospital
Sewage Disposal
Garden Cem
SEABOARD
COAST LINE
BM 115

Name________________________________ Date________________

Instructor ____________________________ Section______________

(a) On the photographs identify areas of cockpit karst (sometimes referred to as polygonal or honeycombe karst), oriented karst (areas of depressions and ridges with a distinct preferred orientation), and mogote karst (areas dominated by tower-like hills of limestone).

(b) Why does oriented karst develop in some areas and not in others?

(c) Look carefully at the Manati photographs and determine what relationship there is between mogote karst and cockpit or oriented karst.

Figure 23.4 Lake Wales, Florida (1:24,000, A.0-1.0 in southwest corner).

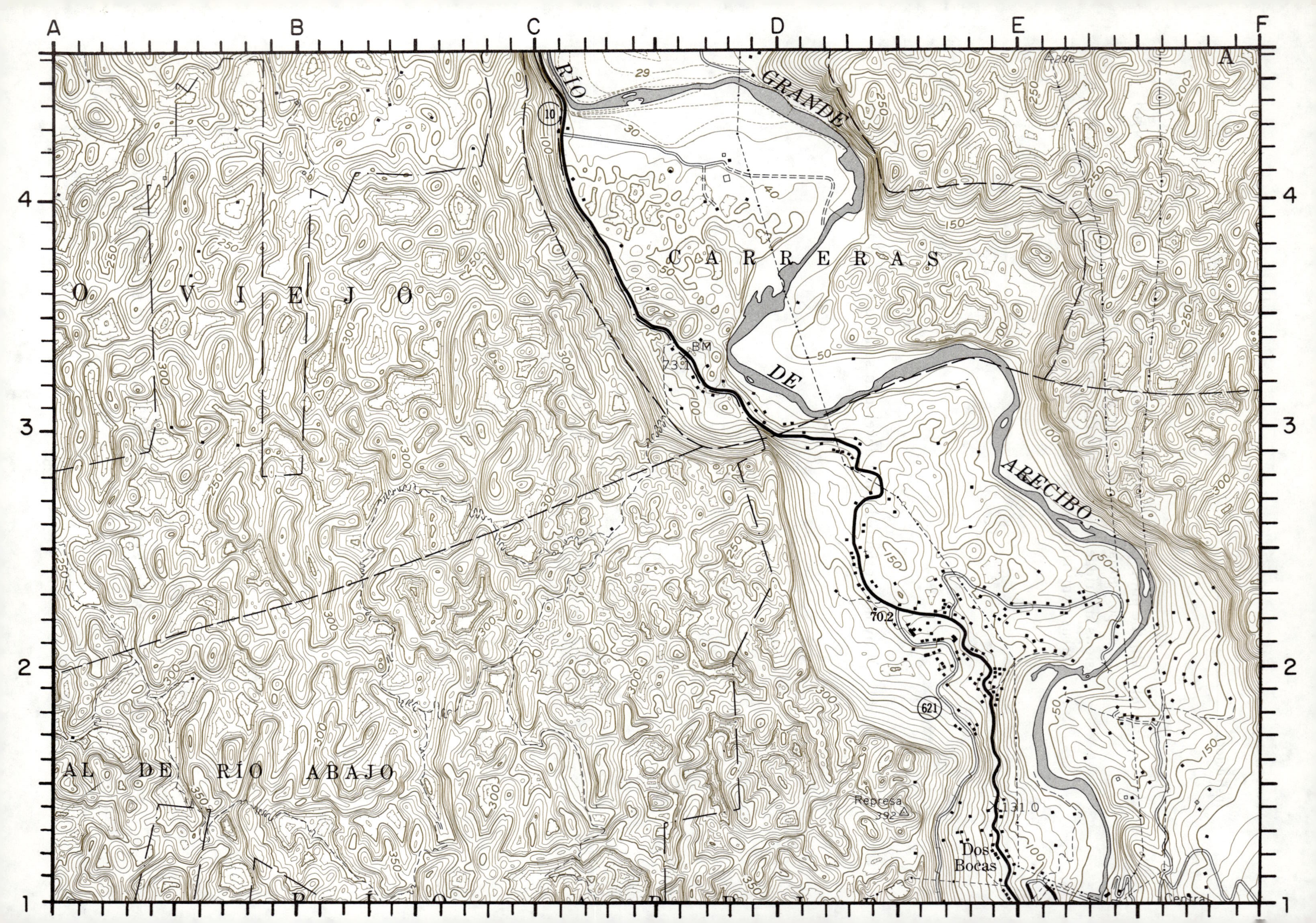

RÍO GRANDE DE ARECIBO
CARRERAS
O VIEJO
AL DE RÍO ABAJO
Dos Bocas
Represa
392
BM
73.1
70.2
131.0
10
621
Central

Name____________________ Date__________

Instructor ____________________ Section__________

(d) With reference to the size of karst features and the intensity of karst development, compare the mogote and cockpit karst of Puerto Rico with the doline karst of Kentucky.

(e) Given that the mean annual precipitation of the Mammoth Cave area is approximately 47 inches and that of north-central Puerto Rico 80 inches, why do you think the karst in these two areas is very different?

Figure 23.5 Utuado, Puerto Rico (1:20,000, A.0-1.0 in southwest corner).

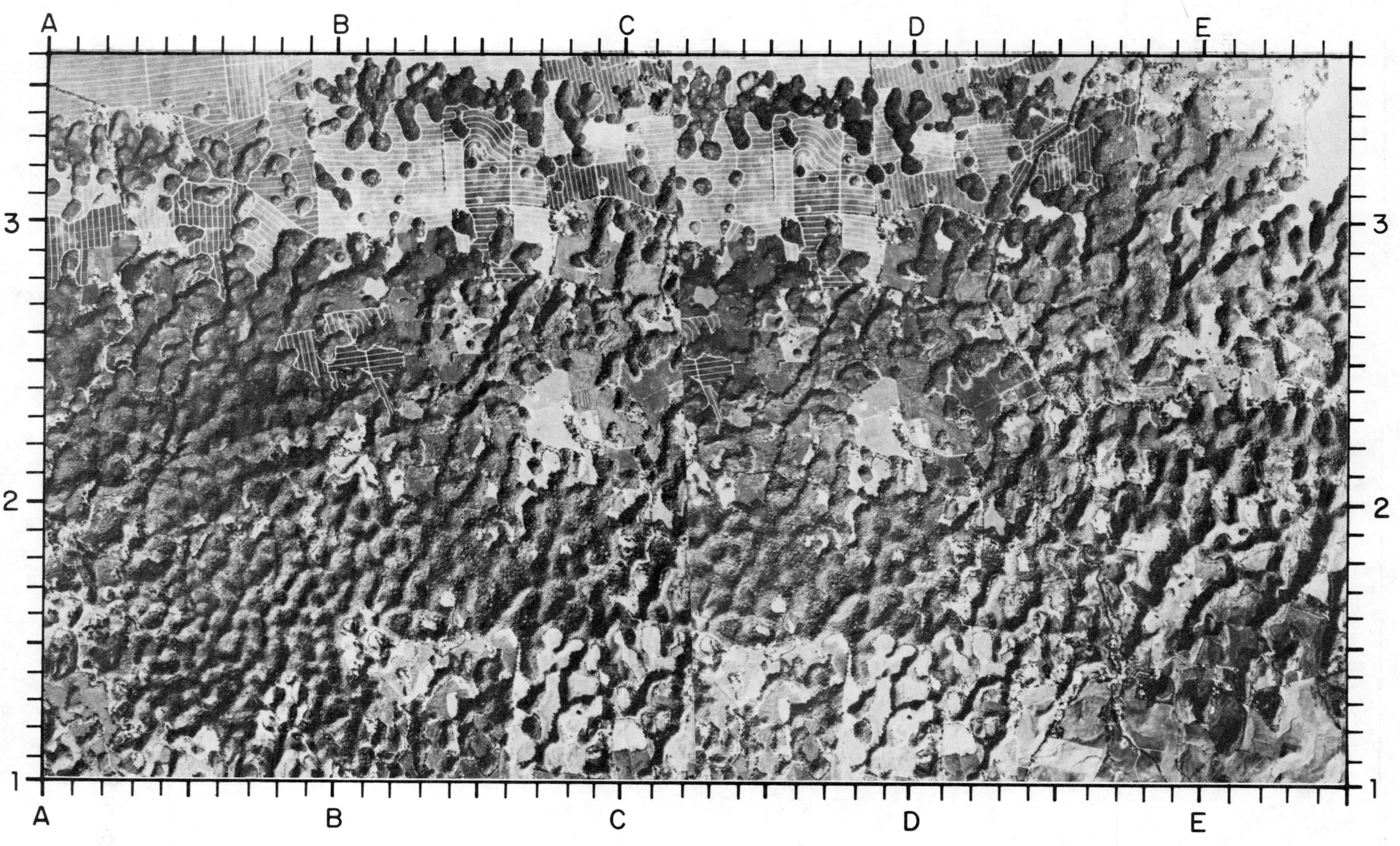

Figure 23.6 Manati, Puerto Rico (1:31,000, A.0-1.0 in southwest corner).

EXERCISE **24**

GLACIERS AND GLACIAL LANDFORMS

Glaciers are masses of ice which under the influence of gravity flow out from the snowfields where they originate. Much of North America and Europe, and parts of northern Asia and southern South America were covered by enormous ice-sheets during the *Pleistocene Epoch*—the time of the 'Great Ice Age' which ended 10,000-15,000 years before present (B.P.) and is thought to have lasted 1 or 2 million years. There were four principal phases of glacier growth during the Pleistocene and these were separated by *interglacial* periods when the climate was at least as warm as it is now (Table 24.1). During glacial events the tremendous volume of water locked up in land ice caused a world-wide or *eustatic* drop in sea level of 300-400 feet. In places the ice sheets were of sufficient thickness that they depressed the crust. During interglacial phases sea level rose and land masses began to rebound.

Table 24.1 Glacial and interglacial phases of the 'Great Ice Age'

AGE (10^3 yrs. B.P.)	European Terminology	N. American Terminology
10-present	Holocene Interglacial?	Holocene Interglacial?
	Würm Glacial	Wisconsin Glacial
100	Riss-Würm Interglacial	Sangamon Interglacial
	Riss Glacial	Illinoian Glacial
600?	Mindel-Riss Interglacial	Yarmouth Interglacial
	Mindel Glacial	Kansan Glacial
2,000?	Günz-Mindel Interglacial	Aftonian Interglacial
	Günz Glacial	Nebraskan Glacial

In many arid and semiarid areas lower temperatures and increased rainfall accompanied glaciation at higher latitudes during what are known as *pluvial events.* Similar conditions to those existing at the present time are thought to have characterized *non-pluvial* or interglacial periods. In the Basin-and-Range Province of the western United States pluvial conditions allowed lakes occupying interior basins to grow in size. The ancient deltas and shorelines of these lakes are clearly visible in many areas today. Perhaps the best known of these Pleistocene lakes is *Lake Bonneville* which occupied the Great Basin. The Great Salt Lake north of Salt Lake City in Utah is the present remnant of this lake.

Today most scientists support the *astronomical theory of climatic* change which suggests that variations in the temperature of the earth's surface are brought about by the relationship in space between the earth and the sun and in particular by the *eccentricity* of the earth's orbit around the sun. The orbit of the earth is an ellipse, the eccentricity of which varies between extremes of almost zero, when the orbit is nearly a circle, to a maximum value when the ellipse reaches its greatest elongation. Peaks in orbital eccentricity are considered to correspond with warm climatic periods at the earth's surface.

Ice-sheets which are continental in size and *ice-caps* which envelope mountain regions may cover millions of square miles and be more than 7,000 feet thick. Greenland and Antarctica contain the only continental ice-sheets in existence today. Mountain peaks that project above broad expanses of ice are known as *nunataks. Cirque* and *valley* glaciers occupy former stream valleys and vary from one to a few hundred square miles in area and from 100-2,000 feet thick. *Alpine* valley glaciers move out of cirque basins as the *snowline* falls while *outlet* valley glaciers are fed by ice-caps and ice-sheets. *Piedmont* glaciers form when a valley glacier passes from a restricted trough to a more open lowland and spreads out. Glacier ice is either temperate or polar. *Temperate ice* is at pressure-melting point throughout its thickness except in winter when the upper layers become too cold. Meltwater is usually abundant in the summer months and circulates freely both within and beneath the ice. *Polar ice* is below the pressure-melting temperature so that even in summer no meltwater penetrates the ice. Cirque and alpine valley glacier ice is temperate while that in ice-sheets and ice-caps is usually polar.

The loss of ice and snow from a glacier as a result of melting, wind action, or sublimation is called *ablation*; its addition is called *accumulation.* At higher elevations on an alpine valley glacier, in the *accumulation zone,* all of the winter snow does not melt during the summer months. The lower limit of this zone is the *equilibrium line* which more or less parallels the *firn line.* At lower elevations is the *ablation zone* in which all of the winter snow melts and also some of the glacier ice. The net loss of snow and ice from the ablation zone is balanced by a flow of ice from the accumulation zone. The *snout* of a valley glacier, which may reach thousands of feet below the firn line marks the position where the net annual ablation just balances net annual accumulation plus ice flow. An increase in accumulation over ablation causes the snout to advance, a decrease causes it to retreat. Ice movement, which is always towards the snout, occurs by a combination of *regelation* which is the melting and immediate refreezing of ice (plastic flow), and *block sliding* which occurs when the sides and base of the glacier are lubricated by meltwater (Figure 24.1).

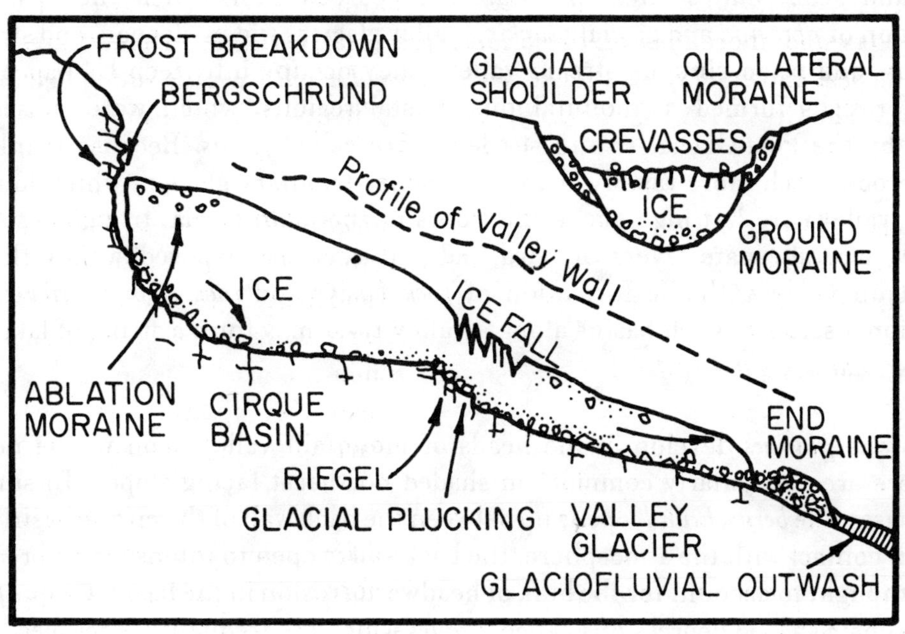

Figure 24.1 Characteristics of the alpine valley glacier.

Glaciers have a surface crust of brittle ice 100-200 feet thick which is carried along on the flowing ice beneath. Fissures or crevasses develop in this upper layer whenever it is stretched. *Marginal crevasses* open because of friction between the glacier and the valley walls. *Transverse crevasses* form across a glacier wherever the bedrock floor is markedly convex. *Longitudinal crevasses* run approximately parallel to the direction of flow of the glacier and open where the ice spreads out laterally. In piedmont glaciers these crevasses may form an almost radial pattern. Where a glacier tumbles across a marked step in the valley profile an *icefall* develops. During the summer months water from melted snow and ice collects into *supraglacial streams* which flow for short distances down the glacier before disappearing into crevasses via deep potholes called *moulins.* This water eventually finds its way to the base of the ice where it forms *subglacial streams* which work their way down the valley and finally emerge at the snout of the glacier.

Rock fragments continually fall from the valley walls above a valley glacier and accumulate to form *lateral moraines.* Debris frozen into the base of the ice or carried along beneath it constitutes *ground moraine* and is pushed up in front of the snout as an arcuate ridge called a *terminal* or *recessional moraine.* In areas where two outlet or alpine valley glaciers coalesce, the inner lateral moraines are carried into the middle of the resulting larger ice tongue as *medial moraine.*

Glaciers heavily charged with angular rock debris are extremely efficient agents of erosion and scour and scratch or *striate* the bedrock over which they pass. By a combination of *abrasion* and glacial *plucking* valley glaciers widen, deepen, and straighten the narrow and often sinuous stream valleys they occupy into deep U-shaped *glacial troughs.* Troughs formed in mountainous coastal regions, which were subsequently drowned by the Post-Glacial rise in sea level, are called *fjords.* Because trunk valley glaciers erode much more rapidly than the smaller tributary glaciers which join them, tributary valleys are left high above the floors of the main glacial troughs as *hanging valleys.* Where alternate layers of hard and soft rock are exposed, valley floors are deepened unevenly with the formation of *rock basins* and *rock steps* or *riegels.* After deglaciation a series of such basins along a valley floor may form a string of lakes called *paternoster lakes.*

Cirque glaciers develop in the heads of mountain valleys and in the northern hemisphere are particularly common on shaded northeast facing slopes. In summer a large crevasse, the *bergschrund,* opens up between the back wall of the cirque basin and the glacier. In contact with the atmosphere, the backwall is open to intense frost breakdown which is thought to account for the bulk of headward erosion in the basin. Cirque glaciers carve armchair-shaped depressions called *cirques* which are frequently occupied by lakes or *tarns* when free of ice. In a mountainous terrain two adjoining cirques may grow until a sharp ridge called an *arête* is all that separates them. An upland region eaten into by cirque erosion from all sides may be reduced to a central peak called a glacial *horn* with arêtes radiating out from it.

Material transported by glacier ice and subsequently deposited may vary from a few feet to several hundred feet in thickness. This *glacial drift* includes both *stratified* deposits laid down by meltwater streams and till, a heterogeneous deposit of clay and boulders which forms *knob and kettle* topography at the former margins of ice-sheets and ice-caps. *Kettle holes* and *kettle ponds* develop when remnant blocks of glacier ice incorporated into the till eventually melt. *Drumlins* are half-egg-shaped mounds of drift which are elongated in the direction of ice movement. The broad *stoss end* faces upglacier, the gently sloping tail downglacier. Drumlins normally occur in large numbers forming what is known as *'basket of eggs' topography.* Their streamlined shapes indicate that they were molded in drift beneath moving ice.

Outwash deposits of glaciofluvial origin are laid down in front of glacier margins by meltwater streams which flow in tunnels beneath the ice. These tunnels eventually fill up with sand and gravel and when the ice retreats these deposits are left as sinuous ridges called *eskers. Ice-dammed lakes* frequently form when a trunk valley glacier obstructs the mouth of a tributary valley. *Deltas* and *terraces* built out into these lakes are formed of *glaciolacustrine deposits* and are called *delta kames* and *kame terraces.*

Name______________________________ Date______________

Instructor______________________________ Section______________

QUESTIONS

24.1 Examine Figure 24.2 showing a portion of the Seward area of Alaska.

(a) Identify and label the following features:

A Alpine Valley Glacier
B Cirque Glacier
C Ice-free Cirque Basin
D Glacial Trough
E Hanging Valley
F Fjord
G Nunatak
H Arête
I Horn
J Ice-dammed Lake
K Tarn
L Meltwater Stream
M Valley Outwash Train

(b) Mark a location on the map (N) where you think there may be a medial moraine.

(c) Calculate the gradient of the upper surface of the Spencer Glacier (C.0-3.6).

(d) Explain the existence of Carmen Lake (E.2-2.9) in the northern portion of Figure 24.2 and briefly outline the events that led to its development.

24.2 The Holy Cross area of Colorado shown in Figure 24.3 does not host glaciers at the present time but there is abundant evidence of former glacier action.

(a) On the map identify the following features:

A Glacial Trough
B Cirque
C Tarn
D Terminal Moraine
E Kettle Hole
F Paternoster Lakes
G Riegel
H Rock Basin

(b) Was the area glaciated by alpine or outlet valley glaciers?

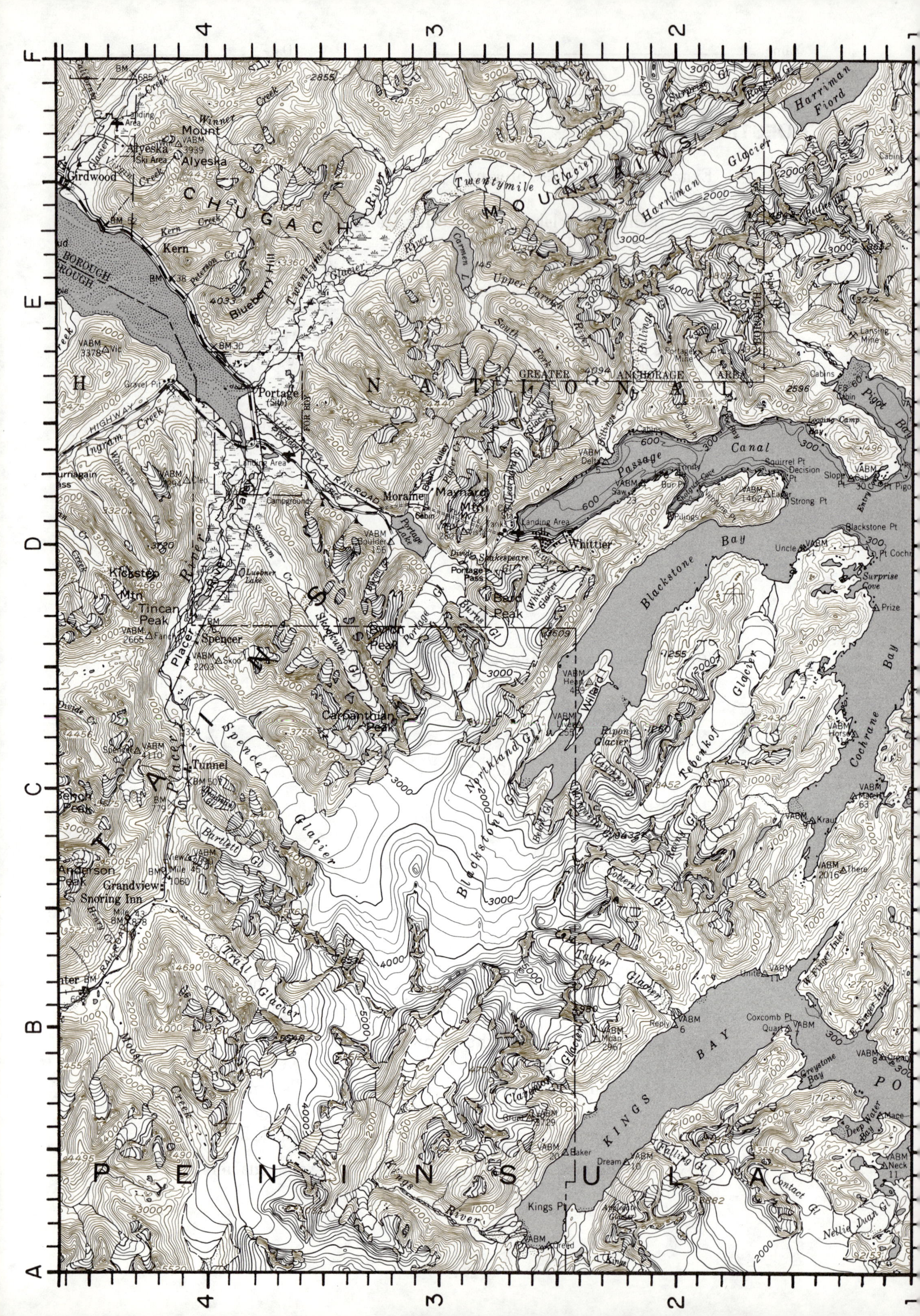

CHUGACH MOUNTAINS
CHUGACH NATIONAL FOREST
KENAI PENINSULA
GREATER ANCHORAGE AREA BOROUGH
Girdwood
Mount Alyeska
Kern
Blueberry Hill
Portage
Whittier
Maynard Mtn
Shakespeare
Portage Pass
Bard Peak
Spencer
Carpathian Peak
Tunnel
Grandview
Snoring Inn
Anderson Peak
Tincan Peak
Kings Bay
Blackstone Bay
Passage Canal
Harriman Fiord
Twentymile Glacier
Harriman Glacier
Spencer Glacier
Blackstone Gl
Trail Glacier
Taylor Glacier
Ripon Glacier
Tebenkof Glacier
Cochrane Bay
Kings Pt
Coxcomb Pt
Blackstone Pt
Decision Pt
Squirrel Pt
Strong Pt
Pigot Bay
Portage Lake
Kings River
Placer River
Twentymile River

Name________________ Date________________

Instructor ________________ Section________________

(c) Explain the form of Lake Fork Valley (A.8-3.5) and describe the events that led to the development of Turquoise Lake (A.5-1.7).

(d) There can be little doubt that an alpine valley glacier formerly occupied Lake Fork Valley. Examine the walls of the valley and estimate the former thickness of ice at the gaging station just west of Turquoise Lake.

24.3 A few glaciers still remain in the Chief Mountain area of Montana (Figure 24.4) but evidence of a former more extensive glaciation is ubiquitous.

(a) Identify the following features on the map:

A	Glacial Trough	E	Horn
B	Cirque Glacier	F	Paternoster Lakes
C	Cirque	G	Riegel
D	Arête	H	Tarn

(b) Explain the shape of McDonald Lake (B.5-3.7) and outline the events that led to its development.

Figure 24.2 Seward, Alaska (1:250,000, A.0-1.0 in southeast corner).

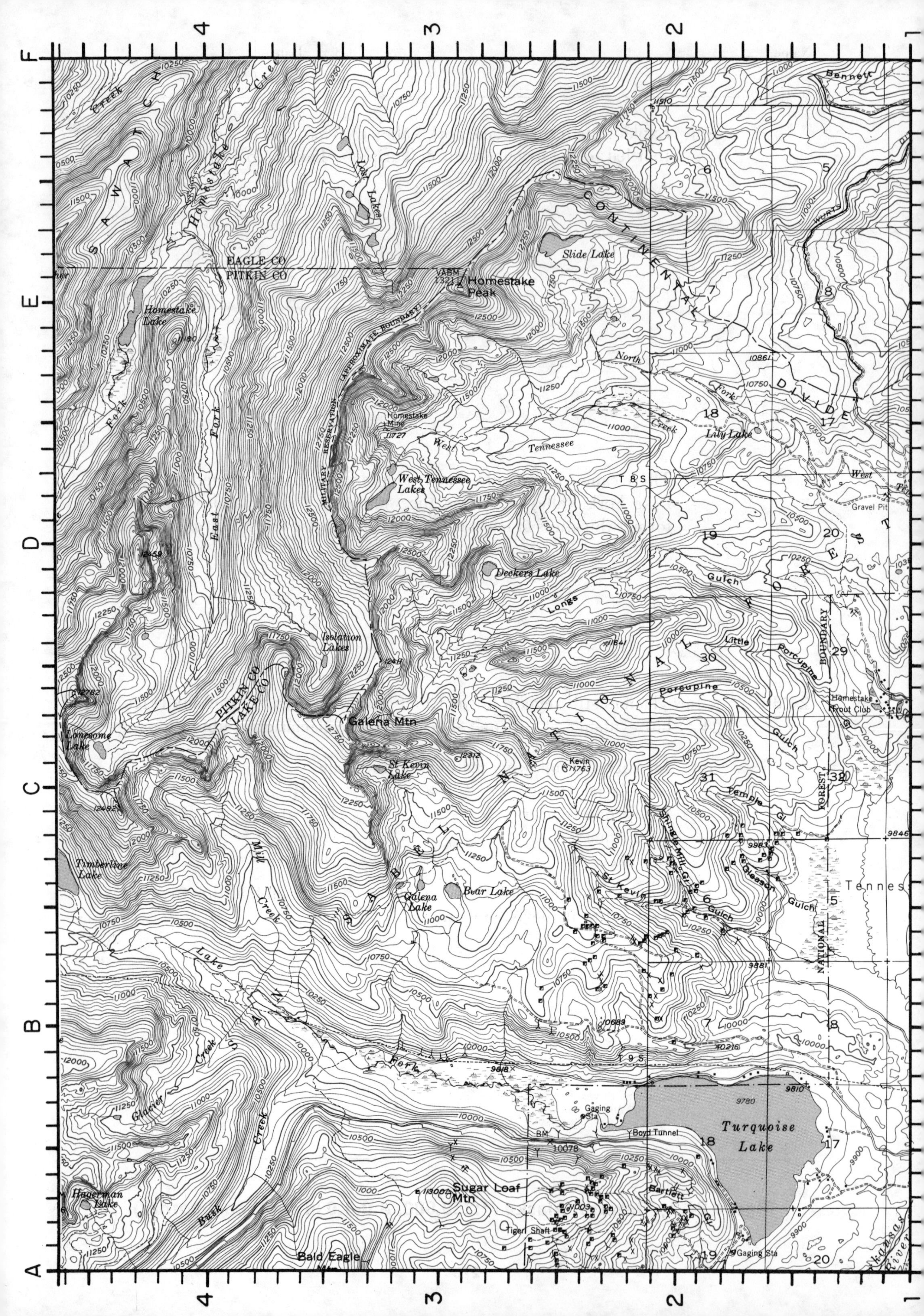
SAWATCH
Homestake Creek
Lost Lakes
EAGLE CO
PITKIN CO
VABM 13211
Homestake Peak
Slide Lake
CONTINENTAL DIVIDE
Homestake Lake
APPROXIMATE BOUNDARY
MILITARY RESERVATION
Homestake Mine
11727
North Fork
West Tennessee Creek
West Tennessee Lakes
Lily Lake
Gravel Pit
T 8 S
East Fork
Deckers Lake
Longs Gulch
Isolation Lakes
Little Porcupine
Porcupine Gulch
BOUNDARY
Homestake Trout Club
PITKIN CO
LAKE CO
Galena Mtn
Lonesome Lake
St Kevin Lake
Kevin 11763
SAN ISABEL NATIONAL FOREST
Temple Gl
Shingle Mill Gl
Gleason Gulch
St Kevin Gulch
Timberline Lake
Mill Creek
Galena Lake
Bear Lake
Tennessee
Lake Fork
T 9 S
Glacier Creek
Busk Creek
Gaging Sta
Boyd Tunnel
Turquoise Lake
BM 10076
Sugar Loaf Mtn
Bartlett Gl
Hagerman Lake
Tiger Shaft
Bald Eagle Mtn
Arkansas River
Bennett
WURTS
NATIONAL FOREST

Name________________ Date________________

Instructor ________________ Section________________

24.4 Examine Figures 24.3 and 24.4 and comment on any preferred directional orientation of cirques and cirque glaciers.

24.5 Figure 24.5 is a view of Taylor Dry Valley looking towards the ice plateau of the Antarctic ice-sheet at 6,000-7,500 feet elevation which is in the background. The upper portion of Taylor Valley is presently occupied by the snout of Taylor Glacier and in front of this is the permanently ice covered Lake Bonney.

(a) Identify and label on the photograph the following features:

A	Piedmont Glacier	F	Hanging Valley
B	Alpine Valley Glacier	G	Nunatak
C	Outlet Valley Glacier	H	Icefall
D	Glacier Trough	I	Transverse Crevasses
E	Cirque	J	Longitudinal Crevasses

(b) Carefully examine the area around the snout of Taylor Glacier and explain the valley-in-valley cross profile of Taylor Valley.

24.6 Examine Figure 24.6:

(a) Commonwealth Glacier is fed by the Antarctic ice-sheet and is an outlet glacier of what type?

(b) On the photograph identify:
A Kame and Kettle Topography
B Meltwater Streams
C Radial Crevasses

Figure 24.3 Holy Cross, Colorado (1:62,500, A.0-1.0 in southeast corner).

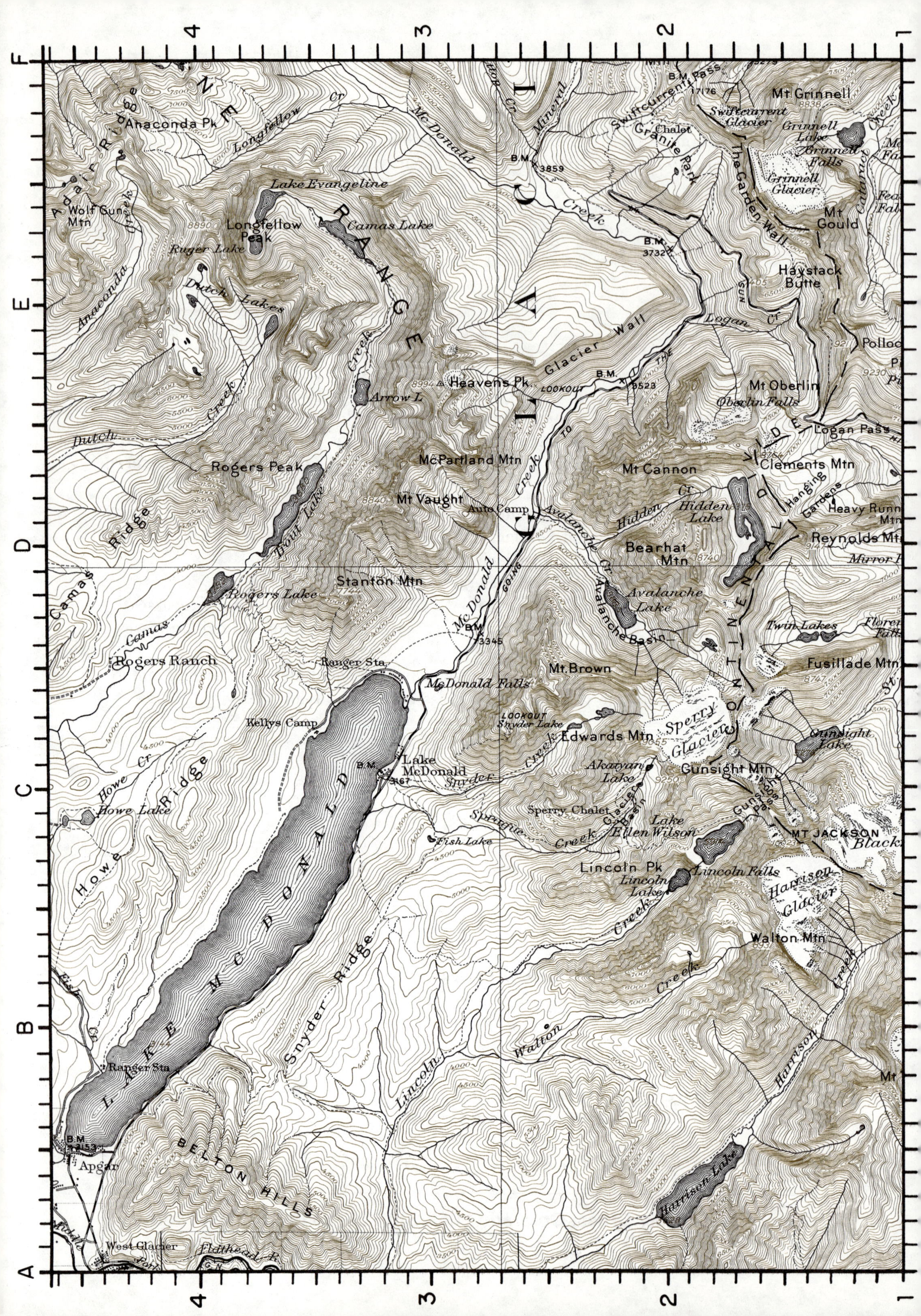

Name ____________________ Date ____________

Instructor ____________________ Section ____________

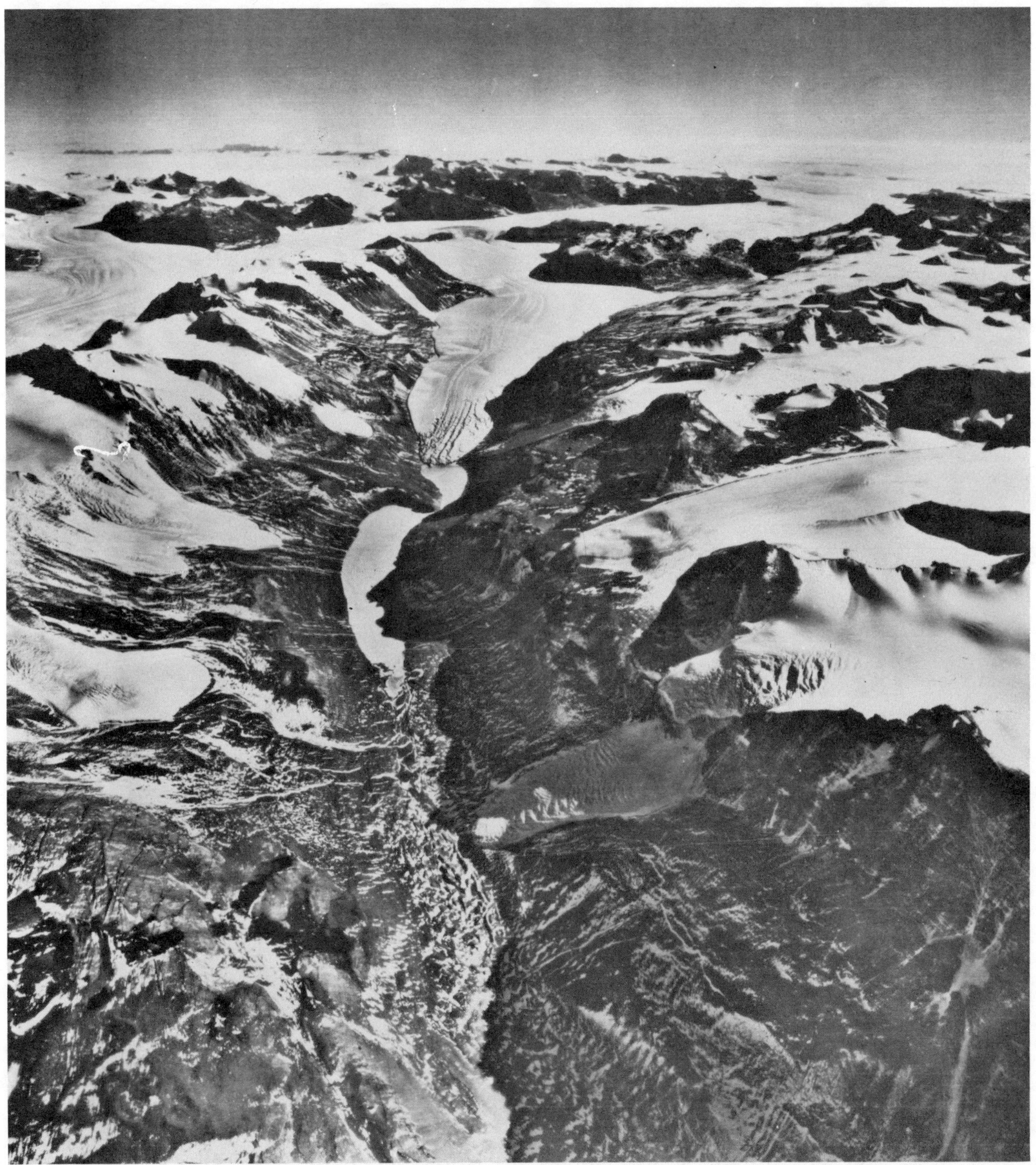

Figure 24.5 Taylor Glacier, Taylor dry valley, Victoria Land, Antarctica.

Figure 24.4 Chief Mountain, Montana (1:125,000, A.0-1.0 in southeast corner).

Figure 24.6 Commonwealth Glacier, Taylor dry valley, Victoria Land, Antarctica.

24.7 Figure 24.7 shows the Crillon Glacier where it flows into Lituya Bay, one of many Alaskan fjords.

(a) On the photographs label the following features:

A	Glacier Snout	F	Icefall
B	Hanging Valley	G	Marginal Crevasses
C	Lateral Moraine	H	Transverse Crevasses
D	Medial Moraine	I	Longitudinal Crevasses
E	Terminal Moraine		

(b) Identify the accumulation and ablation zones of the small valley glacier to the northwest of Crillon Glacier and mark the approximate position of the firn line.

(c) Mark the outlet point of the major subglacial stream which flows from the snout of Crillon Glacier.

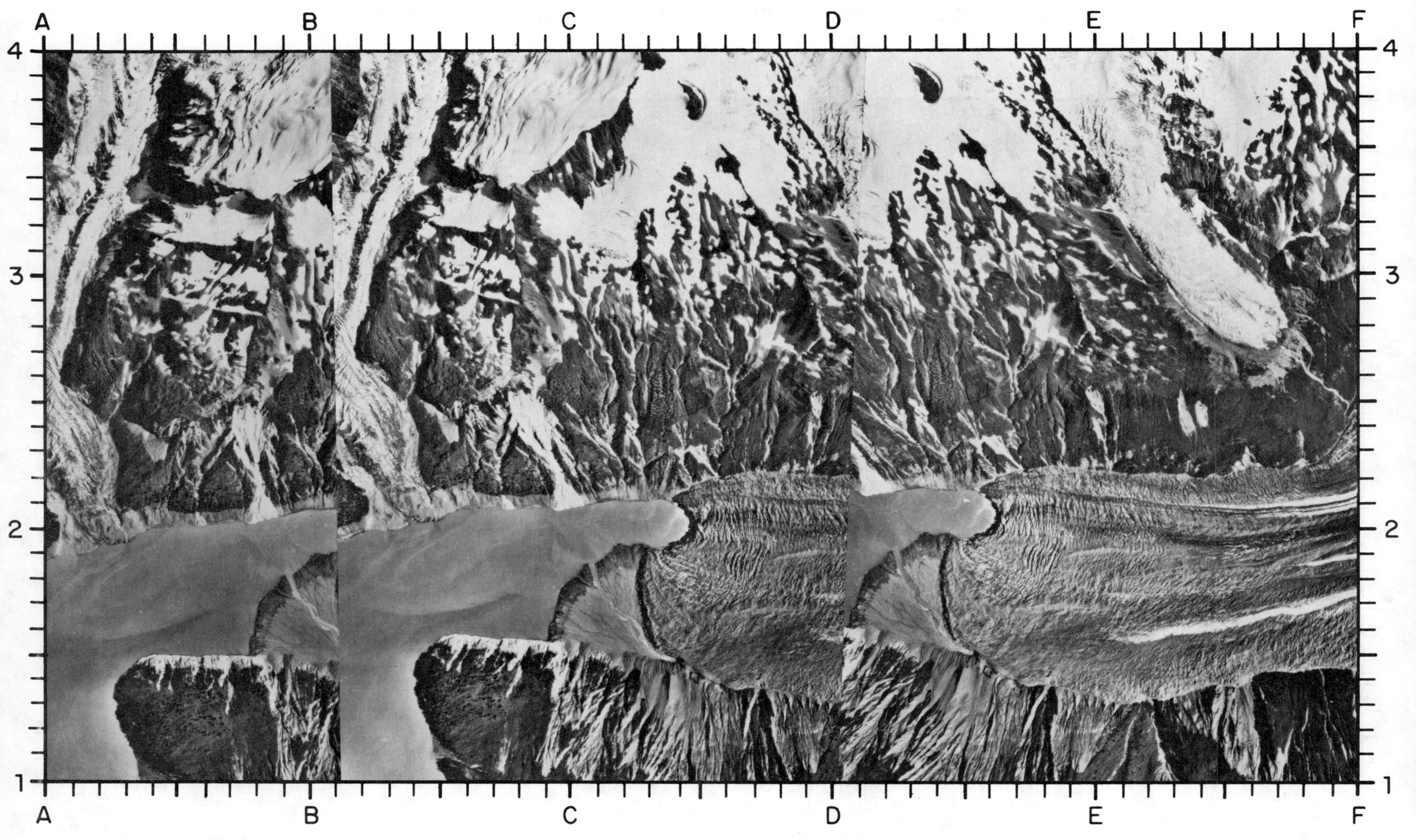

Figure 24.7 Crillon Glacier, Alaska (1:35,500, A.0-1.0 in southwest corner).

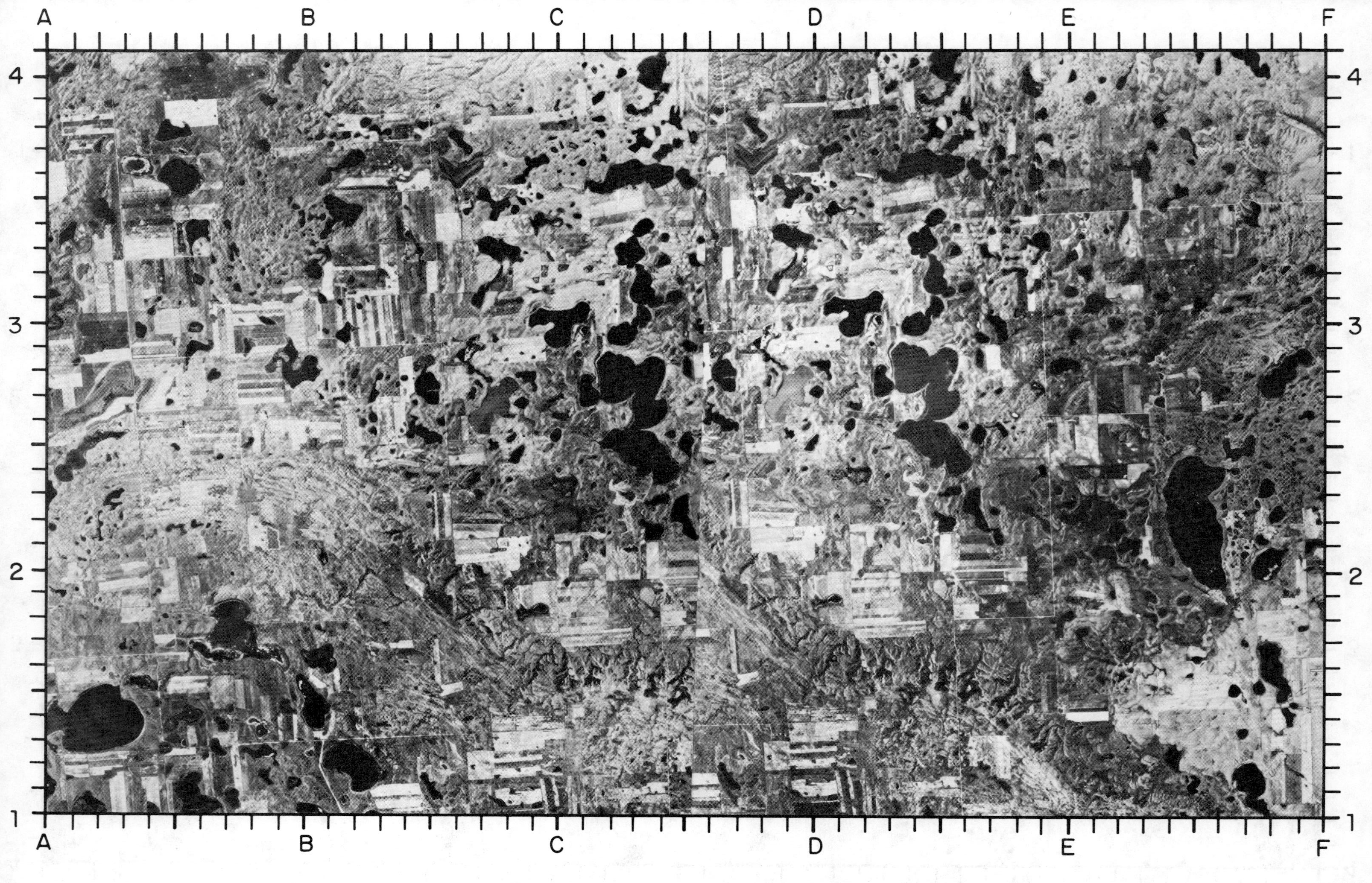

Figure 24.8 Kidder County, North Dakota (1:80,600, A.0-1.0 in southeast corner).

Name ____________________ Date ____________

Instructor ____________________ Section ____________

(d) The northern section of the Crillon Glacier snout is floating and calving has occurred with the formation of icebergs. What is the feature in front of the southern section of the snout and how has it formed?

24.8 Figure 24.8 shows an area of dead ice kame-and-kettle topography in Kidder County, North Dakota. On the photographs identify and label examples of:

A Dry Kettle Hole
B Kettle Lake
C Kame

Also delineate the marked terminal moraine ridge and indicate the flow direction of the ice-sheet that deposited it.

24.9 The southern half of Figure 24.9 shows a field of drumlins in Fond du Lac County, Wisconsin. On the photographs outline a good example of a drumlin and label the stoss end. Which direction was the ice moving when it molded these drumlins beneath it?

24.10 Outline the prominent cirque visible under stereo in Figure 19.6 and identify the terminal moraine at its mouth. Immediately north of the cirque is a steepsided valley cut in limestones and dolomites that was formerly occupied by an alpine valley glacier. Delimit the extent of ground moraine or till on the floor of this valley.

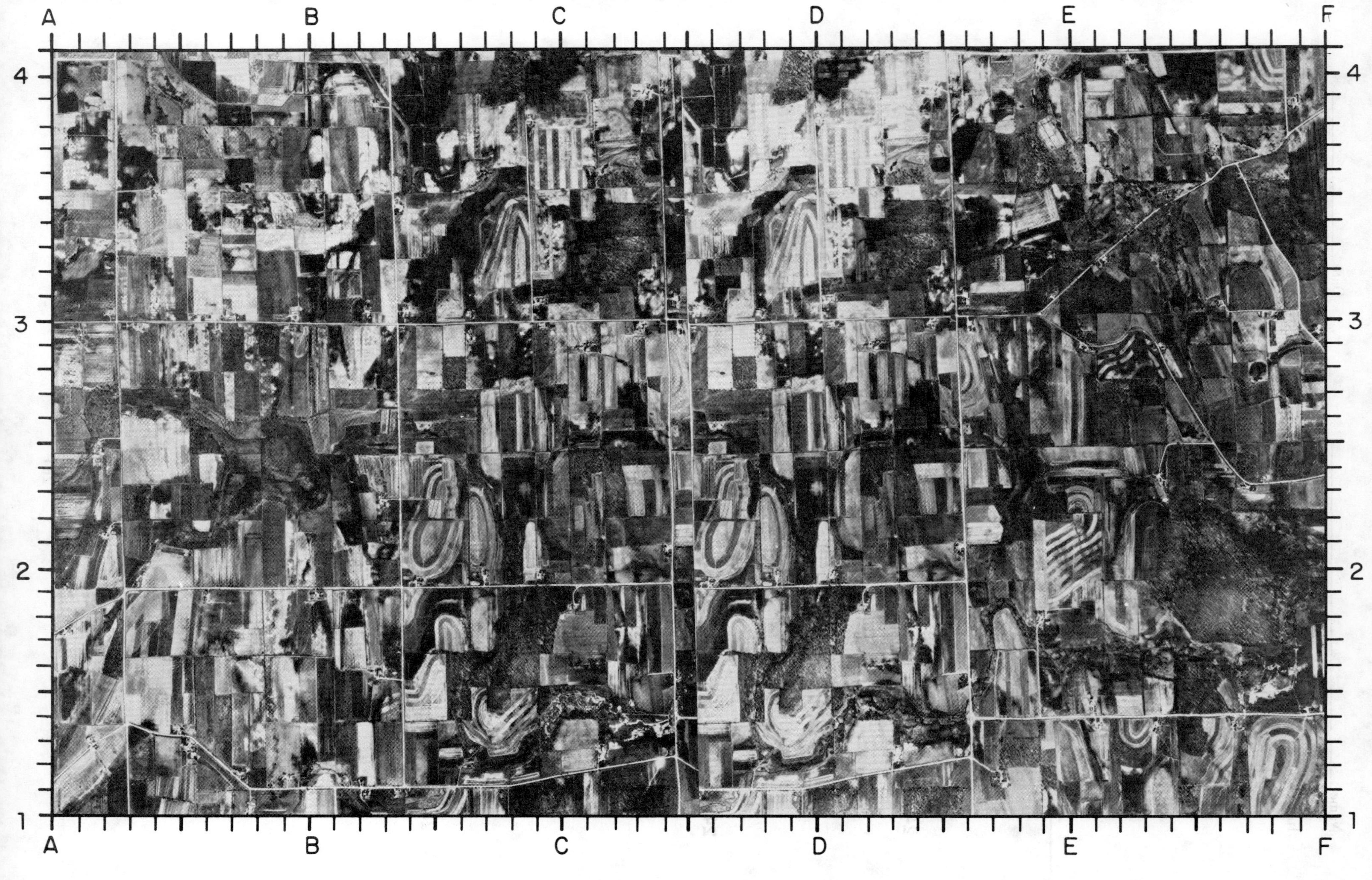

Figure 24.9 Fond du Lac, Wisconsin (1:31,500, A.0-1.0 in southwest corner).

EXERCISE **25**

PERIGLACIAL LANDFORMS

Periglacial regions are those in which the freezing and thawing of water in soils and rock is the dominant geomorphic process. The short summer thaw is frequently insufficient to melt ground frozen during the long winter months so that *permafrost* or permanently frozen ground is often characteristic. The upper surface of permafrost, which varies from a few feet up to more than a thousand feet in thickness, is termed the *permafrost table. Continuous permafrost* extends beneath all topographic features; *discontinuous permafrost* occurs in patches separated by unfrozen zones called *talik* under lakes and rivers; and *sporadic permafrost* occurs in small patches separated by broad areas of unfrozen ground.

For most of the year the ground is frozen solid but during the summer months the surface layer melts. This is called the *active layer* and it may vary from 1-2 inches up to 15 feet in thickness. At the end of the summer *cryostatic* pressures build up in the active layer as the soil refreezes from the surface. Entrapped water moves towards the *freezing front*, freezes and in so doing expands by approximately 9% causing heaving, mixing and sorting of particles within the surface soil layer. Many of the landforms typical of the periglacial environment are due to the seasonal melting and refreezing of the active layer, and can be broadly classified into those produced by the segregation of ice and those produced by thermal contraction.

Polygonal networks of *thermal contraction cracks* develop when the temperature of ice-rich frozen soil drops sufficiently for thermal contraction to take place. Cracks frequently retain water and snow which freezes to form *ice wedges*. Because ice wedges do not melt entirely during the summer months the enclosing sediments are deformed as they attempt to expand and resume their former position. Two types of tundra *ice wedge polygon* relief are differentiated. *Low centered polygons* are typical of low marshy areas and have upstanding rims formed by the upthrust of sediments due to the growth of ice wedges. In poorly drained areas the centers of these polygons and the thermal contraction cracks may contain standing water (Figure 25.1). *High centered polygons* are usually formed by a widening of the peripheral ridges or by vegetation growth. The polygon center is elevated and is surrounded by troughs which may contain water in summer.

When soil in the active layer freezes any water present may segregate into ice lenses. *Ice segregation* is typical of fine-grained materials containing a great deal of soil water. Irregular freezing from the surface down occurs because of lateral variations in heat loss due to differences in *thermal conductivity*. Water in the unfrozen ground moves towards the freezing front as long as a supply of water is maintained and ice lenses may

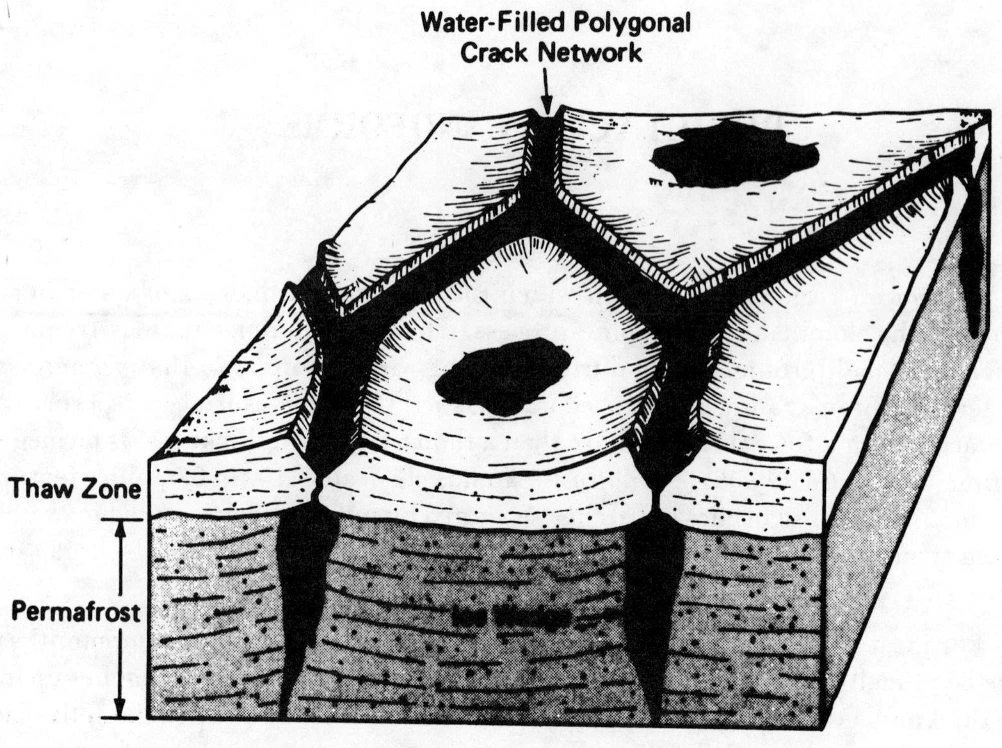

Figure 25.1 Low centered ice-wedge polygons (from Butzer 1976).

develop quickly. In coarser materials with large pore spaces water can not move upwards by capillary action and simply freezes in place with the formation of *pore ice.* Segregated ice lenses vary from a few inches to a hundred or more feet thick. *Pingos,* which may be more than 150 feet high are ice-cored hills which have been domed up from beneath by the growth of ice lenses. They may survive for tens to a few thousand years. Smaller ice-core mounds which are annual or short-lived features are referred to as *frost mounds.* The formation of *needle ice* or *piprake* is also due to ice segregation in the active layer; surface sediments are uplifted as growth occurs.

Patterned ground is formed by a variety of processes including frost sorting, mechanical sorting, contractional cracking and downslope mass movements. The main geometric forms recognized are *circles, polygons, stripes* and *nets* all of which may be sorted or unsorted. Contractional cracking originates polygonal forms while differential heaving is thought responsible for circular forms. Stripes reflect the modifying influence of mass wasting processes upon circular and polygonal forms. *Sorting* of particles occurs because of the repeated freezing and thawing of the ground. When freezing occurs from the surface down the larger particles move upwards, when it occurs from the side they move laterally so that horizontal and vertical layers of different sized particles can result.

Movement occurs because of the greater thermal conductivity of stones as opposed to damp soil. Ice forms beneath a stone before it forms in the nearby soil and continued growth causes the stone to be thrust upwards or sideways towards the cooling front. *Mechanical sorting* also occurs where mounds are produced. Coarser particles migrate under the influence of gravity to form borders of coarse material. *Stone circles* develop in this way (Figure 25.2). When bare rock is subjected to repeated freeze-thaw attack the surface may become a mass of angular fragments which form a *felsenmeer*.

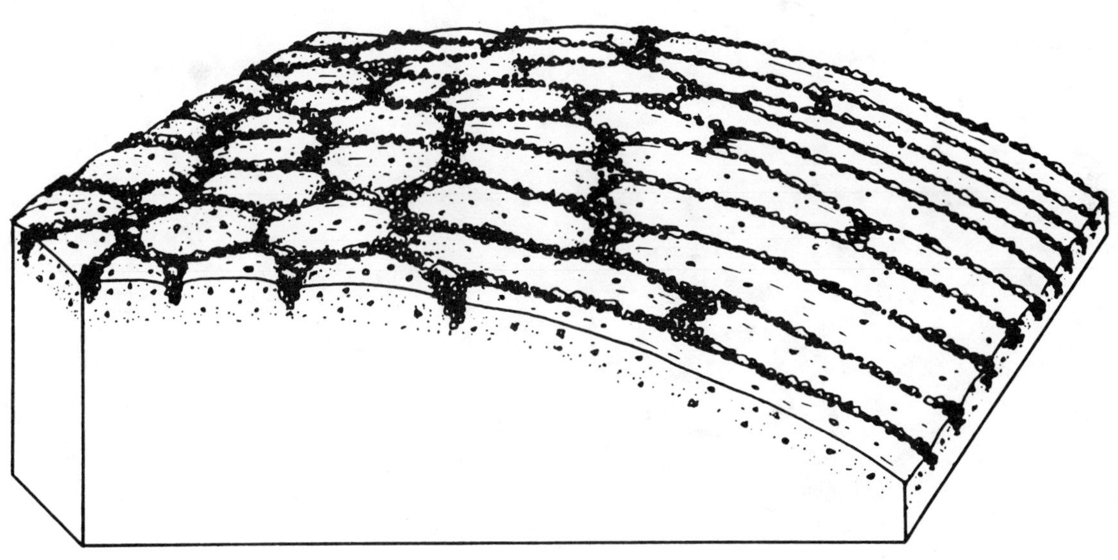

Figure 25.2 Stone circles, garlands and stripes (from Sharpe 1938).

One of the most characteristic landscapes of periglacial regions is *thermokarst* and irregular hummocky terrain formed by the random melting of ground ice, the subsidence of the ground, and the accumulation of the liberated water in the resulting depressions which are known as *thaw lakes*. Thermokarst relief results from an increase in the depth of the active layer usually brought about by an overall amelioration in climate.

Although the landforms generally associated with the periglacial climate are largely the result of stresses set up in the active layer by seasonal freezing and thawing, *fluvial processes* and landforms are nevertheless of considerable importance. The *spring snowmelt* is rapid and because the ground is almost bare of vegetation and because permafrost inhibits vertical movement a high proportion of snowmelt and summer rainfall water becomes *surface runoff*. Streams carry high concentrations of suspended sediment and, where relief allows, may carve steep-sided canyons in bare rock.

Name ____________________ Date ____________

Instructor ____________________ Section ____________

QUESTIONS

25.1 The aerial photographs of Figure 25.3 show periglacial forms developed in stratified silt and sand on the Arctic Coastal Plain near Point Barrow, Alaska. This area falls within the zone of continuous permafrost and is a thermokarst topography.

(a) On the photographs identify and label an example of the following features:

A Thaw lake
B Thaw lake basin still largely frozen
C Low center ice wedge polygons (these can generally be identified where the center contains water)
D High center ice wedge polygons (these have high centers which are separated by troughs)
E Large ice wedge polygons
F Small ice wedge polygons

(b) How have the lake basins in this permafrost region developed?

(c) Why are the shorelines of the thaw lakes serrated?

(d) Given that the scale of the photography is 1:20,000, carefully measure the sizes of the larger and smaller ice wedge polygons on the photographs and then estimate their actual sizes on the ground.

25.2 Remnants of the basaltic plateau to the north of King Hill on the Snake River and the undisturbed upper slopes are characterized by patterned ground phenomena (Figure 19.3).

(a) On the photographs identify and label areas of closely packed circular mounds generally about 50 feet in diameter. The low ground between the mounds is covered by a dark basaltic rubble.

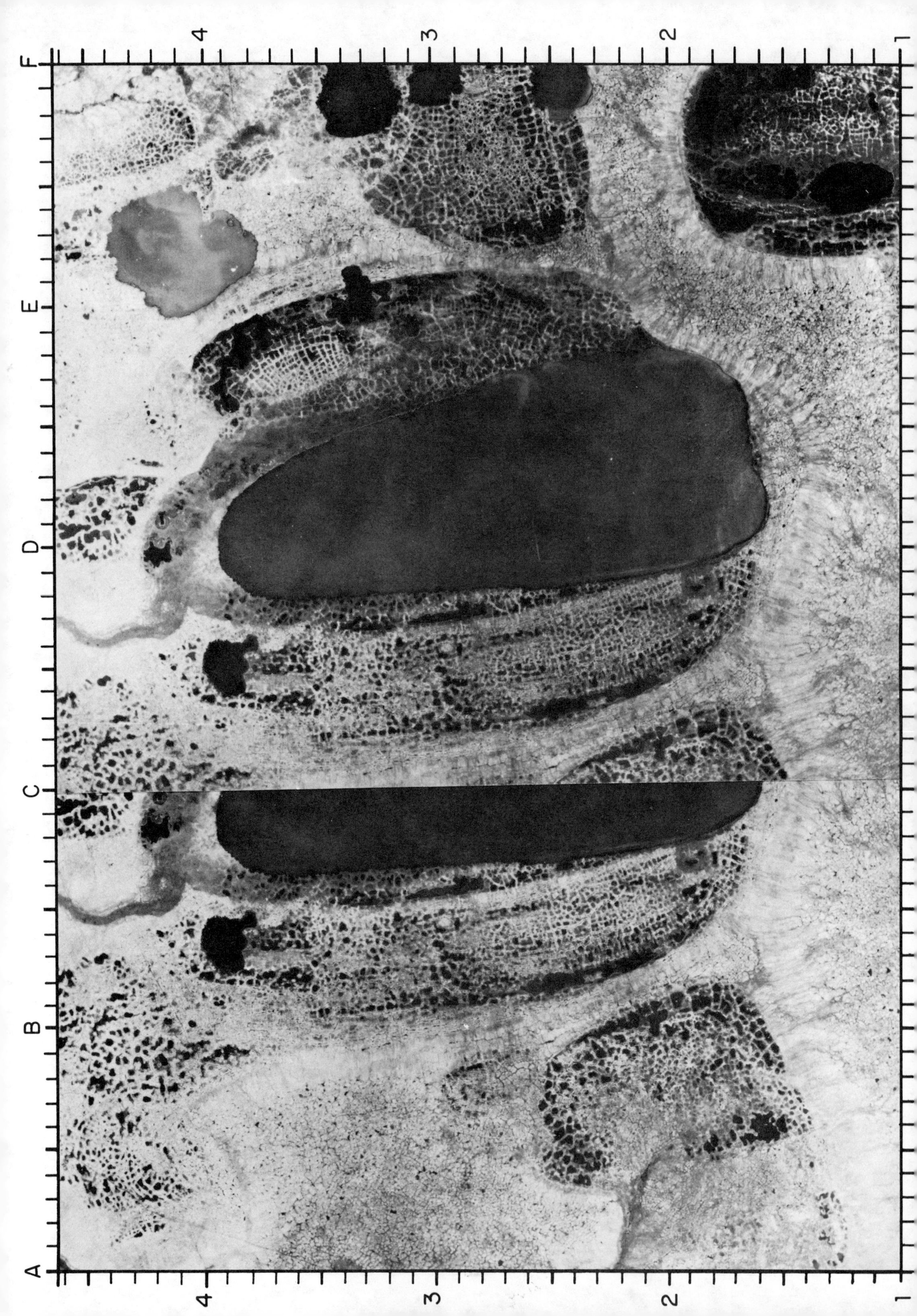

Name____ Date____________________

Instructor ____________________________________ Section__________________

(b) Given that the present climate of this area is not a periglacial one, when did the patterned ground in this region form?

Figure 25.4 Periglacial landforms on Plateau Mountain, southern Canadian Rockies.

25.3 What periglacial features are visible in Figures 25.4, 25.5 and 25.6?

Figure 25.3 Point Barrow, Alaska (1:20,000, A.0-1.0 in southwest corner).

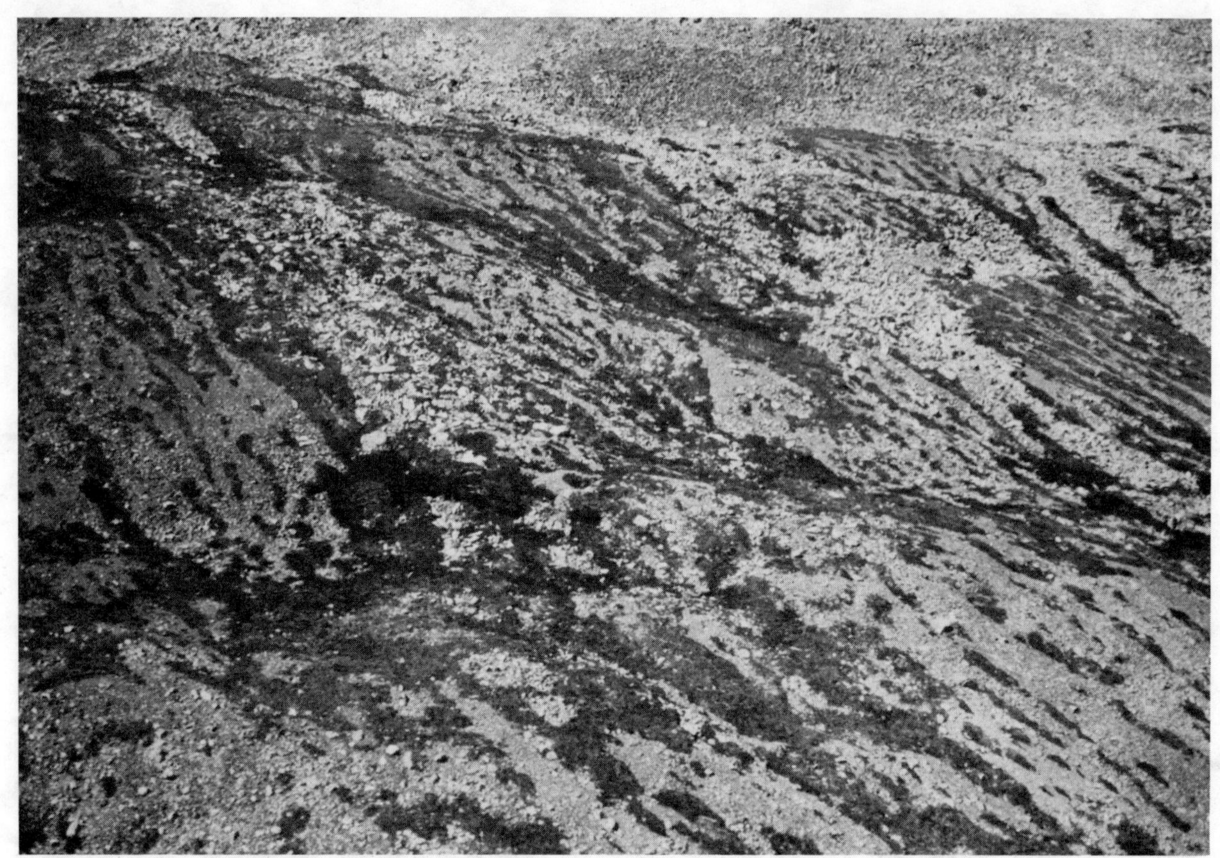

Figure 25.5 Aerial view of tundra zone in the Mackenzie Mountains of Canada.

Figure 25.6 Castleguard Meadow, southern Canadian Rockies.

EXERCISE **26**

COASTAL LANDFORMS

Waves in the deep ocean known as *waves of oscillation* are generated by storms and are the means by which energy is transported. There is little or no actual lateral movement of water out from the storm center. *Wave height* is the vertical distance from *trough* to *crest,* wavelength the horizontal distance between crests measured perpendicular to the *wavefront.* Wave height is controlled by wind velocity and duration, and by *fetch* which is the open stretch of water across which wind can generate waves. When waves move beyond their region of origin they are referred to as *swell waves.* As waves approach the shore they begin to 'feel bottom' which causes an increase in height and steepness. If the *offshore zone* is of variable depth the section of a wave passing over shallow water will be retarded more than the section in deeper water. As a result, the wave front will be bent or *refracted.* Shallow water in front of rocky headlands and deeper water in adjacent bays causes refraction and concentrates wave energy on the headlands which are rapidly eroded. Eventually the crest of a wave may be unsupported and it collapses as a *breaker* or *wave of translation* and there is a rush of water up the beach called the *swash* followed by a return flow known as the *backwash.*

If the wave front steepens gradually until the crest spills over a *spilling breaker* or constructive wave is formed in the *foreshore zone.* Constructive waves are those in which the backwash is weak and does not interfere with the swash of the following breaker. Sediment is swept up the beach and over a long period of time these waves may build a conspicuous *berm* as the coarser material is left stranded at the top of the beach in the *backshore zone* (Figure 26.1). *Plunging breakers* or destructive waves are formed where the sea shallows rapidly. The wave front steepens sharply, the crest curls and finally collapses. In *destructive waves* the backwash is strong and interferes with the swash of the next wave scouring *pools* and *ridges.* Turbulence in the offshore region erodes *submarine troughs* and builds up shallow *bars.* At high tide destructive breakers erode the berm.

The destructive impact of breakers is considerable and is concentrated within a narrow vertical range of 10-20 feet. *Cliffs* and *wave-cut notches* originate where waves undercut coastal rocks and sediments. *Caves* may be excavated along faults and joints. The hydraulic power of wave erosion may connect a sea cave with the surface above via a vertical opening called a *blowhole.* When two caves on opposite sides of a rocky headland unite, a *natural arch* is produced. Collapse of the roof isolates the seaward portion of the headland as a *sea stack* (Figure 26.2). As waves attack rocky coastlines cliffs are undercut and gradually eroded with the formation of a broad, gently sloping *marine-cut platform* which is partially exposed at low tide.

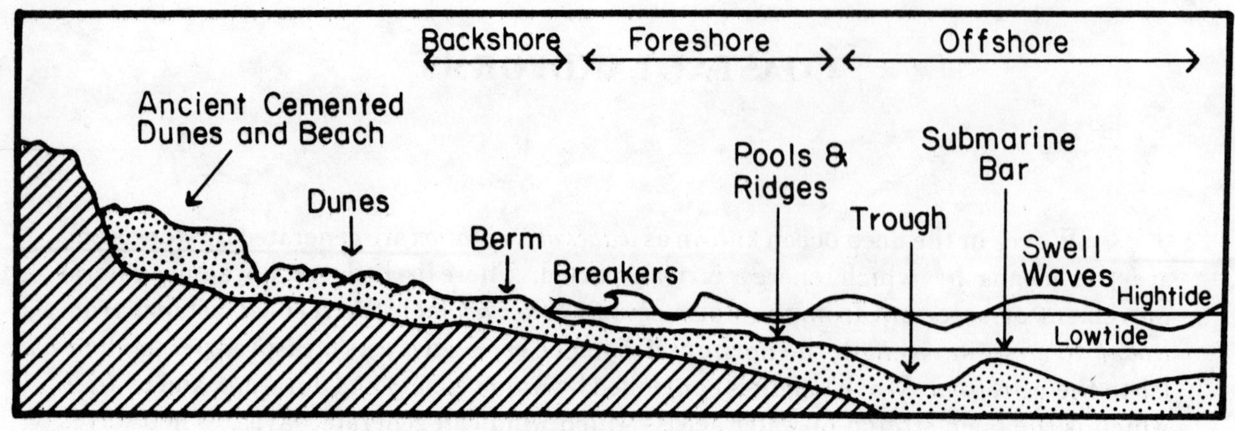

Figure 26.1 The beach environment.

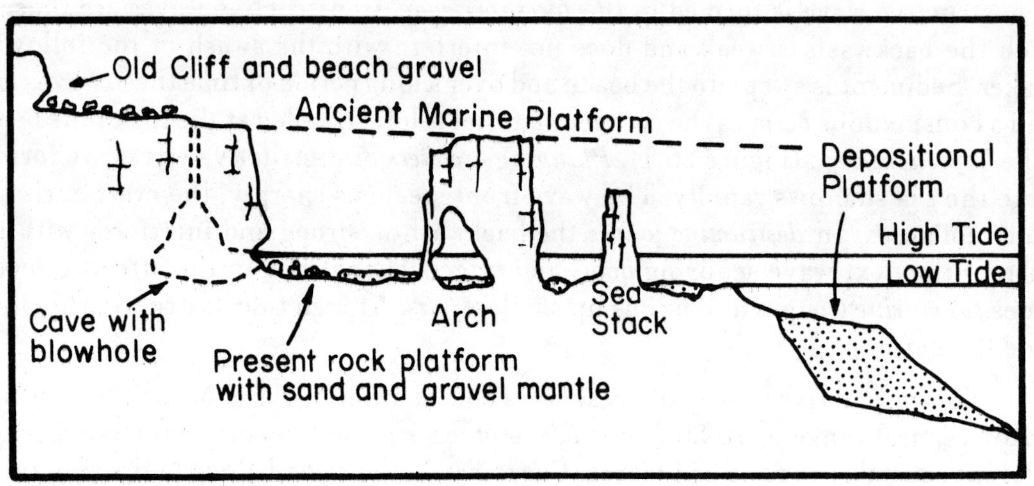

Figure 26.2 Landforms of rocky coastlines.

Longshore transport of sediment is due to *beach drift,* caused by waves striking the shore obliquely, and to *offshore currents.* Although the direction of drift may vary with wave conditions there is generally a net movement of sediment up or down the coast. Where drift is occurring along an *indented shore* spits and bars are constructed. A *spit* is a ridge of sediment built across a bay or *inlet* which terminates in open water. A *cuspate spit* is one that curves towards the coast. A *baymouth bar* or *barrier beach* extends from one headland to another or nearly so and encloses a lagoon. A sand bar connecting an island to the mainland or to another island is called a *tombolo.* The Atlantic and Gulf Coasts of the United States are dominated by *offshore bars* and *barrier islands* separated from the mainland by lagoons (Figure 26.3).

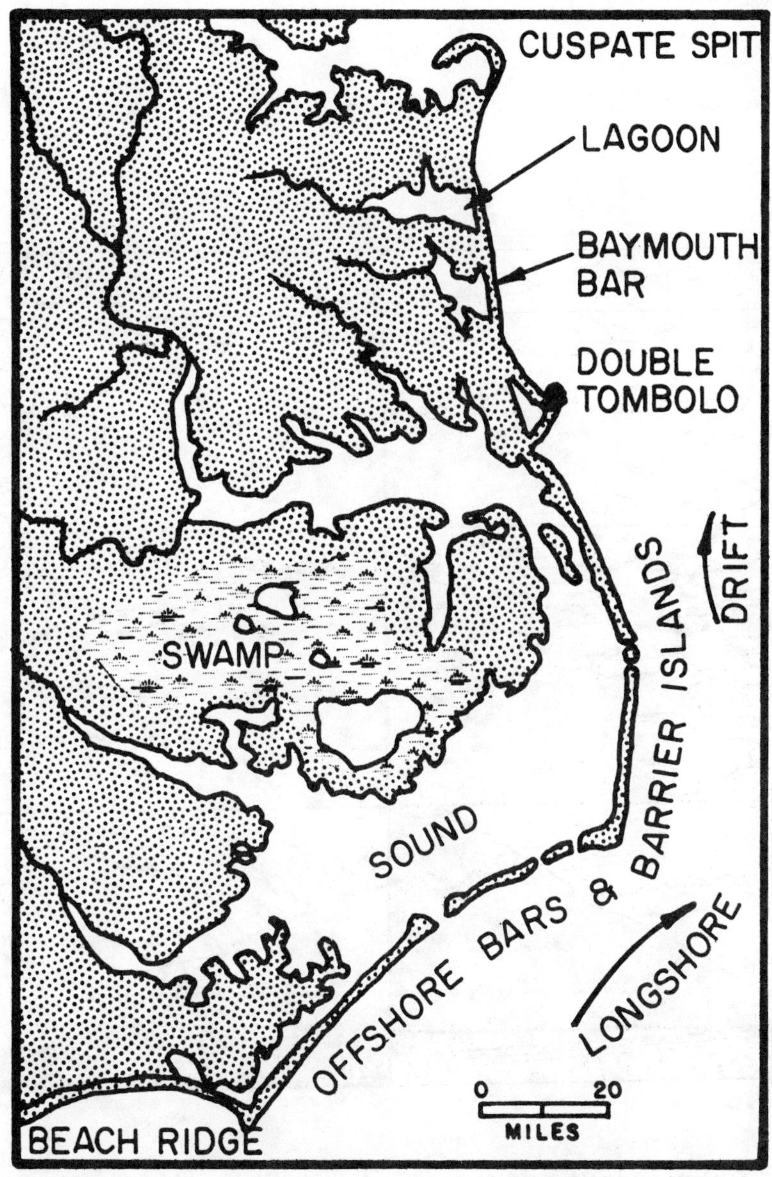

Figure 26.3 Landforms produced by the longshore transport of sediment along a submerged coastline.

Nearly all coastlines have been affected by relative movements between land and sea. Uplift of land due to glacial isostasy, tectonic deformation, and sedimentoisostasy has left marine-cut terraces, cliffs, sea stacks, caves; and ancient coastal dunes, beach sands, and gravels well above present sea level. Coastal regions that have been raised relative to sea level are *coasts of emergence* (Figure 26.4). Growth of huge ice sheets on the continents during the Pleistocene locked great volumes of water on land and caused a world-wide or *eustatic* lowering of sea level of 300-500 feet. The rise in sea level which followed the melting of these ice sheets submerged many of the world's coastal regions which had already adjusted to the lower base level. *Coasts of submergence* are therefore more common than coasts of emergence, and are characterized by bays, estuaries, rias, and fjords.

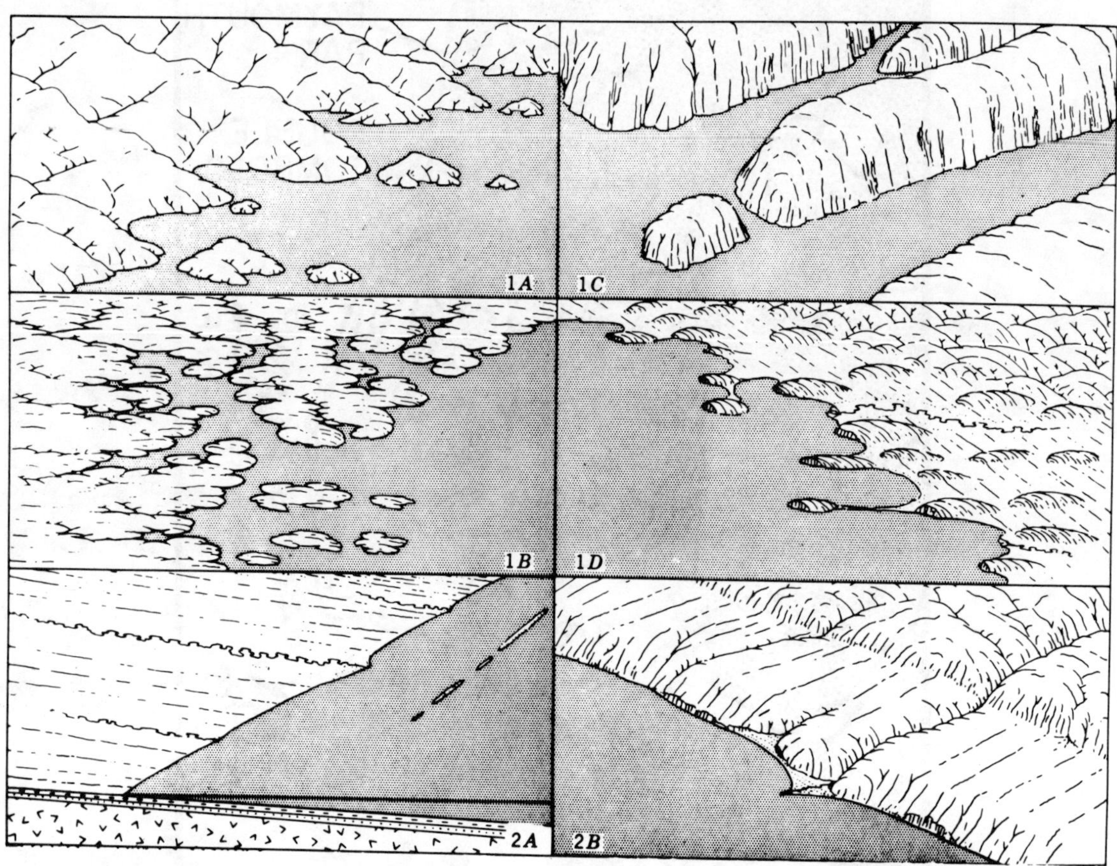

Figure 26.4 Submergent and emergent coastlines. (1A = submerged mountainous coast, 1B = submerged coastal plain, 1C = fjord coast, 1D = submerged drumlin field, 2A = emergent coastal plain, 2B = emergent steeply sloping coastal region) from Strahler 1975.

Name__________ Date__________

Instructor __________ Section__________

QUESTIONS

26.1 Examine Figure 26.5.

(a) Identify and label examples of the following:

A	Offshore Bar	E	Estuary
B	Baymouth Bar	F	Lagoon
C	Spit	G	Tidal Flats or Marsh
D	Inlet or Pass	H	Sand Dunes

(b) How did Laguna Larga (A.7-4.3) originate and what relationship is there between it and Oso Bay (C.2-3.4)?

(c) Outline the events that led to the development of Mustang Island (C.5-2.2) and the tidal flats landwards of it.

CORPUS CHRISTI
CORPUS CHRISTI BAY
NUECES BAY
ARANSAS BAY
Aransas Pass
Ingleside
Portland
Gregory
Taft
Port Aransas
MUSTANG ISLAND
CORPUS CHRISTI SHIP CHANNEL
ENCINAL SHIP CHANNEL
LA QUINTA SHIP CHANNEL
CORPUS CHRISTI AIR STATION
CABANISS FIELD AIR STATION (Auxiliary)
Flour Bluff
Gardendale
ENCINAL PENINSULA
Laguna Vista
LAGUNA LARGA
Chapman Ranch
NUECES CO
KLEBERG CO
ARANSAS COUNTY
NUECES COUNTY
SAN PATRICIO COUNTY
INTRACOASTAL WATERWAY
Corpus Christi Pass (closed)
Newport Pass (closed)
LITTLE DAGGER HILL
Fourmile Hill
BIG HILL

Name____________________ Date__________

Instructor ____________________ Section__________

(d) Why is it necessary to continually dredge a channel between the port of Corpus Christi and the sea?

(e) Could a large tanker drawing 35 feet of water use the Corpus Christi port facilities? Explain your answer.

26.2 Figure 26.6 shows the Provincetown portion of the huge Cape Cod spit which was deposited by longshore transport from the east. The west end of the cape, shown here, was eroded to build the spit to the south of Provincetown. Submerged sandbars are visible offshore which trend roughly at right angles to the beach.

(a) On the map label the following features:
A Complex Recurved Spit
B Tidal Flats
C Active Sand Dunes
D Stabilized Sand Dunes

(b) Outline the stages that led to the development of Long Point (B.0-2.3) and to the different offshore gradients to north and south of Wood End (A.3-3.5).

Figure 26.5 Corpus Christi, Texas (1:250,000, A.0-1.0 in southeast corner).

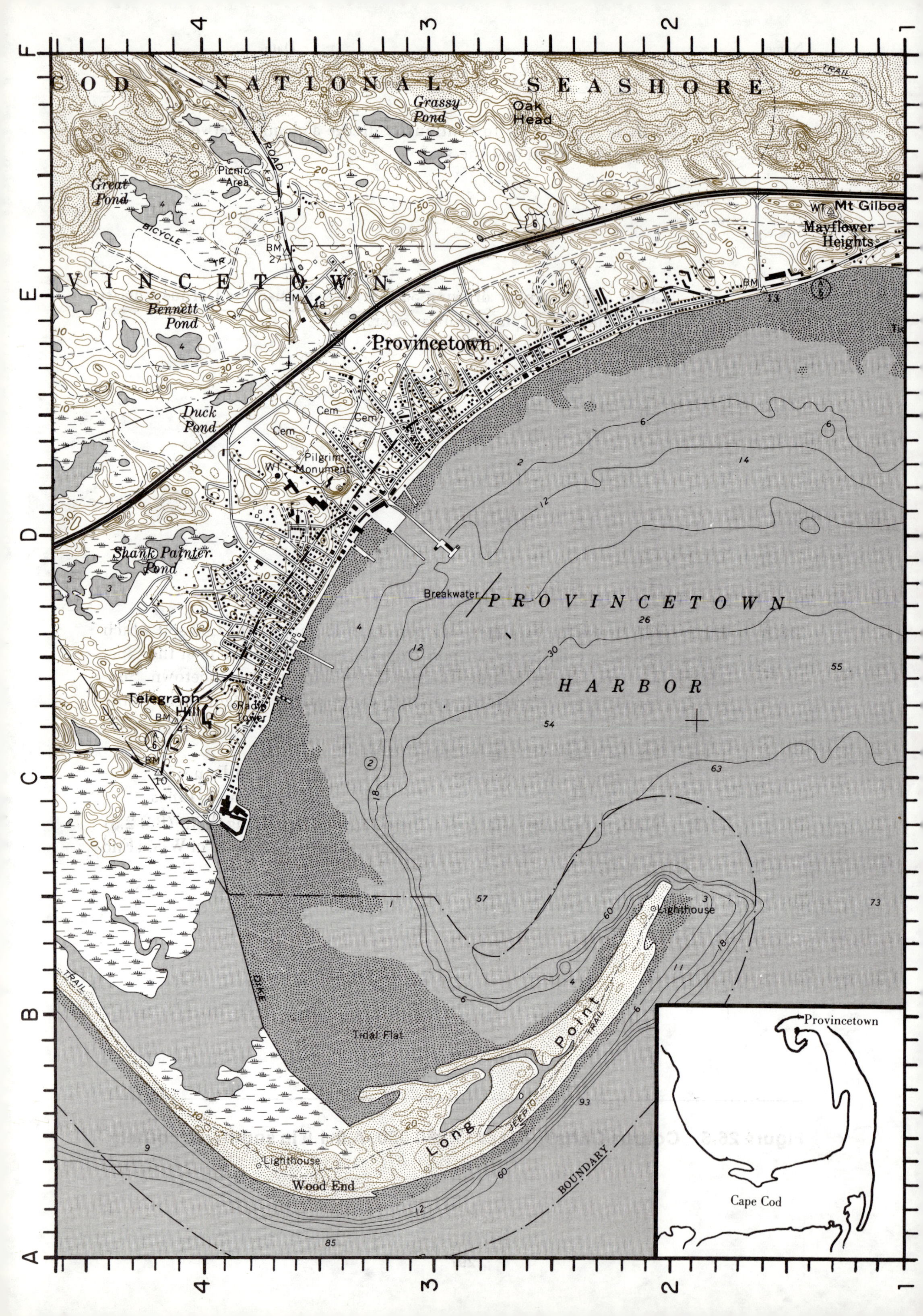

CAPE COD NATIONAL SEASHORE
Grassy Pond
Oak Head
Great Pond
Picnic Area
ROAD
BICYCLE TR
PROVINCETOWN
Bennett Pond
Duck Pond
Provincetown
Mt Gilboa
Mayflower Heights
Cem
Pilgrim Monument
Shank Painter Pond
Breakwater
PROVINCETOWN HARBOR
Telegraph Hill
Radio Tower
Lighthouse
DIKE
Tidal Flat
Long Point
TRAIL
JEEP
BOUNDARY
Wood End
Provincetown
Cape Cod
A
B
C
D
E
F
1
2
3
4

Name____________________ Date____________

Instructor ____________________ Section____________

(c) North of Provincetown is an area of variable relief with numerous small ponds. Explain the origin of this region and of the sand strip on which Provincetown is located.

(d) What evidence is there on the map to indicate that the Pilgrims wintered here before landing at Plymouth Rock on the mainland?

(e) Why is the town of Provincetown and the Pilgrims' resting site located on the south side of the Cape Cod Spit?

26.3 Figure 26.7 shows the southwest portion of San Clemente Island, Los Angeles County, California. The island which is 21 miles long and 4 miles wide and rises more than 2,000 feet above sea level is formed of a series of gently dipping Miocene volcanic rocks.

(a) Identify and label the following features on the photographs:

A Rocky Headland
B Bay
C Ancient Cliff above sea level
D Ancient Marine-cut Rock Terrace above sea level
E Ancient Sea Stack above sea level
F Deep gully systems

(b) Examine the bay immediately east of the headland in the area of stereo coverage. Mark on the photographs the wave crests that are clearly visible and explain why they have been refracted in this way.

Figure 26.6 Provincetown, Massachusetts (1:24,000, A.0-1.0 in southeast corner).

Figure 26.7 Eel Point, San Clemente Island, California (1:26,300, A.0-1.0 in west corner).

Name ____________________ Date ____________

Instructor ____________________ Section ____________

(c) North of the headland is a series of marine-cut rock terraces separated by steps ranging in height from 20-100 feet. Label the present marine terrace 1, an extremely narrow terrace just above sea level 2, and the next four most recent terraces 3,4,5 and 6. On the photographs mark the former cliff shoreline of terrace 3.

(d) Briefly explain the sequence of events that led to the development of the San Clemente Island raised rock terraces.

26.4 Examine the coastal regions shown in Figures 26.5, 26.7, and 24.2 and state which are coastlines of emergence and which are coastlines of submergence.

26.5 Figure 26.8 is a photograph of Marsden Rock on the northeast coast of England.

(a) What kind of coast is shown?

Figure 26.8 Marsden Rock, Durham, England.

(b) What is Marsden Rock?

(c) Explain the pattern of wave refraction in this area.

EXERCISE **27**

ARID AND SEMIARID LANDSCAPES

Arid regions receive less than 10″ of rainfall per year and in any given year may receive no rain whatever. *Semiarid* areas receive considerably more precipitation (10-20″) but still experience long periods of drought. In both areas rainfall is sporadic and of great intensity. Storms tend to be highly localized so that floods frequently dissipate before sizeable channel flow develops. Surface runoff takes the form of *sheetflow* or *rillflow* so that in desert regions drainage lines are poorly developed and in semiarid terrains they are *ephemeral.*

Typical slope-surface material relationships can be seen in a dry, block-faulted area such as occurs in the Basin-and-Range Province of the western United States. Interior drainage basins of tectonic origin surrounded by horst block mountains are preserved because the surface and ground water flow into them is evaporated. Mechanical, and to a lesser extent chemical weathering causes the steep, rocky slopes of the mountain blocks to disintegrate. The larger *detrital* fragments collect as *talus,* the finer material is carried by streams and sheetfloods downslope to form *alluvial fans,* which coalesce into a piedmont slope called a *bahada.* Near the mountains the bahada slopes at 5-15°, further downslope its gradient is reduced to as little as 1-5°. The alluvial veneer of the bahada is only superficially dissected by stream channels. The finest alluvial material is carried into the center of the basin which has a flat floor or *playa* (Figure 27.1). If enough water reaches the center of the basin after heavy rain a temporary *playa lake* may be formed. The dissolved material in the lake water is concentrated by evaporation and deposited when the lake dries. Playa deposits are, therefore, a mixture of silt and *evaporite* materials. Water infiltrates into the coarse gravels of the upper bahada and percolates towards the toe. Wells in this region will often locate a fairly reliable groundwater supply. The gradual dissection of the mountain block, with the extension of *pediments* and the broadening of the bahada leave isolated residuals near the upland margins. These residuals are called *inselbergs.* Mesas and buttes are varieties of inselberg.

Wind can remove dry, incoherent deposits from desert surfaces in a process known as *deflation.* Silt-size particles are carried aloft in *suspension,* sand-size particles are transported by *saltation* within a surface layer of air 5-8 feet deep. Coarser sands and small pebbles *creep* forward along the surface aided by momentum gained from saltating sand particles. Wind removes silt and fine sand from the surface and may create surface *deflation hollows.* Armed with sand grains near the surface the wind is a powerful scouring and abrading agent. Softer rocks are undercut with the formation of *pedestal rocks.* Where alternate hard and soft rock layers are exposed at the surface the softer materials are scoured more rapidly so that the harder layers form positive relief features. The softer rocks of tilted strata may become narrow corridors between overhanging ridges in the harder layers which are called *yardangs.* Deflation of finer material from desert surfaces concentrates rock and coarse sand in a surface lag. Lags of mechanically weathered rock are *hammada* or *rocky deserts,* those of rounded gravel and coarse sand of alluvial origin are *reg* or *pebbly deserts.* About one fifth of the world's desert area is covered by sand forming *sandy desert* or *erg.*

Eolian deposits include both sand and silt or *loess.* Any mound or ridge of sand with a crest or definite summit is called a *dune.* Dunes may occur singly or in fields. A typical dune has a long *windward slope* rising to a crest and a much steeper *leeward slope.* Sand blown over the crest falls into a *wind shadow* and comes to rest at its natural *angle of repose* which for dry sand is 30-35°. Dunes migrate in the direction of the prevailing wind until arrested by vegetation. Five major dune types are recognized (Figure 27.2).

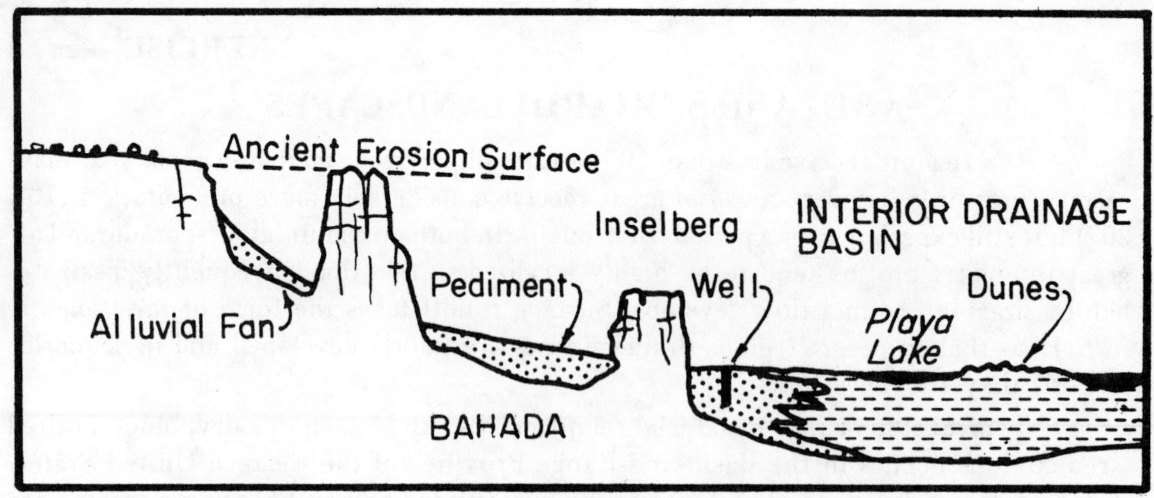

Figure 27.1 Desert landscape model in a block faulted terrain.

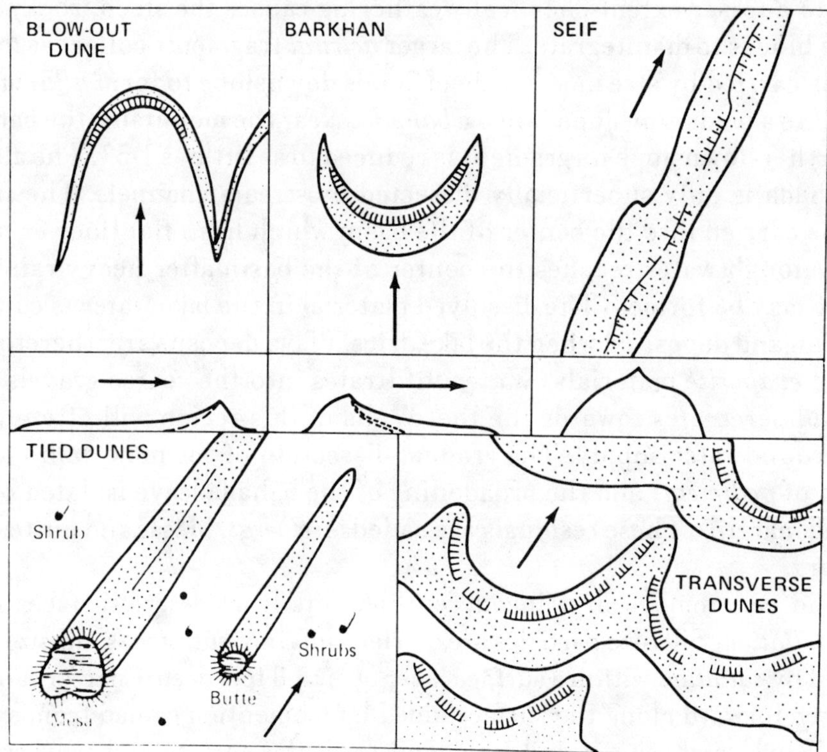

Figure 27.2 Sand dune types. Arrows indicate the effective wind direction (from Butzer, 1976)

Sand drifts or *tied dunes* form in the wind shadow or lee of protruding rocks, bushes or isolated hills. *Parabolic* or blow-out dunes have a sand ridge downwind of a deflation scar the arms pointing upwind. *Transverse dunes* are perpendicular to the prevailing or effective wind direction and form where fields of barkhans merge. Longitudinal or *seif dunes* are sinuous ridges parallel to the effective wind direction. Crescent-shaped *barkhans* have horns which face downwind. They develop from mounds of sand and as the dune migrates the extremities offer less resistance to the wind than the summit region and advance more rapidly extending into wings. The width of a barkhan is commonly 12x height which ranges up to 100 ft. Rates of movement vary from 20 ft/year for the larger dunes to 50 ft/year for the smaller ones. Where the effective wind is occasionally interrupted by strong cross winds which drive in sand from the sides *seif* dunes develop. Vast quantities of dust can be removed from desert regions and deposited elsewhere as *loess,* which is an accumulation of wind-borne dust and silt washed down from the air by rain. In some areas hundreds of feet of loess have accumulated.

Name________________________ Date________________

Instructor ________________________ Section________________

QUESTIONS

27.1 Figure 27.3 shows a portion of the Douglas, Arizona region just north of the United States-Mexico border. The area lies within the Basin-and-Range Province of the western United States where relief is dominated by a series of horst block mountains separated by interior basins developed in graben structures.

(a) What evidence is there on the map to suggest that the Douglas region has an arid or a semiarid climate?

(b) Little Hatchet Mountains are composed of what rock type? Explain your answer.

(c) Construct a 25-mile long east to west topographic profile from Antelope Pass in the northern section of the Peloncillo Mountains (A.9-3.9) eastwards towards the Little Hatchet Mountains (E.7-3.6). The profile should pass through Tank Mountain. Use a horizontal scale of 1:250,000 and a vertical scale of 1 inch equals 2,000 feet. On the section label the following features.

A Eroded Horst Block Mountain
B Downfaulted Basin or Graben
C Interior Drainage Basin
D Playa
E Playa Lake
F Pediment or Bahada
G Alluvial Fan
H Inselberg

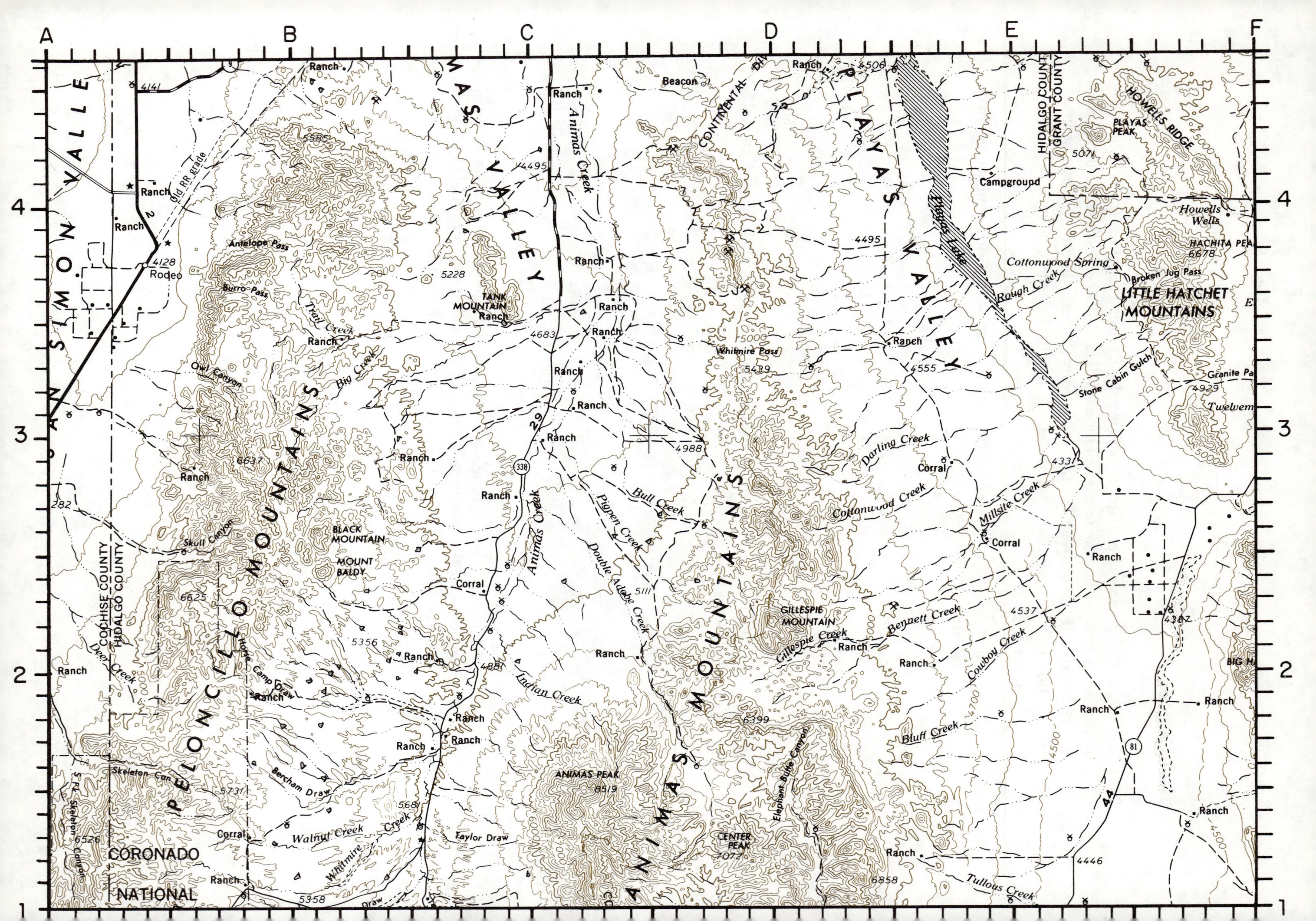

A
B
C
D
E
F
4
3
2
1
SAN SIMON VALLEY
ANIMAS VALLEY
PLAYAS VALLEY
PELONCILLO MOUNTAINS
ANIMAS MOUNTAINS
LITTLE HATCHET MOUNTAINS
HOWELLS RIDGE
PLAYAS PEAK
HACHITA PEAK
6678
Howells Wells
Broken Jug Pass
Cottonwood Spring
Rough Creek
Playas Lake
Campground
HIDALGO COUNTY
GRANT COUNTY
COCHISE COUNTY
HIDALGO COUNTY
CONTINENTAL DIVIDE
Beacon
Rodeo
4128
4141
Old RR grade
Antelope Pass
Burro Pass
Trail Creek
Big Creek
Owl Canyon
Skull Canyon
Deer Creek
Horse Camp Draw
Bercham Draw
Walnut Creek
Whitmire Creek
Skeleton Cyn
S Fk Skeleton Canyon
CORONADO
NATIONAL
TANK MOUNTAIN
BLACK MOUNTAIN
MOUNT BALDY
Animas Creek
Whitmire Pass
Bull Creek
Pigpen Creek
Double Adobe Creek
Indian Creek
Taylor Draw
ANIMAS PEAK
8519
CENTER PEAK
7072
Elephant Butte Canyon
GILLESPIE MOUNTAIN
Gillespie Creek
Bennett Creek
Cowboy Creek
Bluff Creek
Tullous Creek
Darling Creek
Cottonwood Creek
Millsite Creek
Stone Cabin Gulch
Granite Pa
Twelvem
BIG H
Ranch
Corral
5585
5228
4495
4683
5000
5439
4555
4506
5071
4929
4331
4988
5111
4537
4382
6637
6625
5356
4881
6399
5731
5358
6526
6858
4446
4500
282
29
338
81
9
2
44

Name________________________ Date______________

Instructor ____________________ Section______________

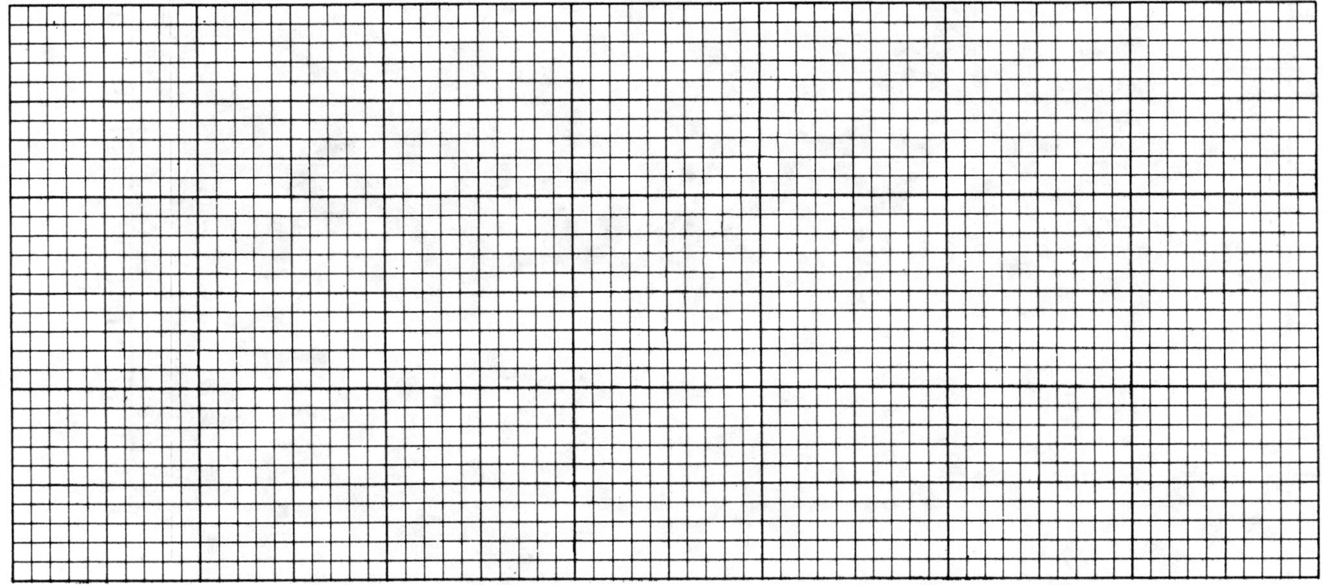

(d) With particular reference to the origin of Playas Lake (D.8-3.8) describe the drainage characteristics of the Douglas area. Why are streams identified with a dashed rather than a solid blue line?

27.2 Figure 27.4 shows part of the northwest side of the Sacaton Mountains in Arizona. The region is dominated by a dense network of dry river channels, outlined by desert shrubs, which are developed in an alluvial cover.

(a) Why are the desert shrubs restricted to dry stream beds?

(b) On the photographs label examples of the following:

A Braided Stream
B Parallel Drainage
C Pediment
D Inselberg

27.3 Figure 27.5 shows the Comb Ridge region of Arizona. Comb Ridge (D.1-1.0) is a north-facing cliff capped by Navajo Sandstone that dips gently to the southeast. In the north Little Capitan Valley (D.0-2.4) separates the ridge from the dip slope on the De Chelly Sandstone (D.5-4.0).

(a) Identify and label on the photographs good examples of the following:

A Mesa
B Butte
C Transverse Dune
D Barkhan
E Tied Dune
F Longitudinal Dune
G Blow-out Dune
H Ephemeral Lake in Dunes

Figure 27.3 Douglas, Arizona (1:250,000, A.0-1.0 in southwest corner).

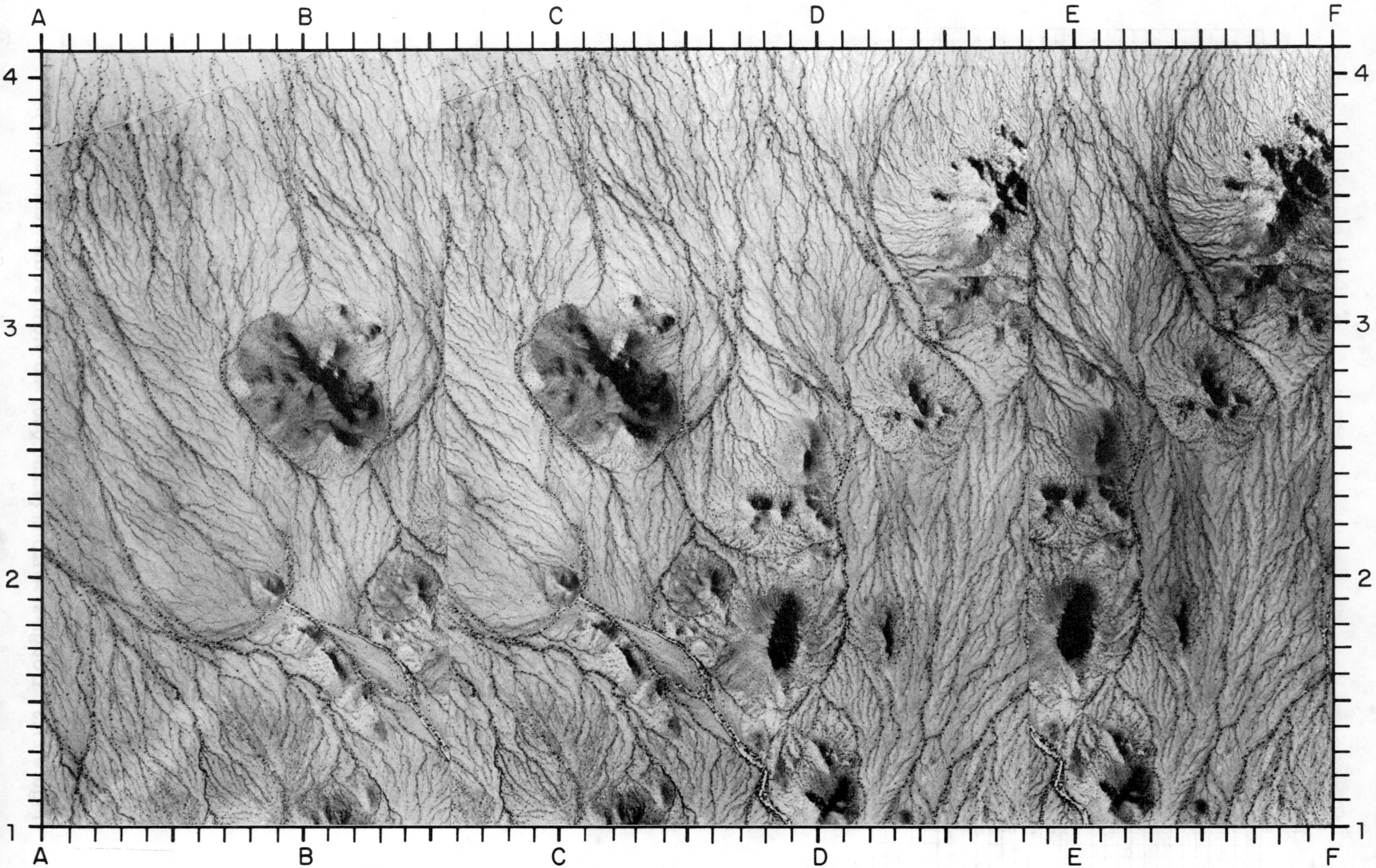

Figure 27.4 **Sacaton Mountains piedmont, Arizona (1:26,800, A.0-1.0 in southeast corner).**

Name ______________________________ Date ______________

Instructor ______________________________ Section ______________

(b) What is the direction of the prevailing wind?

(c) A series of washes drain into Little Capitan Valley from the north. How have the sand dunes affected these drainage routes?

(d) What is the source of the sand in this area?

27.4 Figure 27.6 shows the sand dunes of the Kharan Kalat region of western Pakistan. The crests of the major dunes run northwest to southeast. Small white patches of salt indicate that locally a pond sometimes forms at the foot of a dune (e.g. D.5-3.6).

(a) What kinds of dunes dominate this area and what is the spacing between crests?

(b) On a suitable dune mark the crest, slip-off slope and the shallow windward slope, and indicate the direction of the prevailing wind.

(c) Calculate the relative height of one of the dunes near to the center (principal point) of one of the photographs by assuming that the angle of rest of sand grains on the slip-off slope is approximately 35°.

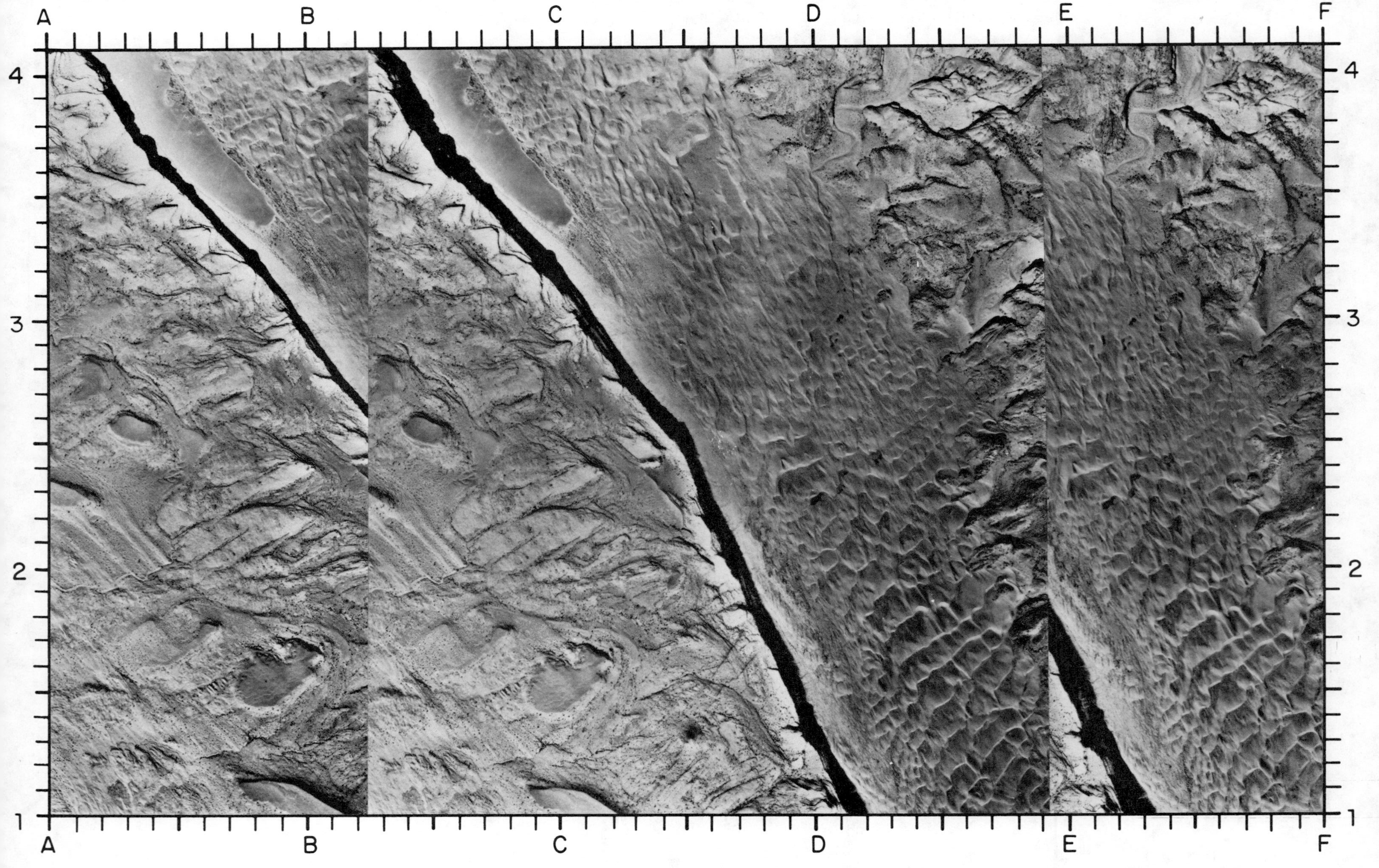

Figure 27.5 Comb Ridge, Arizona (1:26,600, A.0-1.0 in southeast corner).

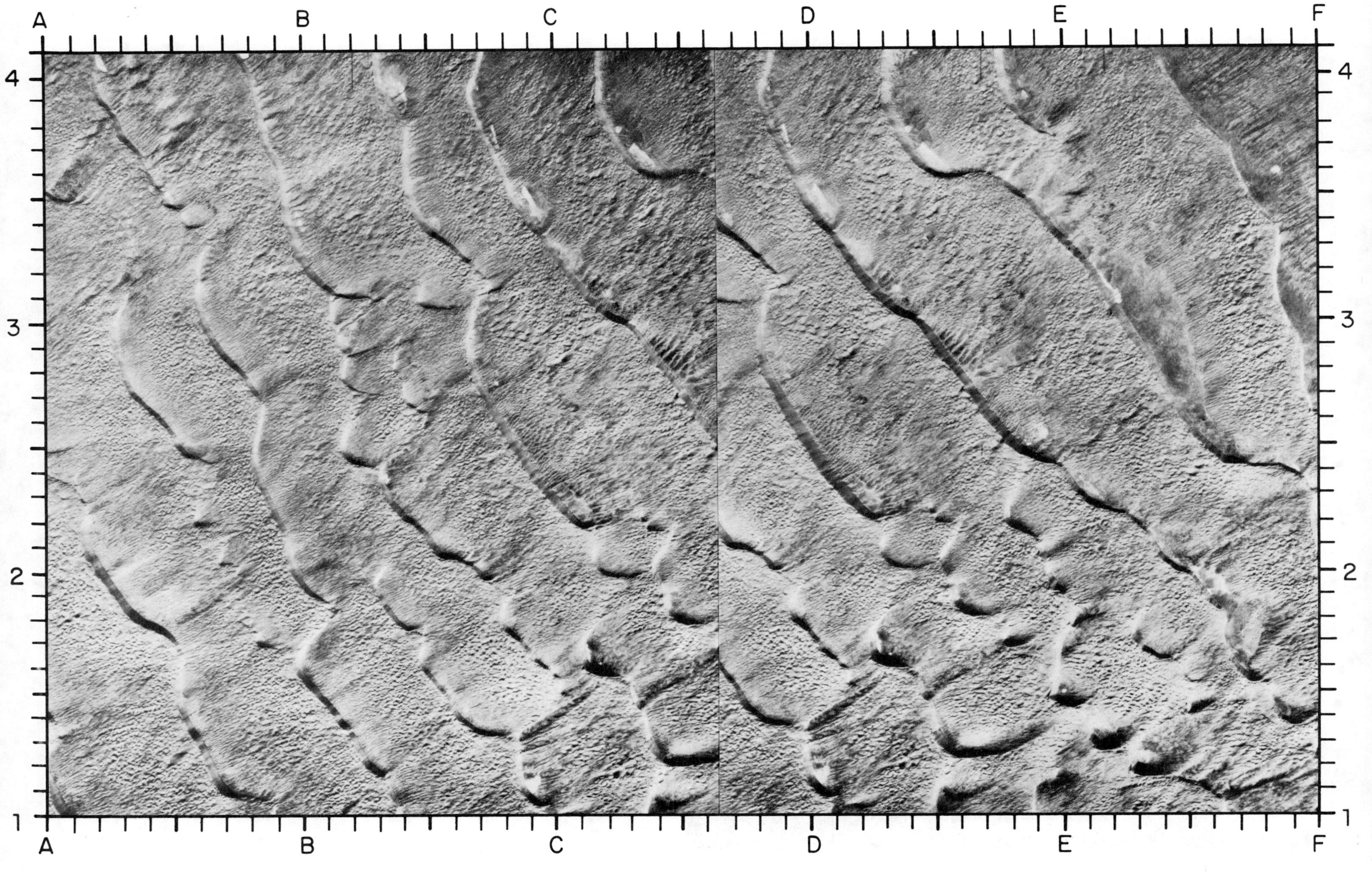

Figure 27.6 Kharan Kalat, Pakistan (1:53,200, A.0-1.0 in southwest corner).

Figure 27.7 Mojave Desert, southern California

27.5 Figure 27.7 shows the Twenty Nine Palms region of the Mojave Desert in southern California. In the photograph label the following features:

A Alluvial fans	C Bahada
B Fault block mountains	D Hammada

EXERCISE 28

THE PHYSICAL PROPERTIES AND ERODIBILITY OF SOILS

The study of soils is the science of pedology. The pedologist is interested in the appearance of the soil, its mode of formation, its physical, chemical and biological composition, and its classification and distribution. Soil is composed of solids, liquids, and gases. Assessment of soil for many purposes is based on evaluation of these properties. While a true soil or *solum* is composed of both mineral and organic particles, the underlying material is usually wholly mineral matter. Factors of soil formation include parent material, organisms, climate, topography, and time. A rock in a part of the landscape, when acted upon by atmospheric processes (particularly temperature and precipitation) disintegrates into loose unconsolidated inorganic debris (the parent material of a soil). Soil formation from rock is indicated in the following equation.

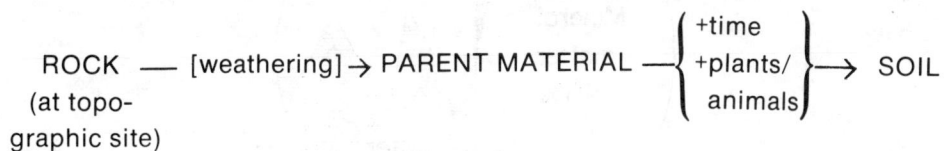

$$\text{ROCK (at topographic site)} \text{ — [weathering]} \rightarrow \text{PARENT MATERIAL} \text{ —} \left\{ \begin{array}{c} \text{+time} \\ \text{+plants/} \\ \text{animals} \end{array} \right\} \rightarrow \text{SOIL}$$

Soil structure is an important physical characteristic of any soil. The structure is brought about by the individual particles of sand, silt or clay aggregating together in larger units known as *peds*. The peds have been described as the 'architecture' of the soil, and the spaces around them act as channels to conduct water through the soil. Soil structure readily falls into 6 categories: structureless (each soil particle independent of all others), massive (entire soil mass clings together, no lines of weakness), crumb or granular, platy, blocky, and prismatic (Fig. 28.1).

Figure 28.1 Soil structures formed by the aggregation of sand, silt and clay particles: (a) prismatic; (b) columnar; (c) angular blocky; (d) subangular blocky; (e) platy; (f) crumb or granular.

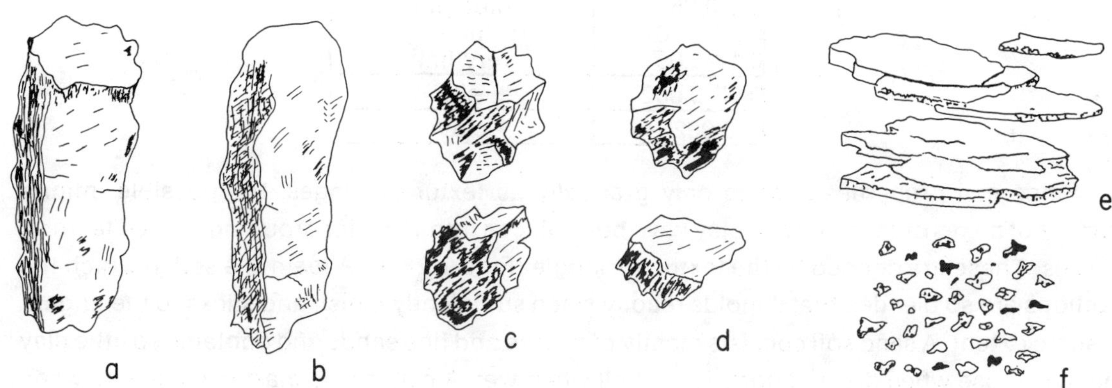

A typical topsoil contains 45% mineral matter, 25% water, 25% air, and 5% organic matter (Fig. 28.2). Soils are developed in material which has already been weathered from solid rocks. This weathered mantle or regolith can be as deep as 50 m in the humid tropics. One physical characteristic of the soil solids is their texture. Soil particles grade continuously in size from > 2 mm in diameter (coarse fragments) to < 0.002 mm in diameter (clay). For convenience of definition and measurement this continuum is divided into particle-size groups or *separates* called sand, silt, and clay. The particle-size composition of a soil can then be described in terms of its composition of each of these three separates.

Figure 28.2 Volume composition of a typical topsoil; amounts are approximate as the percentage of certain constituents, e.g. air and water, is constantly varying.

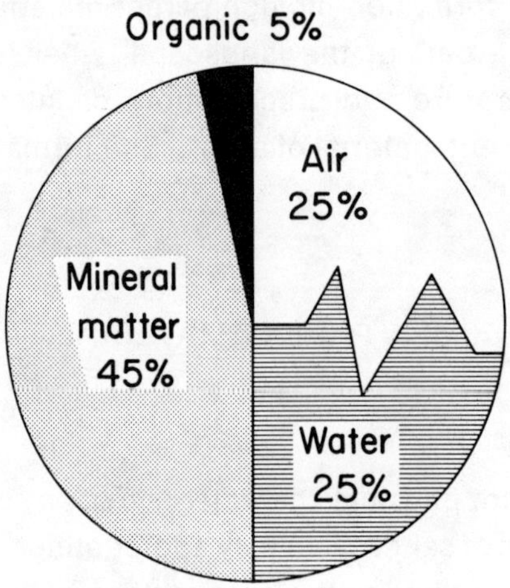

Table 28.1 Particle-size Classification System of the U.S. Department of Agriculture

USDA	
size, mm	**name**
> 2.0	coarse fragments
2.0-0.05	sand
2.0-1.0	very coarse
1.0-0.5	coarse
0.5-0.25	medium
0.25-0.10	fine
0.10-0.05	very fine
0.05-0.002	silt
< 0.002	clay

Since soil properties change only gradually as texture changes, the possible infinite number of mixes of sand, silt and clay have been placed into 12 major groupings called textural classes. These are defined by the textural triangle (Figure 28.3). A loam is a soil in which the fractions are so blended that it molds readily when sufficiently moist, and sticks to the fingers to some extent. A sand soil consists mostly of coarse and fine sands and contains so little clay that it is loose when dry and not sticky at all when wet. A clay soil is plastic and sticky when moistened sufficiently and gives a polished surface on rubbing. A silt soil has the smooth, soapy feel of silt. These texture classes are used for the description of texture within a profile but there is a need for broader groupings of soil textures when describing the whole profile or groups of similar soils. The broader groupings are shown in Fig. 28.4. These different soil textures have properties which affect the management and economic use of the soil.

Figures 28.3 and 28.4: Soil texture classes, U.S.D.A. The three sides represent base lines for sand, silt and clay with the apices opposite representing 100 percent of each constituent. Percentages can be read off to give the textural name for any soil sample. **(28.4)** Broad soil texture groupings.

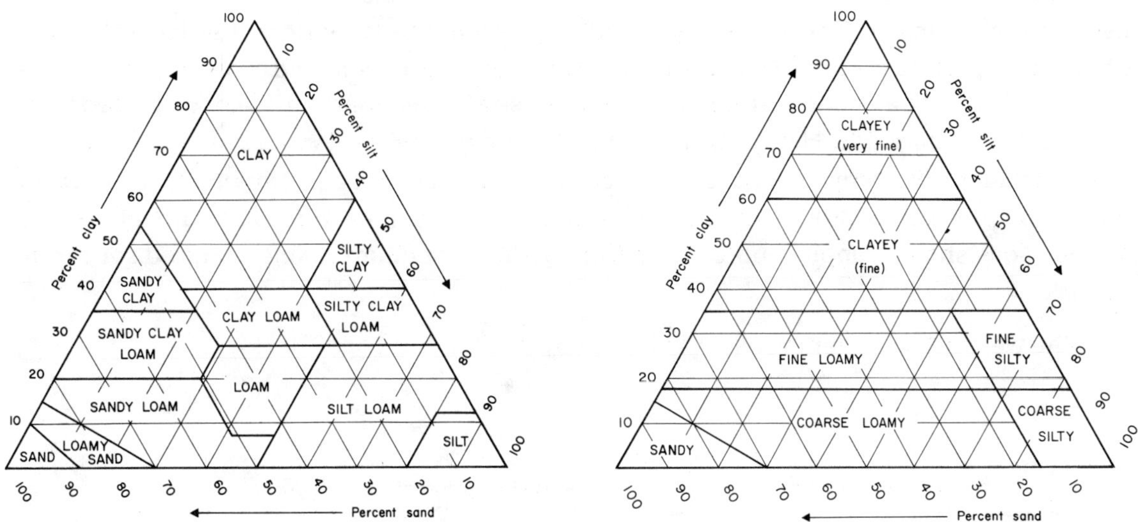

Soil water fills all the pore spaces and fissures in a soil. If a soil drains freely, all of the water contained in the larger pores and cavities is removed—this is the gravitational soil water. Soil that has lost its gravitational water is said to be at *field capacity*. In this state considerable amounts of water are still held in the finer pores and by capillary action. If the soil continues to lose moisture from these reserves of capillary water, the point is reached when plants can not obtain enough water to continue transpiration. Wilting then takes place from which the plant does not recover; this is the *permanent wilting point*. The amount of water held in the soil between field capacity and permanent wilting point is the *available water capacity* of the soil. This amount varies with soil texture and structure.

Size and number of voids influence water movement, aeration, and storage capacity of plant-available water. It is desirable, therefore, to have information about the soil pore space. If the pores and fissures of a soil are filled with water, fresh oxygen can not easily diffuse into the soil. The oxygen present is soon used so that anaerobic conditions are produced. It is in these conditions that the growth of most plants is inhibited and the process known as gleying is brought about. The amount of pore space in a soil can be calculated if the bulk density and particle density of soil are known.

The density of a soil is its mass per unit volume,

$$D = W/V$$

where D = density (g/cm^3), W = weight (g), and V = volume (cm^3).

The distinction between particle and bulk density is in the volume term. For bulk density (BD) the total soil volume V_t, including all voids or pores, is used. In particle density (PD) only the volume of the soil solids V_s, excluding pore volume, is used. Since the weight W (the weight of water-free soil dried in an oven to 105°-110° C) is the same in both cases the difference in the V term allows calculation of pore space.

$$\% \text{ volume of solids} = 100 \frac{BD}{PD}$$

$$\% \text{ PS (pore space)} = 100 \, (1 - {}^{BD}/_{PD})$$

Bulk density is determined by obtaining a cylinder of undisturbed soil of known volume using a core sampler. The soil is dried and weighed in the laboratory. Particle density is obtained by placing the dried soil from the core sampler into a graduated cylinder containing a known volume of water and stirring vigorously to free air. The volume of water displaced is the volume of the soil solids. High bulk density ($>$ 1.6 g/cm^3) means low porosity—a situation in topsoils that results from compaction. This causes restricted root penetration, impeded water infiltration and percolation, and low soil aeration—all detrimental to plant growth. Sandy surface soils show a range in pore space from 35-50%, medium- to fine-textured soils from 40-60%.

The weight % of water in soil (P_w) is given by

$$P_w = \frac{W_{sm} - W_{sd}}{W_{sd}} \times 100 = \frac{W_{H_2O}}{W_{sd}} \times 100$$

where W_{sm} = wt. of soil moist (g), W_{sd} = wt. of soil dry (g), and W_{H_2O} = wt. of water in sample (g).

The volume % of water (assuming that the density of water is 1 g/cm^3) is given by

$$P_v = P_w \times BD.$$

Compared with atmospheric air, soil air is usually saturated with water vapor and is richer in CO_2 (Table 28.2) because of production of CO_2 in the soil by root respiration and litter decay.

Table 28.2 Composition of Atmospheric and Soil Air

	Oxygen	Carbon Dioxide	Nitrogen
Soil Air (%)	20.65	0.25	79.2
Atmospheric Air (%)	20.97	0.03	79.0

The soil water will dissolve any soluble constituents that may be present and as such contribute to the *soil solution* which is the medium whereby plants are supplied with nutrients. The concentration of hydrogen ions (H^+) in solution is indicated by the pH scale in which neutrality is pH = 7, values below this are acid and above it alkaline. In humid regions the normal range is pH = 5-7, in arid regions it is pH = 7-9 (Table 28.3).

TABLE 28.3 Soil Acidity and Alkalinity.

pH	4.0 4.5	5.0	5.5	6.0 6.5	6.7 7.0	8.0		9.0	10.0 11.0
Acidity	Very strongly acid	Strongly acid	Moderately acid	Slightly acid	Neutral	Weakly alkaline	Alkaline	Strongly alkaline	Excess-ively alkaline
Soil groups	Podzols	Gray-brown podzolic soils Tundra soils		Brown forest soils Prairie soils Latosols	Tropical black earths	Chestnut and brown soils		Black alkali soils	

Erosion occurs naturally in all environments. However, when erosion exceeds a normal rate it is referred to as accelerated erosion. Soil erosion is a major and serious aspect of man's role in changing his own environment. Although there are many techniques available for attempting to reduce the intensity of the problem, it still appears intractable. The erosion of croplands by wind and water is one of man's biggest environmental problems. Although construction, urbanization, mining and other such activities are often significant in accelerating erosion of the soil, the prime causes are deforestation and agriculture. One serious consequence of accelerated erosion is the sedimentation that takes place in reservoirs, thereby shortening their lives and reducing their capacity. Forests protect the underlying soil from the direct effects of rainfall, runoff is generally reduced, tree roots bind the soil, and the litter layer protects the ground from rainsplash. With the removal of forest for agriculture soil loss will increase. The rates of erosion that result will be particularly high if the ground is left bare, though under some crops the increase will be less marked. The method of ploughing, the time of planting, the nature of the crop, and the size of fields will all have an influence on the severity of erosion. In general, the greater the proportion of a river basin that is deforested the greater will be the sediment yield per unit area. In the USA the rate of sediment yield appears to double for every 20% loss in forest cover.

The "universal soil-loss equation" developed by workers in the Soil and Water Conservation Research Division, Agricultural Research Service, U.S. Dept. of Agriculture, includes factors that affect soil erosion from a cultivated area. The equation can be used to predict soil erosion from farmland. The equation is:

$$A = R\,K\,L\,S\,C\,P,$$

where A is the computed soil loss per unit area, R = rainfall, K = soil erodibility, L = slope length, S = slope gradient, C = crop management (vegetative cover), and P = erosion control practice. Working together these factors determine how much water enters the soil, how much runs off, and the manner and rate of its removal.

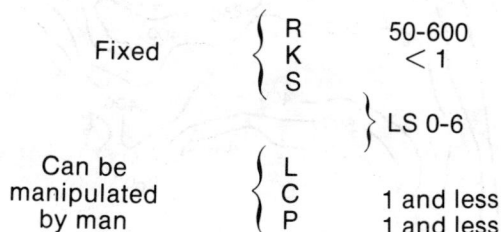

Fixed	R	50-600
	K	< 1
	S	LS 0-6 (S and L together)
Can be manipulated by man	L	
	C	1 and less
	P	1 and less

The rainfall factor R is a value expressing the capability of the locally expected rainfall to erode soil from an unprotected (fallow) field. It includes information on the average energy of storms and their intensity. Rainfall erosion indexes have been developed for various parts of the U.S.A. (Figure 28.5). The approximate range of average annual indexes for a large part of the USA ranges from < 50 in the western semi-arid plains to > 600 in the Gulf States of the southeast.

Physical properties of the soil greatly influence the rate at which different soils erode. Some of the more important ones are: texture, size and stability of soil structure, type of clay, soil permeability and infiltration, organic matter content, and soil depth. The factor (K) in the soil-loss equation is in tons of soil loss per acre per unit of rainfall erosion index for a slope of 9% and length of 72.6 ft. It expresses the loss from continuous cultivated fallow without the influence of crop cover or management. K factors for selected soils are given in Table 28.4.

Solution of the soil-loss equation is made easier by combining the equations for length and percentage of slope (Figure 28.6). The family of curves in this Figure has been prepared for use in solution of the soil-loss equation. Steepness of slope and length of slope increase the rate of erosion and the soil-loss ratio, which varies from 0-6.

The base value for the soil-loss equation is average annual soil loss from cultivated, continuous fallow. This base value is quantitatively expressed for specific field conditions by the product of the terms R, K, L and S. When a field is cropped, the soil loss is reduced because some protection from erosion is now provided. Fallow offers little or no protection. Thus for use in the soil-loss equation, the cropping management factor (C) is the expected ratio of soil loss from cropped land to the corresponding value of soil loss for the same land under fallow. Typical cropping-management values for some crop sequences are given in Table 28.5.

Figure 28.5 Average annual rainfall erosion index or rainfall factor (R) for states east of the Rocky Mountains. (From Wischmeier, W.H. and D.D. Smith. Agricultural Handbook 282. United States Government Printing Office, Washington, D.C. (1965))

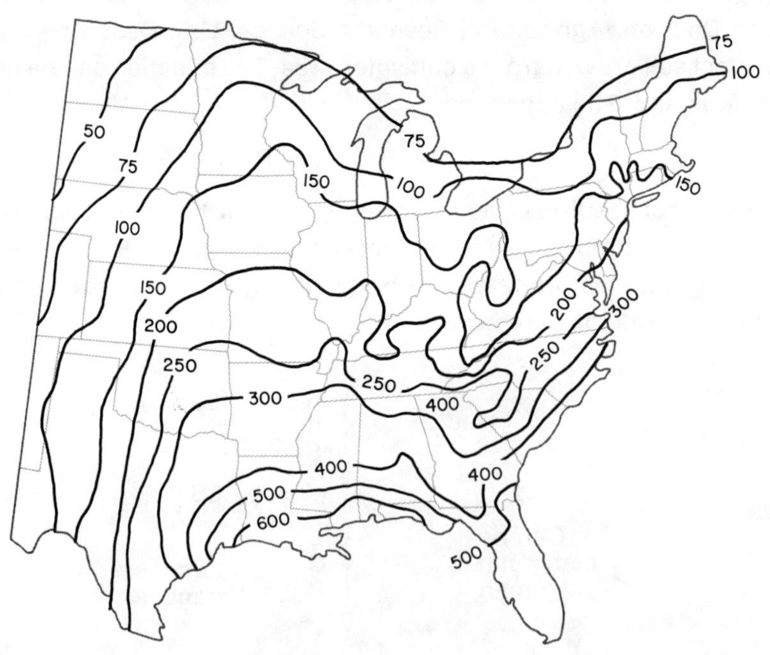

Table 28.4 Universal Soil Loss Equation K Factors for Selected Soils (after FAO, United Nations, 1965).

Location	Soil type	Factor (*K*) (t/acre)
Bethany, Mo.	Shelby silt loam	0.41
Castana, Ia.	Ida silt loam	0.26
Clarinda, Ia.	Marshall silt loam	0.33
Hays, Kan.	Colby silty clay loam	0.32
La Crosse, Wis.	Fayette silt loam	0.37
Madison, Wis.	Dodge silt loam	0.50
McCredie, Mo.	Mexico silt loam	0.28
Zanesville, Ohio	Muskingum silt loam	0.48

Table 28.5 Cropping-management Values (C) for Some Crop Sequences (after FAO, United Nations, 1965).

CROP SEQUENCE AND MANAGEMENT*	C
Crl	0.365
Crr-Wrr	0.286
Crl-Crl-Crr-W-H	0.262
Crl-Crl-Crr, cc-O-H	0.197
Crl-Crr-W-H	0.184
Crl-Crr, cc-O-W-H2yr	0.146
Crl-Orr-W-H	0.144
Crr-Crr-W-H2yr	0.137
Crr-W-H	0.118
Crr-W-H2yr	0.102
Crr-W-H3yr	0.078
Crl-O-H2yr	0.063
O-H3yr	0.011
W-H3yr	0.011

*rl = residue left
rr = residue removed
cc = cover crop
C = corn
W = wheat
O = oats
H = hay

Studies have shown that contouring tillage conservation practices (factor P) appears to produce its maximum average effect on medium slopes (2-7%). Here, soil losses with contouring are about one half of that occurring with up-and-down-hill farming. In strip-cropping, meadow strips alternate with grain strips and slow down the flow of runoff water and catch soil eroded from the grain or cultivated strips. Erosion from strip-cropped fields averages from 45-25% of that expected from up-and-down tillage (Table 28.6).

Soil-loss tolerance is the amount of soil (tons/acre) that can be lost and still maintain a high level of productivity over a long period of time. This value varies from 1-5 tons/acre.

Table 28.6 Practice Factor (P) Values for Contouring and Contour Strip-cropping (after FAO, United Nations, 1965)

Percentage slope	Practice factor (*P*) values	
	Contouring	Contour strip-cropping
1.1 - 2.0	0.60	0.30
2.1 - 7.0	0.50	0.25
7.1 - 12.0	0.60	0.30
12.1 - 18.0	0.80	0.40
8.1 - 24.0	0.90	0.45

Figure 28.6 Length and Steepness of Slope Factor (LS) Universal Soil-Loss Equation. (From Wischmeier, W.H. and D.D. Smith. Agricultural Handbook 282. United States Government Printing Office, Washington D.C. 1965)

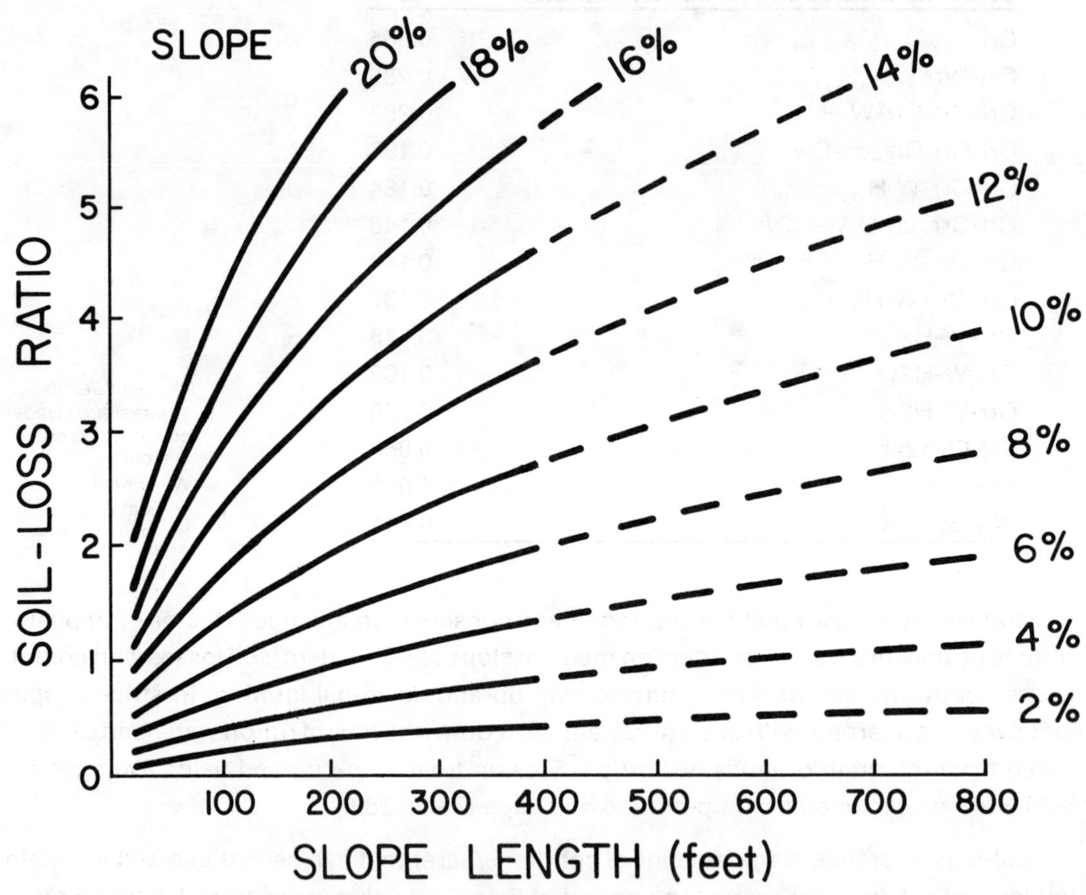

Name ______________________ Date ______________

Instructor ______________________ Section ______________

QUESTIONS

28.1 Examine Figure 28.7 showing the profiles of two soils. Determine the soil structure that is characteristic of each horizon. Use Figure 28.1 in deciding upon your answer. Note that two structures are apparent in the B horizon of soil (i).

Horizon	Structure: Soil (i)	Structure: Soil (ii)
A		
E		
B upper		
lower		
C		

Figure 28.7 Soil Horizons and Structure

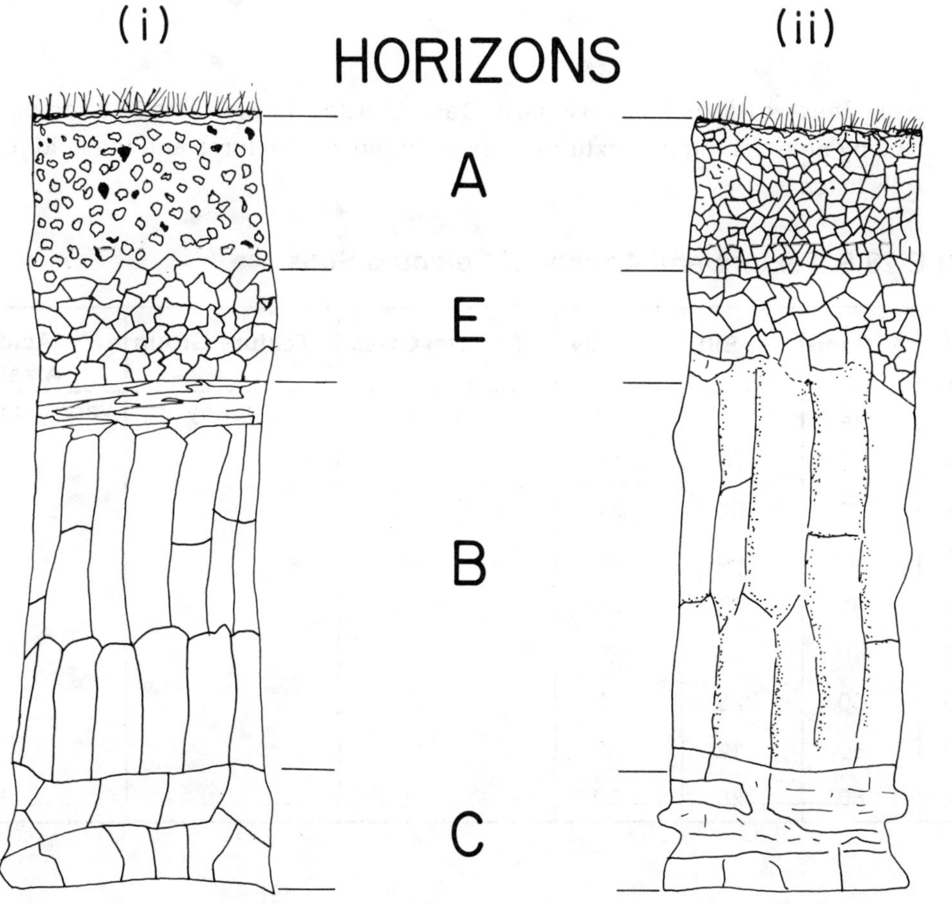

28.2 Using the textural triangle in Fig. 28.3 record the compositional limits (minimum and maximum amounts) for sand, silt, and clay for the texture classes listed in Table 28.7.

Table 28.7 Compositional Limits for Soil Texture Classes

	Textural Limits		
Textural Class	**% Sand**	**% Silt**	**% Clay**
sandy clay			
silty clay	0-20	40-60	40-60
clay loam			
loam			
loamy sand			
sandy loam			
sandy clay loam			
silt loam			
silt			
sand			
clay			
silty clay loam			

28.3 Using the textural triangles in Figs. 28.3 and 28.4 and data in Table 28.3 complete Table 28.8 by determining the textural classes, broad soil textures and the acidities of the soils listed.

TABLE 28.8 Texture and Acidity of Selected Soils

pH	% Sand	% Silt	% Clay	Textural Class	Texture Group	Acidity/ Alkalinity
4.5	92	5	3	sand	sandy	very strongly acid
5.2	20	70	10			
6.8	7	62	31			
8.8		40	60			
10.8	31	33	36			
8.3	40	40	20			
6.3	60	25	15			
4.8	60	12	28			
5.6	60	20	20			

Name ______________________ Date ______________

Instructor ______________________ Section ______________

28.4 Several cylinders of soil were extracted with a coring device and analyzed in the laboratory. From the results given in Table 28.9 estimate bulk density (BD), particle density (PD), % pore space, weight % of water (P_w) and volume % of water (P_v).

Table 28.9 Physical characteristics of selected soils.

Volume of core samples (cm^3)	Weight of wet sample (g)	Weight of oven-dried sample (g)	Volume of solids in a 25 g sample (cm^3)	Bulk Density (g/cm^3)	Particle Density (g/cm^3)	% Pore Space	Weight % of water	Volume % of water
50	92.5	72.5	9.4	1.45	2.66	45.5	27.5	40%
50	95.3	83.2	9.0					
50	84.5	78.2	9.2					
50	77.3	65.0	9.7					
50	68.1	64.2	9.5					

28.5 Plot on graph paper the line representing % PS = 100 (1 - $^{BD}/_{PD}$). Assume PD = 2.65 g/cm^3. Plot BD from 1.0 to 2.0 in units of one-tenth on the horizontal axis, and plot % PS on the vertical axis (0 - 100%). What happens to % PS as BD increases and why?

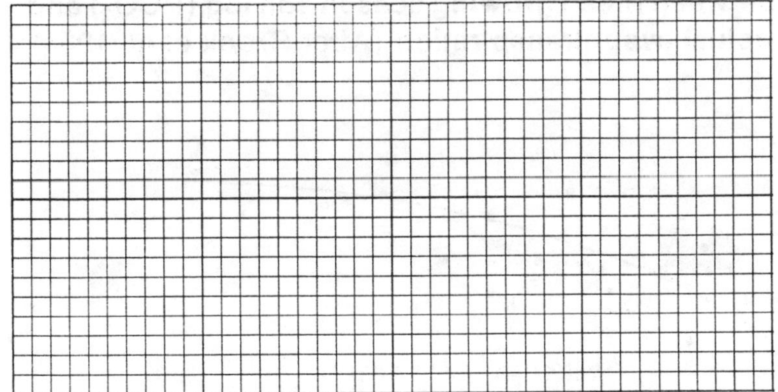

28.6 The maximum plant-available water in a soil is 10% (P_w). The soil BD is 1.30 g/cm^3. First derive P_v and then estimate how much plant available water in inches (and mm) can be stored in 18 inches (457 mm) of soil? Show your calculations.

28.7 A soil at field capacity has P_w = 25%. Its BD and initial water content (P_w) are 1.3 g/cm^3 and 15%. How much water (in inches) would be required to wet the soil to a depth of 18 inches?

28.8 At field capacity, the top 18 inches of soil (BD = 1.3) has a P_w = 23%. If the soil is presently at the permanent wilting point (P_w = 15%) to what depth would a 1.5 inch rainfall wet the soil if no runoff occurs?

28.9 Examine Figure 28.8 which shows the relationship between mean growing season soil CO_2 and mean annual actual evapotranspiration. Look at the climates and vegetation characteristics of the sites used to model soil CO_2 and explain why soil CO_2 increases with increasing AET.

Figure 28.8 Relationship between mean growing season soil log (PCO_2) and mean annual actual evapotranspiration (After Brook et al. 1983)

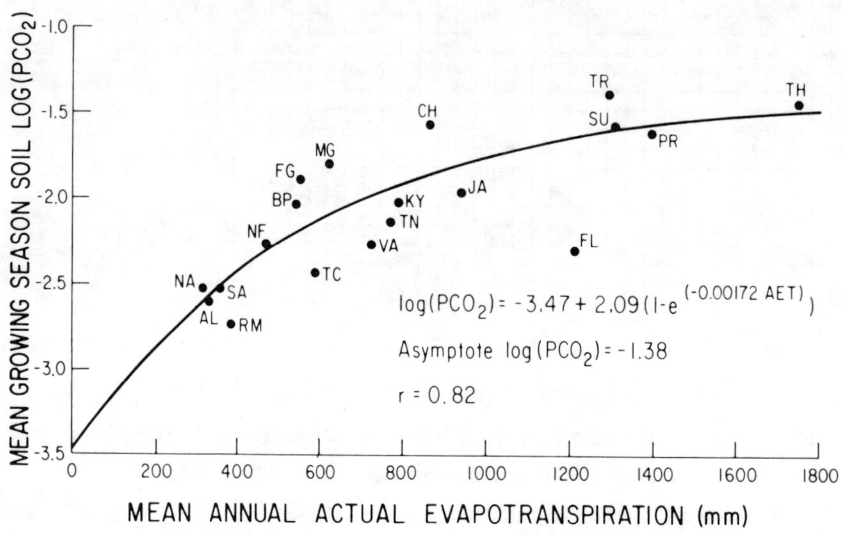

NA =	Nahanni, Canada	FL =	south Florida, USA
SA =	Saskatchewan, Canada	FG =	Frankfurt-Main, W. Germany
RM =	Rocky Mountains, Canada	MG =	Mullenbach, W. Germany
NF =	Newfoundland, Canada	JA =	Jamaica
BP =	Bruce Peninsula, Canada	TR =	Trinidad
TC =	Trout Creek, Ontario, Canada	PR =	Puerto Rico
AL =	Alaska, USA	CH =	Yunan, China
VA =	Reston, Virginia, USA	SU =	Sulawesi
TN =	Sinking Cove, Tennessee, USA	TH =	Phangnga, Thailand
KY =	Mammoth Cave, Kentucky, USA		

Name ______________________ Date ______________

Instructor ______________________ Section ______________

28.10 Examine Figure 28.9 showing a strong correlation between soil pH and (D - S) + 1000. D is the average annual soil water deficit, and S the average annual soil water surplus. Why do you think that soil pH might be dependent upon the balance between deficit and surplus (D - S)?

Figure 28.9 Relationship between soil A-horizon pH and water balance (D-S) 1000 (After Folkoff et al. 1981)

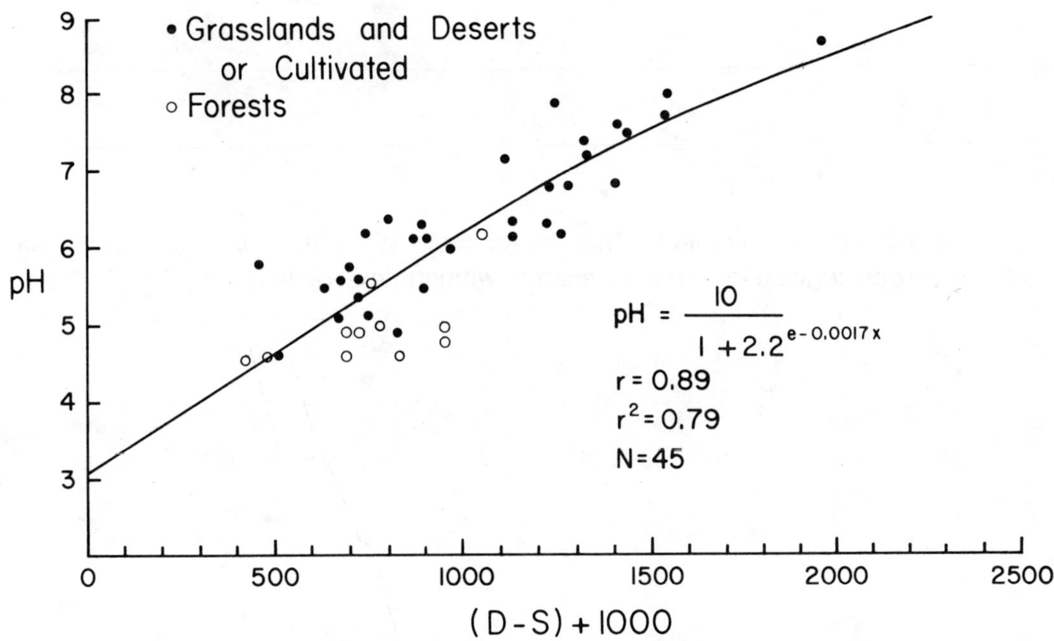

28.11 The base value for the soil-loss equation is average annual soil loss from cultivated, continuous fallow. This base value is given by the product of R, K, L and S.

(a) Estimate base erosion for the Dodge silt loam near Madison, Wisconsin (Table 28.4) for a slope steepness of 8% and a slope length of 200 feet. Determine R from Figure 28.5 and LS from Figure 28.6. Your answer will be in tons/acre/yr.

(b) Obviously the situation above describes the worst possible soil loss conditions. What would the soil loss be with contour ploughing and a crop sequence of Crl (Tables 28.5 and 28.6)?

(c) What would the soil loss be with contour strip-cropping conservation practices and a Crl crop sequence?

(d) Is this value of soil erosion within the tolerance of 1-5 tons/acre/yr? If not, what would you suggest to bring soil erosion within tolerable limits?

EXERCISE **29**

SOIL HORIZONATION, MAPPING AND CLASSIFICATION

The succession of horizons exposed when a vertical cut is made through the soil to the parent material comprises the soil profile. Where present, superficial organic horizons are designated O, surface eluvial horizons by A and E, subsurface illuvial horizons by B, the parent material by C, and the unweathered rock by R (Fig. 29.1). The term *solum* is used to refer to the A and B horizons since they are the horizons truly developed through soil formation. The A horizon is the mineral soil horizon formed at or near the surface and is characterized by the presence of humified organic material. The E horizon underlies the A, it is lighter in color and contains less organic matter, sesquioxides of iron and/or clay than the horizon beneath. The B horizon is normally differentiated from adjacent horizons by color and structure. It is characterized by illuvial concentrations of silicate clay, iron, aluminum, or humus.

Figure 29.1 Main horizon types in a soil profile

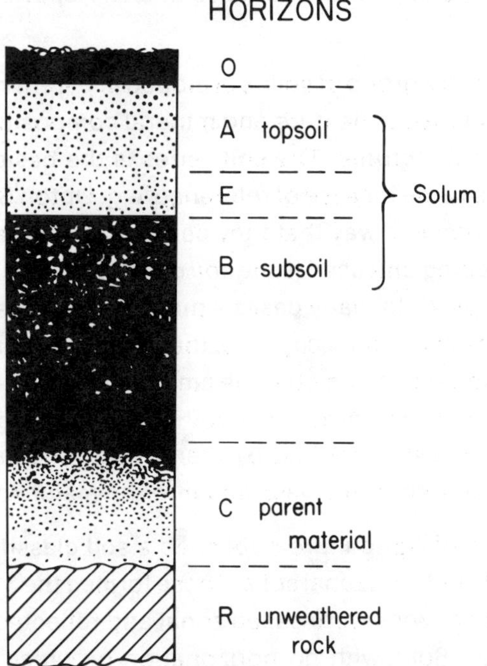

Organic matter in soils is mainly decomposed by soil micro-organisms and also by larger animals including earthworms, insects, and some mammals. Many of these larger animals perform a dual function because they also help to mix the soil.

Various processes operate in the soil to cause the development of horizons. The process of *translocation* is the movement of particles in suspension (*eluviation*) or in solution (*leaching*) to lower levels in the soil. Because of this many soils develop a relatively porous surface layer due to downward movement of the finest particles. The deposition of translocated materials at lower levels is *illuviation*, and this can produce layers rich in finer particles.

Laterization involves the solution and removal of silicon to produce the red iron-rich soils that are widely encountered in the permanently or seasonally wet parts of the tropics. *Podzolization* involves the breakdown of clay and the eluviation of iron, other soluble material, and humus with silica left behind as a residue. This eluviation results in a bleached A horizon. In dry climates soluble material, principally calcium carbonate, is leached from the A horizon after rain. Where the water evaporates from the B horizon $CaCo_3$ is deposited. This accumulation of $CaCO_3$ in the B horizon is *calcification*. In closed basins in dry areas there is a large seasonal input of salts which are eventually, through evaporation, precipitated in the soil. This process is *salinization*. Persistent waterlogging in low-lying areas causes accumulation of organic material and the formation of reduced iron (instead of oxidized iron) giving the soil a blue-gray color; this is *gleization*.

Detailed soil surveys of the U.S.A. are published on a county basis by the U.S. Department of Agriculture. The soil survey report consists of a group of soil maps in the back of the report and a discussion of soil properties within the county. Detailed, large-scale soil maps are usually printed on aerial-photobase maps. Each soil area is surrounded by a boundary line and identified by a symbol. Areas marked with the same symbol are the same soil. Each survey describes the soils in the county in terms of their locations, profile characteristics, relation to each other, and suitability for various management practices. The report also contains information about soil morphology, soil genesis, soil conservation, and soil productivity, and presents data on soil *series, complexes,* and *associations.* Reports include a county-sized map of soil associations.

A soil series is defined as a group of soils developed in parent materials having the same or similar lithological characteristics (texture and mineral composition) and the same or similar succession of horizons in the profile. The soil series is the basic mapping unit used in soil survey. It is usually given a name for ease of reference, e.g. Upton Series. Sometimes, however, two series occur in an intermixed way that they can not be separated for mapping purposes. When this occurs the mapping unit used is the *soil complex*—also designated by a local name (e.g. Upton-Simona Complex). In many cases a number of different soil series occur consistently in adjacent positions in the landscape, i.e. they are always associated with one another. Because of this, several soil series can often be amalgamated for ease of description as a *soil association.* In this way soils can be grouped together in a meaningful way to make a map of an area. Soils in a county are also classified by identifying fundamental processes operating within the profiles. Individual soils are classified into the major soil groups.

In 1938 the U.S. Dept. of Agriculture published a soil classification system utilizing the concepts of zonal, azonal, and intrazonal soils. These formed the 3 main orders of soils. Mature soils with well developed horizons were called zonal and reflected the world zones of climate and vegetation (Fig. 29.2). Soils with no horizonation or which had not developed typical profiles were azonal soils. Intrazonal soils were those whose characteristics were determined by local non-climatic conditions of drainage or parent material.

The zonal soils were divided into two major groups. Soils in humid areas subject to laterization and podzolization processes, which cause them to accumulate iron and aluminum at some level in the soil profile, were called *pedalfers*. In drier areas calcification becomes important producing a group of soils known as *pedocals*. Each of the main soil orders were divided into suborders and great soil groups. The great soil groups were subdivided into families and the families into series (Table 29.1).

Figure 29.2 World Distribution of Zonal Soils (Source: H. Robinson, 1972: 94)

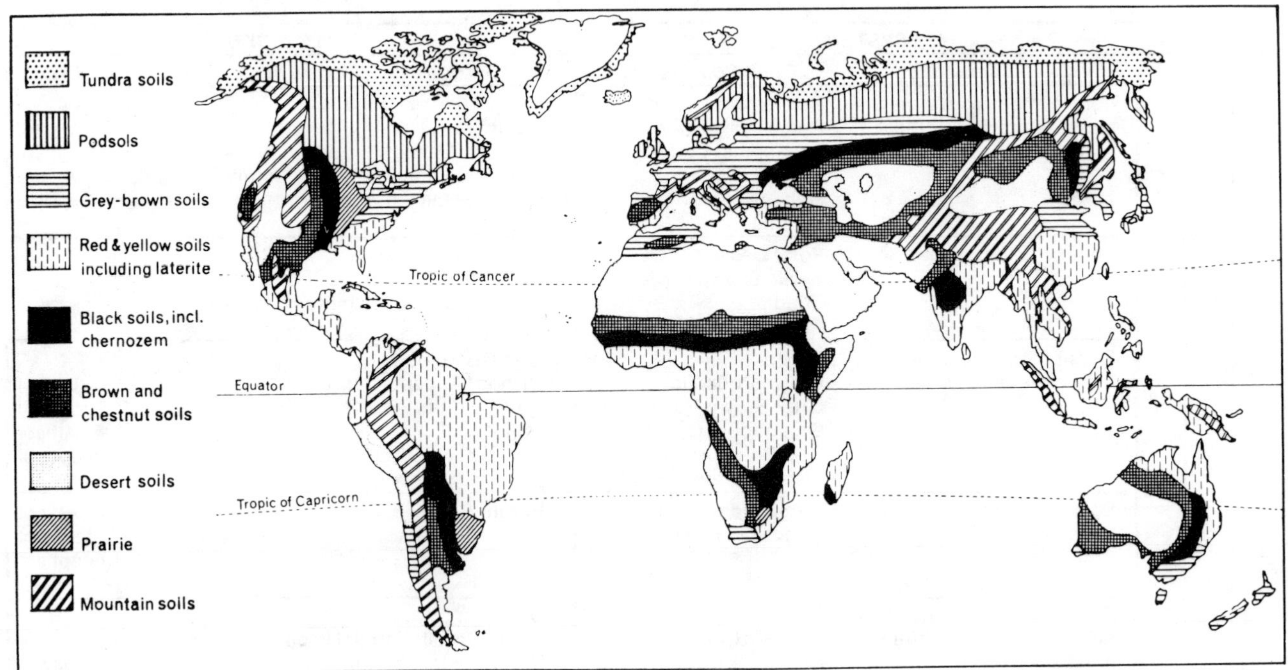

In 1960 a new soil classification, the United States Comprehensive Soil Classification System (or 7th Approximation) was introduced. In this classification there are ten soil orders, each subdivided into several suborders. The suborders are divided into great groups and then into families and soil series (Table 29.1). Entisols are soils with poor horizonation, histosols are composed primarily of plant material. Vertisols are soils in which horizon development is impeded by the churning effects of repeated changes in their volume due to alternating wetting and drying. Inceptisols are young soils in humid regions which have suffered some leaching but which show no clear illuvial horizon. Aridisols have low organic matter, are thin, stony and calcification and salinization may have occurred. Mollisols have dark, humus-rich A horizons and develop under semiarid to subhumid grasslands. Spodosols are soils in which a leached and eluviated light-colored A horizon overlies an illuvial B horizon. Alfisols have yellowish-brown partially leached A horizons, causing the upper soil to be colored by iron and aluminum compounds. Ultisols are similar to Alfisols, but are more thoroughly leached and therefore have redder A horizons. Oxisols are even more thoroughly leached than Altisols and are the soils of the wet tropics.

Table 29.1 1938 U.S. Soil Classification System and Approximate Equivalents in the 1960 Comprehensive Classification (after Birkeland P.W., 1974)

1938 SYSTEM				1960 SYSTEM	
Order		*Suborder*	*Great Soil Groups*	*Suborder*	*Order*
ZONAL	*Pedalfer*	Soils of forested warm-temperature and tropical regions	Red-brown Lateritic Red-Yellow Podzolic Low-humic Latosol Latosol Humic Latosol Laterite	Humult, Udalf, Udult Udult Tropept, Ustox Tropept, Humult, Andept, Ustult Humult, Humox, Andept Orthox	Oxisol Ultisol Alfisol Inceptisol
ZONAL	*Pedalfer*	Soils of cold regions	Polar Desert Arctic Brown Tundra Alpine Turf	Aquept, Ochrept, Umbrept, Andept	Inceptisol
ZONAL	*Pedalfer*	Soils of forested cool-temperate regions	Podzol Brown Podzolic Gray-Brown Podzolic Gray Wooded Sols Bruns Acides Western Brown Forest	Orthod, Humod Orthod, Andept, Ochrept Udalf, Udult Boralf Ochrept, Umbrept	Ultisol Spodosol Alfisol Inceptisol
ZONAL	*Pedalfer*	Soils of forest-grassland transition	Degraded Chernozem Noncalcic Brown	Boralf, Boroll Xeralf, Ochrept	Alfisol Mollisol Inceptisol
ZONAL	*Pedalfer* / *Pedocal*	Dark-colored soils of semiarid, subhumid, and humid grasslands	Reddish Prairie Prairie (Brunizem) Chernozem Reddish Chestnut Chestnut	Ustoll Udoll, Boroll, Xeroll, Ustoll Boroll, Ustoll, Xeroll Ustalf, Ustoll Xeroll, Ustoll, Boroll	Mollisol Alfisol
ZONAL	*Pedocal*	Light-colored soils of arid regions	Reddish Brown Brown Sierozem Red Desert Desert Polar Desert	Ustalf, Orthid, Argid Ustoll, Xeroll, Argid, Orthid, Boroll Argid, Orthid Argid, Orthid Argid, Orthid Argid, Orthid	Aridisol Mollisol Alfisol
INTRAZONAL		Hydromorphic soils in areas of imperfect drainage or high water table	Humic Gley Low-Humic Gley Alpine Meadow Bog Half Bog Planosol Ground-Water Podzol Ground-Water Laterite	Aquoll, Aquept, Aquult, Aqualf Aquult, Aquent, Aquept, Aqualf Aquod, Aquoll, Umbrept Suborders of Histosol Aquept, Aquoll, Aqualf Aqualf, Alboll Aquod Aquult, Udult, Usult	Inceptisol Mollisol Alfisol Spodosol Ultisol Entisol
INTRAZONAL		Halomorphic soils (saline and alkali) in areas of imperfect drainage in arid and coastal areas	Solonchak Solonetz Soloth	Orthid, Aquept Natric great groups of Alfisol, Mollisol, Aridisol Natric subgroups of Mollisol and Alfisol	Inceptisol Aridisol Mollisol Alfisol
INTRAZONAL		Calcimorphic soils formed from calcareous parent materials	Brown Forest Rendzina	Ochrept, Xeroll, Udoll Rendoll	Inceptisol, Mollisol
AZONAL			Lithosol Regosol Alluvial		Entisol Inceptisol Mollisol

Name ______________________ Date ______________

Instructor ______________________ Section ______________

QUESTIONS

29.1 The Thornthwaite moisture index (I_m) is given by the equation:

$$I_m = 100\left(\frac{P}{PET} - 1\right)$$

where P is the annual precipitation and PET the annual potential evapotranspiration. The value of I_m is positive at places where P is greater than PET and negative where P is less than PET.

Examine Figures 29.3 and 29.4 and answer the following questions.

(a) What correlation exists between the Thornthwaite moisture index and the boundary between pedocal and pedalfer soils?

Figure 29.3 Regional variations in the Thornthwaite Moisture Index in the Coterminous USA (After Muller and Oberlander, 1978)

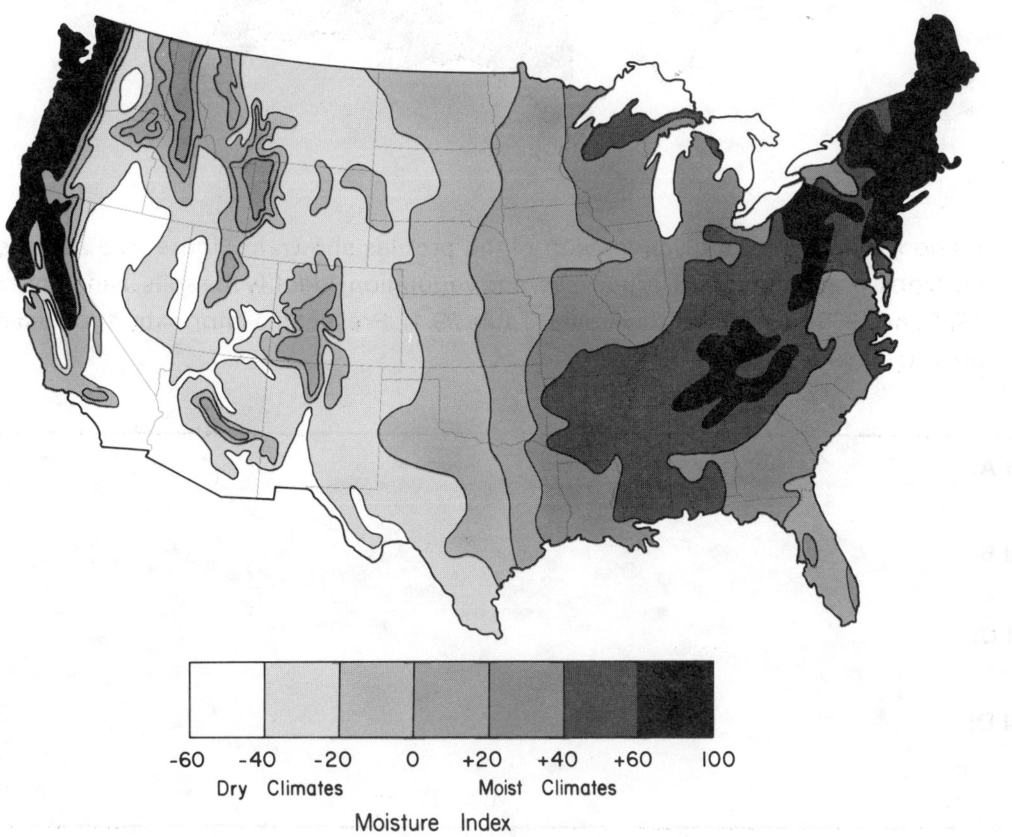

(b) Of what importance is the balance between precipitation and potential evapotranspiration in the formation of these two major soil categories?

Figure 29.4 Pedocal and Pedalfer Soils of the Coterminous U.S.A.

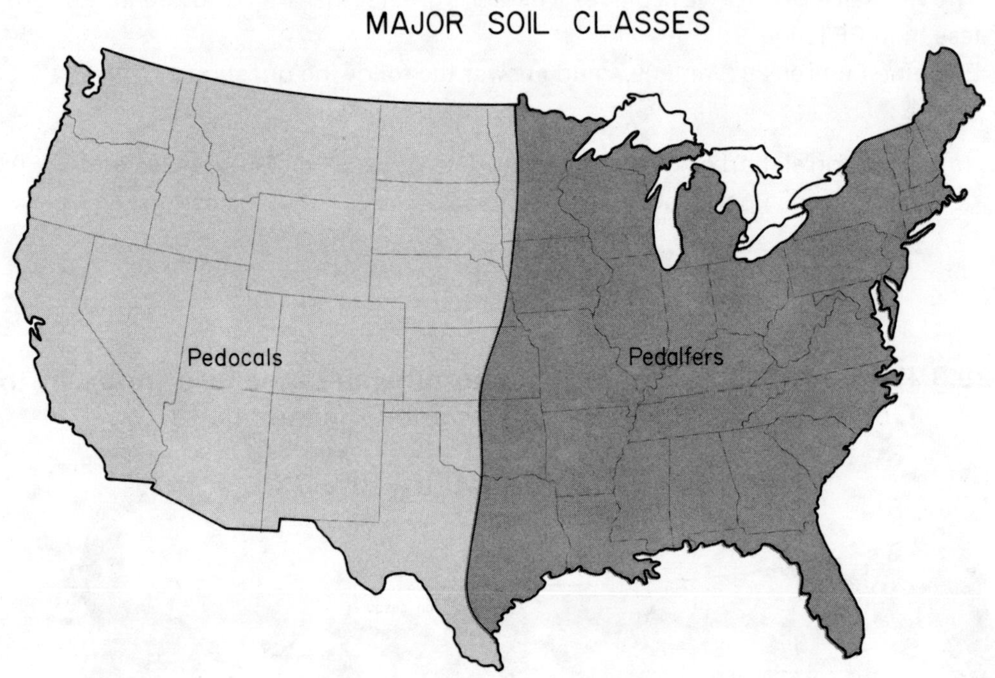

29.2 Using the information given in each of the profiles shown in Figure 29.5 about soil horizons, predominant soil processes, and vegetation, identify the soils using both the 1938 and 1960 classification systems (Table 29.1). Present the information you used in arriving at your answers.

Soil A:

Soil B:

Soil C:

Soil D:

Name ______________________ Date ______________

Instructor ______________________ Section ______________

Figure 29.5 Diagrammatic Profiles of Four Common Soils (After H. Robinson, 1972)

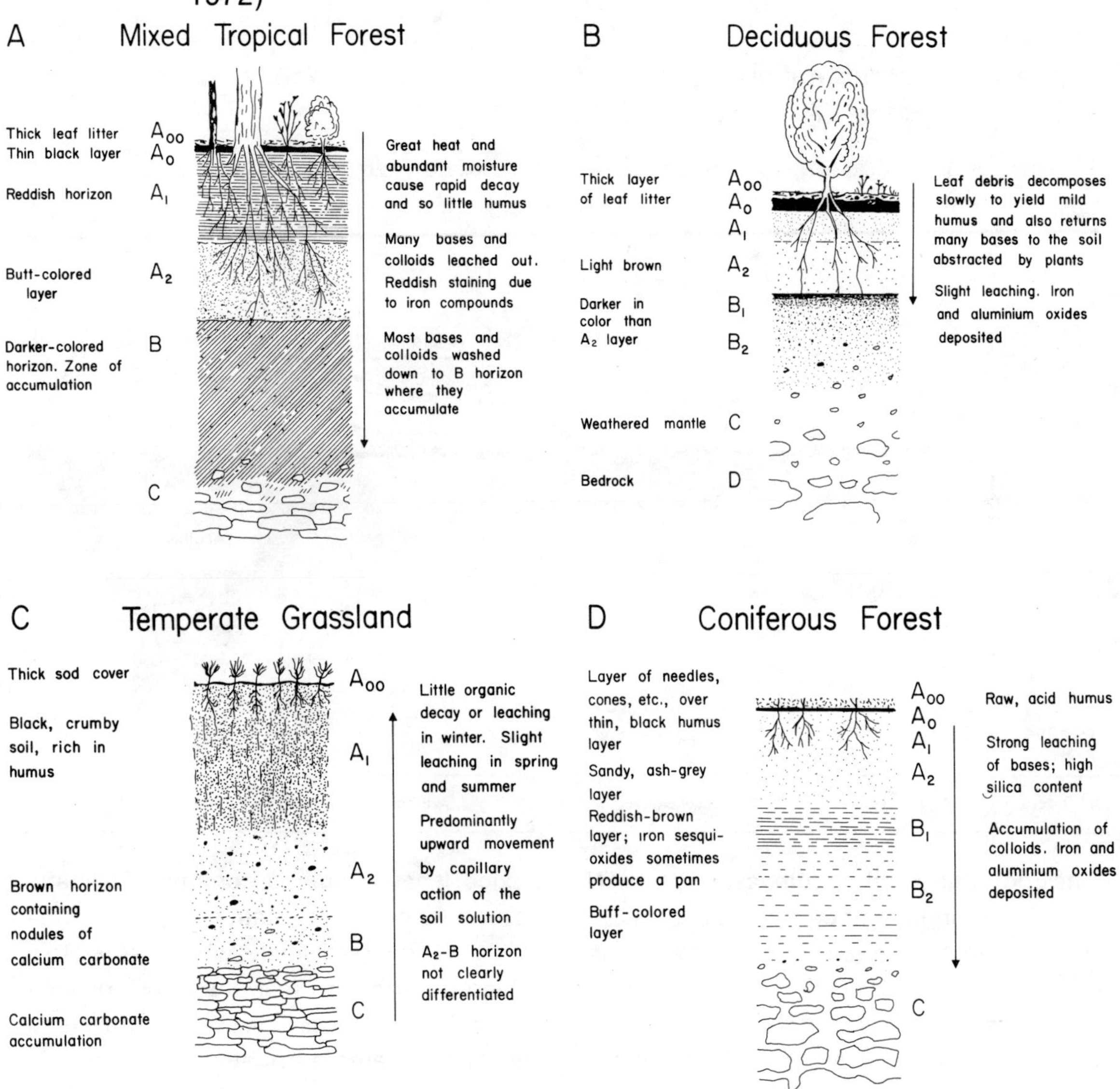

29.3 Use the information presented in Figure 29.6 to answer the following questions.

(a) Explain why Inceptisols are characteristic of Polar Desert and Tundra environments, why Spodosols, Altisols, and Mollisols are characteristic of the Taiga Zone, Aridisols of the Semidesert and Desert zones, and Ultisols and Oxisols of the Tropical Forest and Savanna zones.

(b) What dominant processes of soil formation are occurring in the Taiga, Semidesert and Desert, and Tropical Forest zones?

Figure 29.6 Latitudinal distribution and formative influences of the major soil orders (After Strakhov, N.M. 1967)

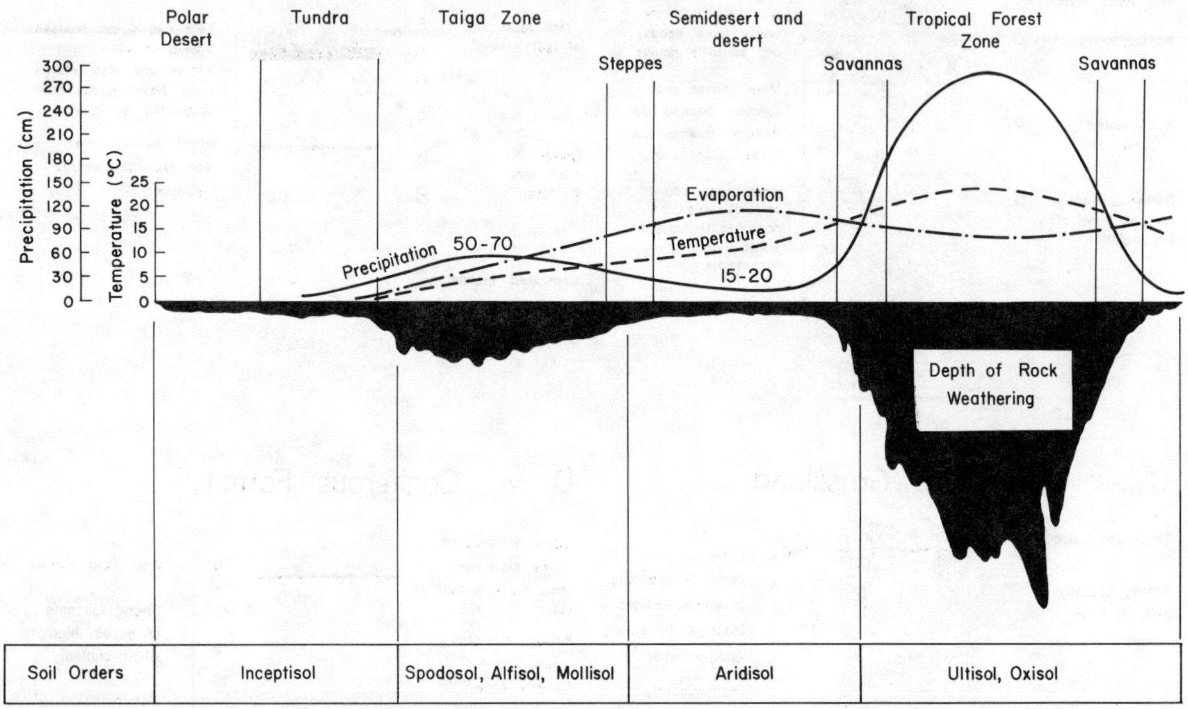

The Eddy area is in the southeastern part of New Mexico. It has a semiarid, continental climate. The mean annual temperature is 26.1° C (79° F) and mean annual precipitation is 31.5 cm (12.4 inches). Most of the soils in the area are developed in a low-rainfall climate, which has led to a limited amount of leaching of salt and other minerals. In most areas the depth of lime accumulation is generally shallow or very shallow. About 97% of the area is used for grassland with ranching the main occupation. Approximately 3% of the area is irrigated.

In the Soil Survey report of the Eddy area there is a general soil map of the soil associations of the region (Fig. 29.7). A soil association is a landscape that has a distinctive proportional pattern of soils and normally consists of one or more major soils and one or more minor soils. The soils in one association may occur in another but in a different pattern. Areas where the soil material is so rocky or shallow that it cannot be classified by soil series are also mapped as part of the associations. These areas are given descriptive names such as Limestone Rock Land and are called land types. Seven soil associations are recognized in the Eddy area. They are listed and described briefly in the legend of Fig. 29.7.

29.4 The Simona-Pajarito soil association occupies approximately 14% of the Eddy area. The main soil series, the Simona and Pajarito, occupy 45% and 40% of the association respectively. The remaining 15% of the association is characterized by Largo, Potter, and Bippus soils, and by three land types. (Fig. 29.8).

Name ______________________ Date ______________

Instructor ______________________ Section ______________

(a) Briefly list the clear relationships evident in Figure 29.8 between soil series and land types, and geological and topographical characteristics.

(b) Which soil series are developed on (i) the shallowest, and (ii) the steepest slopes of the association?

(c) Two of the soil series are developed in stream deposits. Which ones?

29.5 Examine the geology and soil association maps of the Eddy area (Figs. 29.9 and 29.7).

(a) What are the two dominant soil associations (in terms of area) to the east of the Pecos River?

(b) What are the two dominant soil associations west of the Pecos River?

(c) In terms of geology, explain why the dominant soil associations west of the Pecos River differ from those to the east of it.

(d) What soil association is common along the Pecos River and what deposits are these soils developed in?

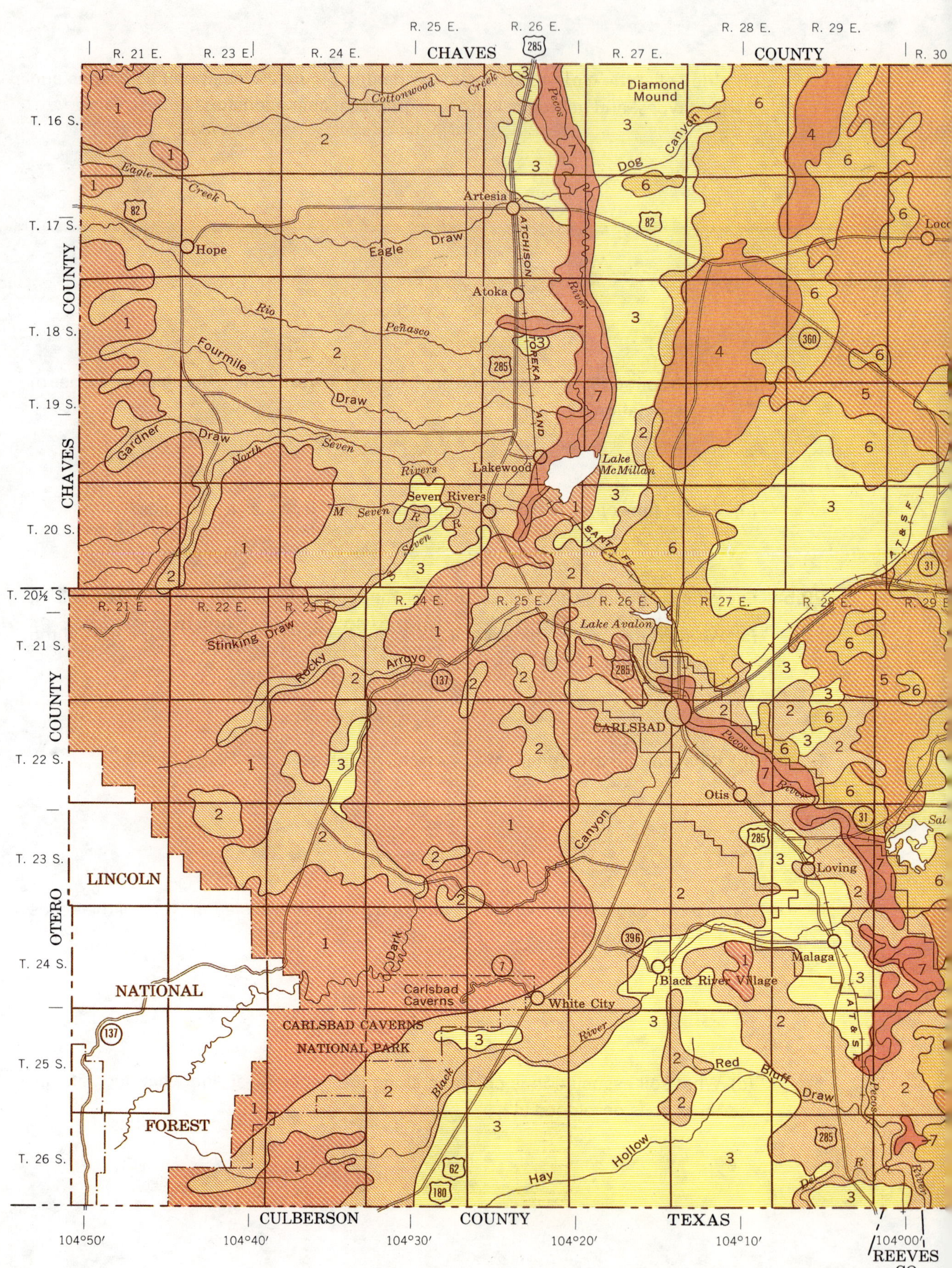

CHAVES
COUNTY
CHAVES
COUNTY
LINCOLN
NATIONAL
FOREST
OTERO
CULBERSON
COUNTY
TEXAS
REEVES
CO
CARLSBAD CAVERNS
NATIONAL PARK
Artesia
Hope
Atoka
Lakewood
Seven Rivers
Lake McMillan
Lake Avalon
CARLSBAD
Otis
Loving
Malaga
Black River Village
White City
Carlsbad Caverns
Diamond Mound
104°50′
104°40′
104°30′
104°20′
104°10′
104°00′

Name________________________________ Date______________

Instructor ___________________________ Section____________

(e) Two soil associations appear to derive their principal characteristics from bedrock. Name these associations and explain why soils associated with them tend to be very shallow.

(f) Name the soil associations of the Eddy area that have developed in alluvial materials.

Figure 29.7 Map of soil associations in the Eddy area of New Mexico (Source: U.S.D.A., *Soil Survey, Eddy Area, New Mexico*, 1971)

Scale 1:506880

1 0 1 2 3 4 5 6 7 8 Miles

SOIL ASSOCIATIONS

1 Limestone rock land-Ector association: Rock land and very shallow, stony and rocky, loamy soils over limestone; on hills and mountains

2 Reagan-Upton association: Loamy, deep soils and soils that are shallow to caliche; from old alluvium

3 Reeves-Gypsum land-Cottonwood association: Loamy soils that are very shallow to moderately deep over gypsum beds, and Gypsum land

4 Kimbrough-Stegall association: Loamy soils that are very shallow to moderately deep to caliche; from old alluvium

5 Kermit-Berino association: Sandy, deep soils from wind-worked mixed sand deposits

6 Simona-Pajarito association: Sandy, deep soils and soils that are shallow to caliche; from wind-worked deposits

7 Arno-Harkey-Anthony association: Loamy, deep soils from recent mixed alluvium

29.6 Even though calcification is one of the major soil-forming processes in the area, some of the soils, such as those of the Berino Series (part of the Kermit-Berino association), are mature, with clayey subsoil and deeply leached salts. If the present semiarid climate in the Eddy area is not conducive to leaching how would you explain the leaching of the Berino soils?

Figure 29.8 The typical relationships between soil series and land types in the Simona-Pajarito association (Source: U.S.D.A., *Soil Survey, Eddy Area, New Mexico,* 1971)

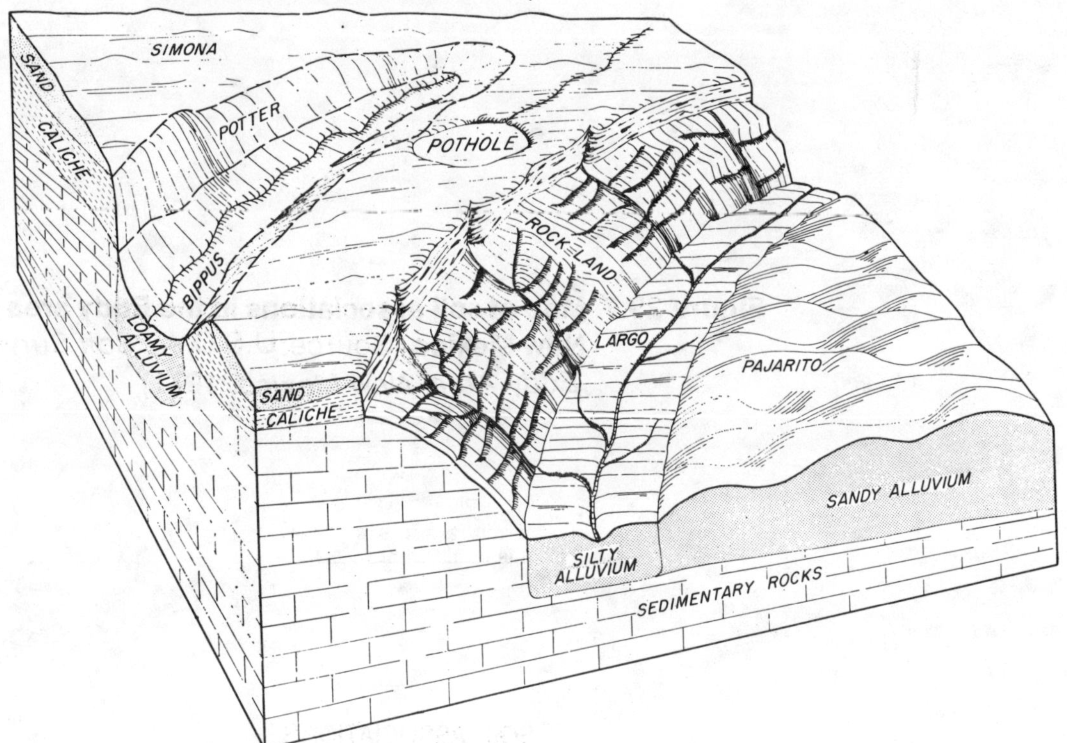

29.7 Three of the ten soil orders of the 7th Approximation are represented in the Eddy area (Table 29.2).

(a) What are entisols, aridisols, and mollisols?

(b) Why do aridisols appear to be the most common soil type in the Eddy area?

Name ______________________ Date ______________

Instructor ______________________ Section ______________

(c) Using Table 29.2, classify the soil series of the Simona-Pajarito association into soil orders and then by referring to Figure 29.7, explain why these soils are classified this way.

Table 29.2 Soil Orders Represented in the Eddy Area

SOIL ORDERS	SOIL SERIES IN THE EDDY AREA
ENTISOLS	Anthony, Arno, Cottonwood, Harkey, Kermit, Largo, Likes.
ARIDISOLS	Atoka, Berino, Cacique, Karro, Mobeetie, Pajarito, Potter, Reagon, Reeves, Russler, Simona, Tonuco, Upton, Wink
MOLLISOLS	Bippus, Dev, Ector, Kimbrough, Pima, Stegall

29.8 The seven soil associations of the Eddy area can probably be grouped into four principal categories in terms of the parent materials in which they have developed. What are these four groups?

Figure 29.9 Generalized geologic map of the Eddy Area, New Mexico (Source: U.S.D.A., *Soil Survey, Eddy Area, New Mexico*, 1971)

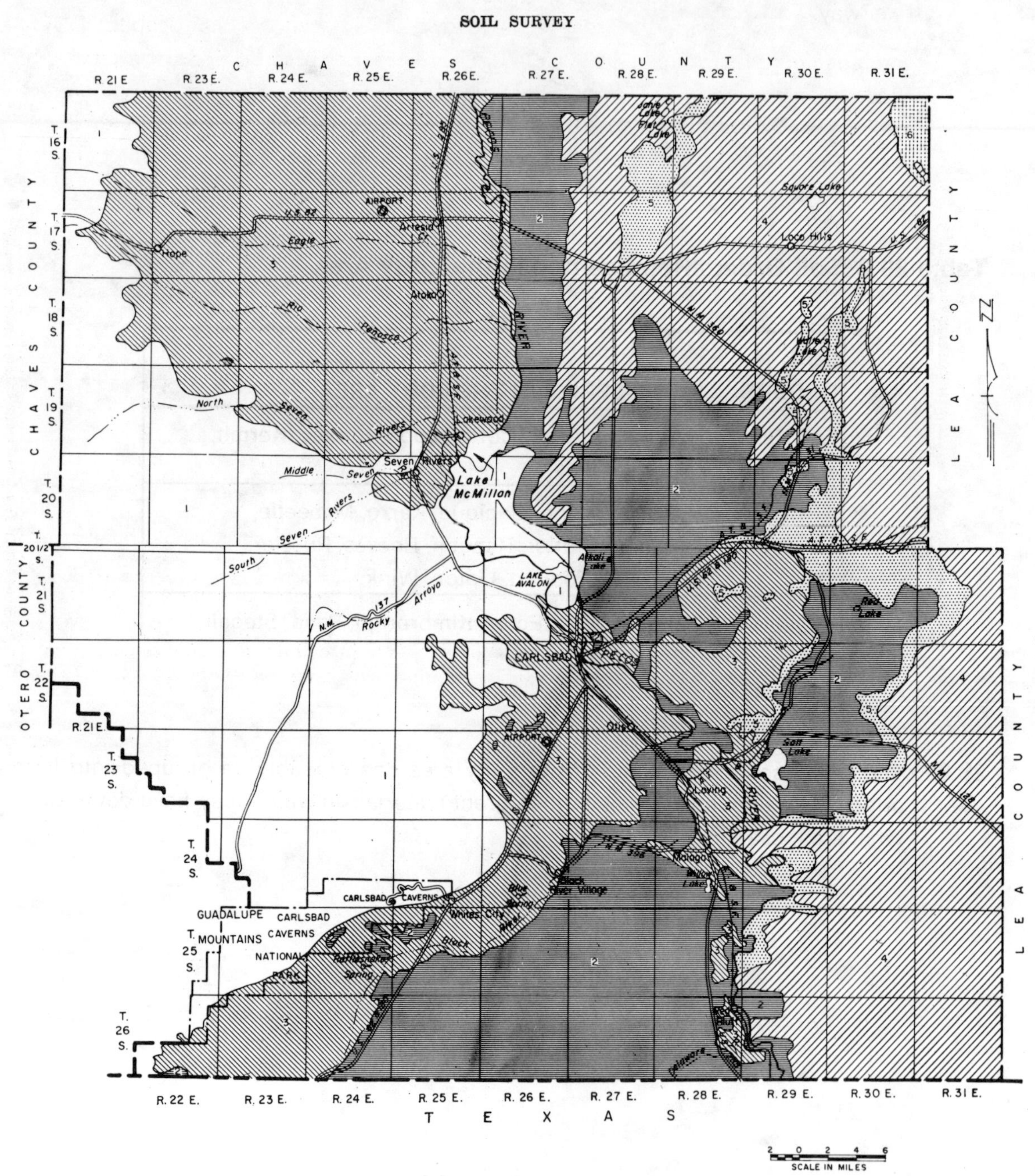

1. Rocks of Permian age, primarily carbonatic.
2. Rocks of Permian age, primarily gypsiferous.
3. Loamy deposits of Quaternary age.
4. Sandy deposits of Quaternary age.
5. Rocks of Triassic age.
6. Rocks of Tertiary age.

Name ______________________ Date ______________

Instructor ______________________ Section ______________

29.9 A map of soil associations is useful to people who want a general idea of the soils in an area, who want to compare different parts of an area, or who want to know the location of large tracts of land that are suitable for a certain kind of farming or land use. Usually the Soil Survey reports for areas in the U.S.A. have a section on soil capability and management. In the capability system, soils are grouped at three levels: the capability class, subclass, and unit. The broadest groups, the capability classes, are designated by Roman numerals I through VIII. The numerals indicate progressively greater limitations and narrower choices for practical use. The eight capability classes are described below:

Class I soils have few limitations that restrict their use. (None of the soils of the Eddy area are in class I.)

Class II soils have moderate limitations that reduce the choice of plants or that require moderate conservation practices.

Class III soils have severe limitations that reduce the choice of plants, require special conservation practices, or both.

Class IV soils have very severe limitations that reduce the choice of plants, require very careful management, or both.

Class V soils are not likely to erode but have other limitations, impractical to remove, that limit their use largely to pasture, range, woodland, or wildlife. (None of the soils of the Eddy area are in class V.)

Class VI soils have severe limitations that make them generally unsuited to cultivation and limit their use largely to pasture or range, woodland, or wildlife.

Class VII soils have very severe limitations that make them unsuited to cultivation and that restrict their use largely to pasture or range, woodland, or wildlife.

Class VIII soils and landforms have limitations that preclude their use for commercial plants and restrict their use to recreation, wildlife, or water supply, or to esthetic purposes.

(a) Which soil association in the Eddy area could provide the best cropland, if irrigated? Where is this soil association located?

(b) Into which capability class would you place the irrigable soils of the Eddy area?

(c) Into which two capability classes would you place the majority of soils that are present in the Eddy area? Explain your answer.

(d) Within many of the soil associations of the Eddy area there are active sand dunes and rock land units. To which capability class do these areas belong and why?

EXERCISE **30**

VEGETATION STRUCTURE AND PLANT AND ANIMAL DISTRIBUTIONS

Biogeography is the study of the geographical aspects of plants and animals, especially their present and past distributions. *Phytogeography* is the geography of plants, *zoogeography* the geography of animals. Vegetation consists of different species of plants which occur in groups known as *populations.* All of the plant populations in a given area form a plant *community.* The community and the physical habitat function together as an *ecosystem.* Gradients of communities and environments—or gradients of ecosystems—are called *ecoclines.* Most plant communities are influenced by one species or group of species called *dominants.* A dominant plant is one that because of its coverage above ground or because of the root space it occupies below ground modifies the immediate environment and determines conditions available to other members of the community.

Most plant communities are *stratified* with different species extending to different heights above the ground. In a forest, for example, the trees with the upper foliage in full sunlight form the *canopy* or uppermost layer. Beneath the canopy there is a lower layer of smaller trees using some of the remaining light. This lower tree stratum contains younger individuals of the canopy tree species and mature individuals of smaller tree species. Beneath this layer are shrubs adapted to photosynthesize using the weaker light within the forest. The shrubs further reduce the amount of light reaching herbs on the forest floor. Shrublands and grasslands are also a mixture of species of different heights.

When vegetation is disturbed naturally (e.g. by fire) or by man it does not immediately re-establish itself. A sequence or *succession* of plant communities occurs, one replacing another until a stable community is established. The general trend in succession is towards taller and more diverse vegetation. The resulting stable plant community is the *climax* or *natural vegetation* if it evolves in response to the regional climate and without the interference of man.

The earth's many different plant communities have been grouped into twelve major types called *plant formations* or *biomes* (when the characteristic animal population is also being referred to). The major biomes of the world are tundra, northern coniferous forest, temperate deciduous forest, temperate grassland, broad-leaved evergreen forest, desert, tropical savanna, tropical deciduous forest, tropical scrub-forest, tropical rain forest, and mountain (Figure 30-1). In a general way the biomes of the world reflect the spatial variability of regional climates and zonal soils (also a response to regional climates). Rainfall appears to determine the great plant formations of forest, grass, and scrub, while temperature appears to control the detail within each formation (e.g. broad-leaved deciduous forest or coniferous forest). Biomes in areas which are climatically favorable to life support many plant and animal species (e.g. tropical rain forest), those in areas of less favorable climate (e.g. tundra, desert) support fewer species.

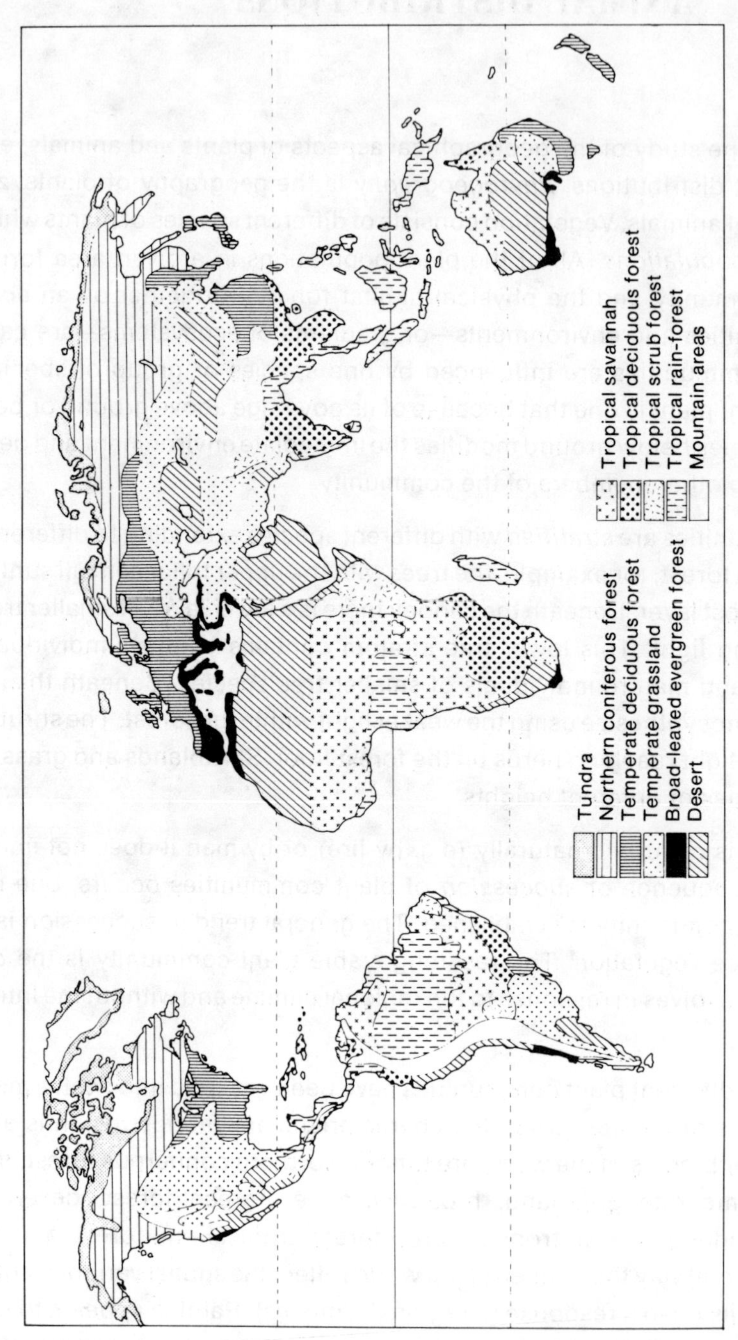

Figure 30.1 The Biomes of the Earth (Source H Robinson 1972: 438)

Biomes are seldom sharply separated from one another. They grade into one another gradually through a transition zone called an *ecotone*. Along a gradient from a favorable environment to a climatically extreme environment (e.g. desert or tundra) there is normally a decrease in plant biomass, productivity, and the percentage of the ground surface covered. In addition, there is generally a decrease in the heights of the dominant plants and a reduction of vertical stratification. *Biomass* is the amount of organic material at the earth's surface in dry g/m^2. There are two measures of plant productivity. *Gross primary productivity* in $g/m^2/yr$ is the rate at which organic matter is created by photosynthesis, and *net primary productivity* in $g/m^2/yr$ is gross primary productivity minus plant *respiration.* In arid climates there is a positive, almost linear relationship between plant productivity and annual precipitation. In humid areas there is a moisture level beyond which productivity shows virtually no increase with increased precipitation. Temperature also influences plant productivity, which increases along the temperature gradient from the Arctic to the Tropics.

The *tolerance range* of a species is the range of environmental conditions under which it will grow. Any species has an upper and lower limit of tolerance as a result of its light, heat, and moisture requirements. The *ecological range* of a species, which may be more restricted than the tolerance range, is the range of environmental conditions within which a species actually grows. The *geographical range* is the geographical limit of the ecological range. The distributions of species have also been affected by geological factors such as continental drift, by the mobility of species in various stages of evolution, and by man's activities.

Moisture and temperature are perhaps the most important environmental factors that influence the distributions of species. Moisture is essential to plant growth, and according to the amount of water they require, plants can be divided into: *hydrophytes* if they live in water or in very damp regions; *xerophytes* if they are adapted to dry conditions; *mesophytes* if they require moist but not wet conditions; and *tropophytes* if they are adapted to seasonal wet and dry conditions.

Heat supplies energy for plant growth. Plants can survive within very specific temperature limits that vary for each plant species. *Megatherms* are plants favoring warm regions, *microtherms* those favoring cold regions, and *mesotherms* are plants adapted to intermediate temperature conditions. There appear to be four main temperature controls influencing the geographical distributions of plant species: (a) control by minimum temperature, in which case there is a correlation between the geographical range of a species and isotherms for the average temperature of the coldest month or the average monthly minimum temperatures; (b) control by insufficient heat with the geographical range coincident with the isotherm for the warmest month; (c) limitation by excessive heat; and (d) limitation due to insufficiency of temporary cold (e.g., plants may need a cold season for germination or flowering). The distribution of the loblolly pine in the southeastern U.S.A., for example, seems to be controlled to the north and west by low temperatures and limited soil moisture. The upper elevational limits of many desert plants are controlled largely by temperature and often by frost. Upper elevational limits of trees are often related to mean or maximum temperatures during the warm months of the growing season.

Man has had a profound effect on the distributions of plants and animals. Whole landscapes have been modified as a result of vegetation changes brought about by activities such as fire, grazing, deforestation, desertification, and pollution. Man has introduced plants and animals into new areas either deliberately or accidentally and has modified the animal kingdom by domestication. One of the most important results of man's interference with animals has been the extinction of many species.

Name ______________________ Date ____________

Instructor ______________________ Section ____________

QUESTIONS:

30.1 In Figure 30.2,

(a) Identify and estimate the heights of the following layers: dominant canopy trees; subdominant trees; shrubs; and herbs.

(b) What species of trees are present in the dominant canopy layer?

(c) What species of trees characterize the subdominant layer?

Figure 30.2 Structure of a Typical Eastern Deciduous Forest

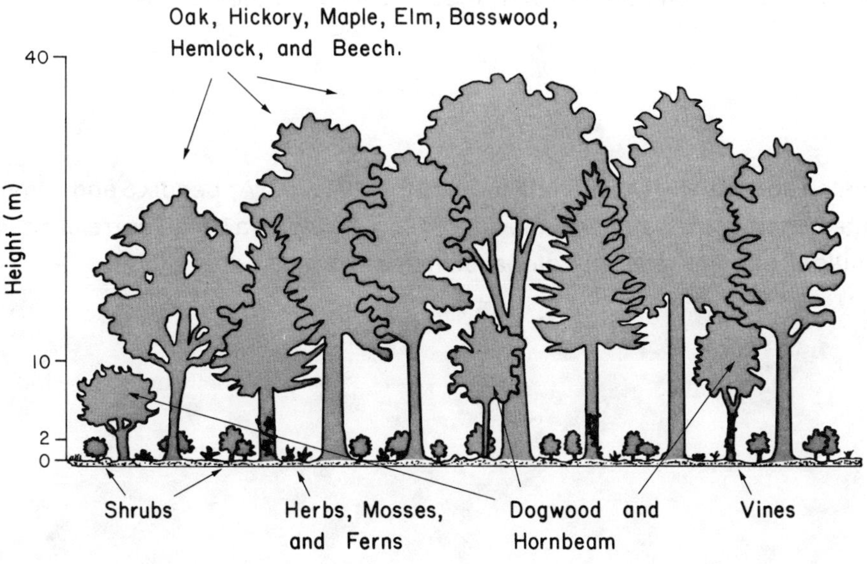

Figure 30.3 Stratification and Light Extinction in a Forest (After Whittaker, 1975)

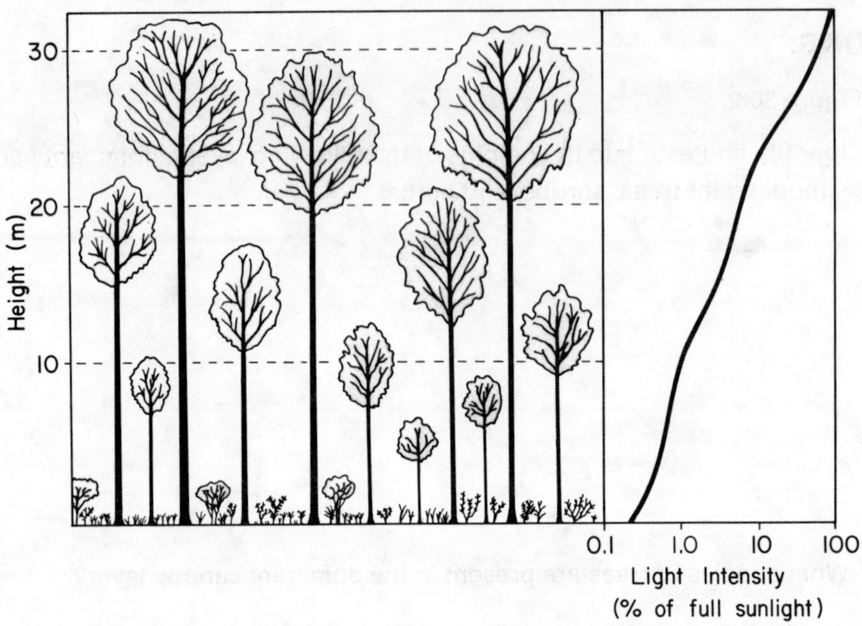

30.2 Examine Figure 30.3

(a) Layer the forest profile by identifying herb, shrub, canopy and other strata. Give the average height above ground (m) of each level.

(b) Estimate the amount of light (as a percentage of that reaching the canopy) reaching each of the layers you identify. What do your results suggest about plants near the forest floor compared to, say, the trees forming the canopy?

30.3 Complete Table 30.1 by examining Figures 30.1, 29.2, and Appendix 5 and comment on the statement that the pattern of the major biomes shows a broad correlation with the distributions of zonal soils and global climates.

Name ______________________________ Date ______________

Instructor ______________________________ Section ______________

Table 30.1 Biomes, Zonal Soils, and Global Climates

BIOMES	AREAL EXAMPLES	ZONAL SOILS	GLOBAL CLIMATES
Northern Coniferous Forest	Northern Canada and Alaska		
	Northern U.S.S.R.		
Desert	North Africa		
	Australia		
Tundra			
Tropical Rain Forest			
Tropical Deciduous Forest			

30.4 Using Figure 30.1, identify two examples of groups of biomes that are elongated generally east-west, and two examples of groups of biomes that are elongated north-south. Explain your examples in terms of broad climatic and/or topographical characteristics.

30.5 Examine Figs. 29.3 and 30.1. Present examples of and explain the considerable degree of correlation between Thornthwaite's moisture index (I_m) and the biomes of the U.S.A.

30.6 Examine Fig. 30.4 and answer the following questions.

(a) What happens to net primary productivity (NPP) as mean annual precipitation increases?

Figure 30.4 Net Primary Productivity Above and Below Ground in Terms of Mean Annual Precipitation (From Lieth, 1973)

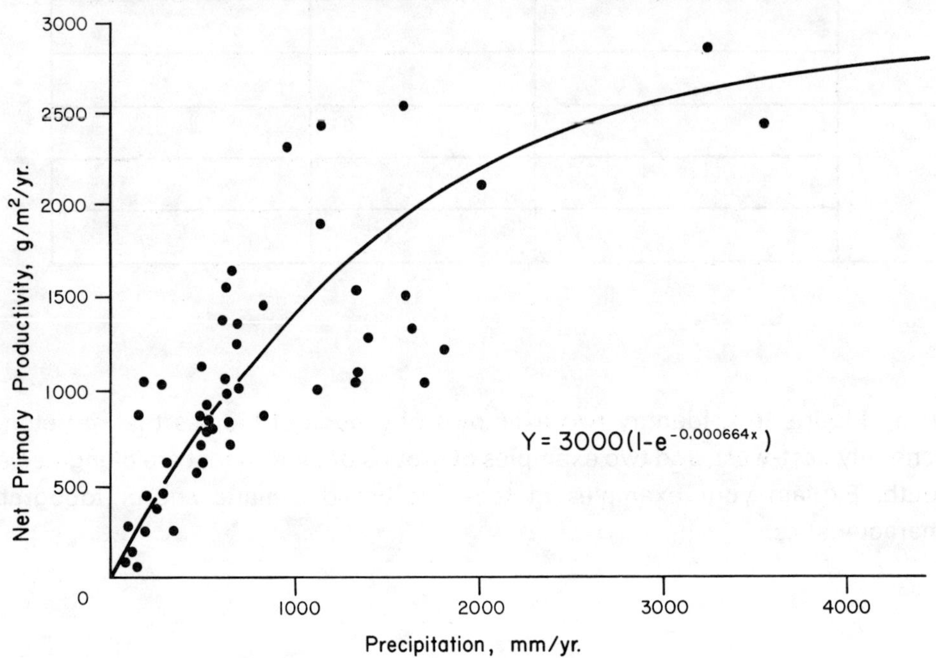

(b) What is the rate of change of NPP (per 100 mm of precipitation) between 0 and 1,000 mm, and between 3,000 and 4,000 mm of annual precipitation?

Name________________________________ Date________________

Instructor ____________________________ Section_______________

(c) Why does NPP increase less between 3,000 and 4,000 mm than between 0 and 1,000 mm of precipitation?

30.7 Examine Fig. 30.5 and answer the following questions.

(a) What happens to NPP as mean annual temperature increases?

(b) Compare the rates of change of NPP (per °C) between -13° and 0°C, between 0 and 15°C, and between 25 and 30°C. Why are the rates from -13 to 0°C and from 25 to 30°C lower than they are from 0 to 15°C?

Figure 30.5 Net Primary Productivity Above and Below Ground in Terms of Mean Annual Temperature (After Lieth, 1973)

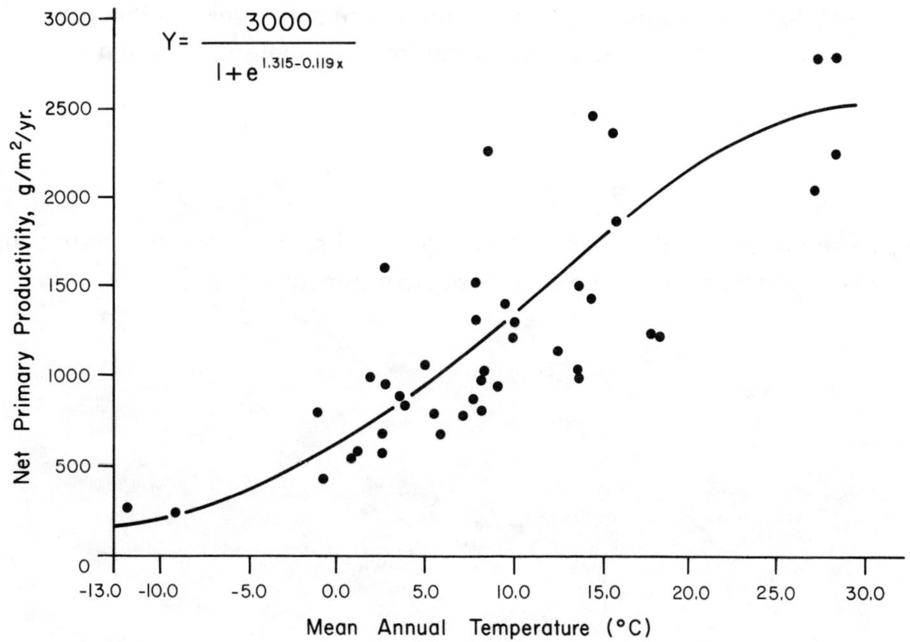

30.8 Using Figures 30.4 and 30.5 and information in your answers to questions 30.6 and 30.7, explain why NPP and biomass increase from desert scrub through temperate deciduous forest to tropical rain forest (Table 30.2).

Table 30.2 Net Primary Production and Biomass for Selected Continental Habitats

Habitat	Net primary production per unit area (gm/m²/yr)	Mean biomass per unit area (kg/m²)
Tropical rain forest	2000	44.00
Tropical seasonal forest	1500	36.00
Temperate evergreen forest	1300	36.00
Temperate deciduous forest	1200	30.00
Boreal forest	800	20.00
Savanna	700	4.00
Woodland and Shrubland	600	6.80
Temperate grassland	500	1.60
Tundra and alpine meadow	144	0.67
Desert shrub	71	0.67

30.9 Figs. 30.6 and 30.7 are profile diagrams of ecoclines from the southern Appalachians to Arizona and from Venezuela to the Arctic of northern Canada.

(a) Using the map of global climates (Appendix 5) determine whether the ecoclines of Fig. 30.6 and Fig. 30.7 fall along moisture or temperature gradients.

(b) Describe and explain changes in vegetation species, height, stratification, biomass, and ground cover along these environmental gradients.

Name ____________________ Date ____________

Instructor ____________________ Section ____________

Figure 30.6 Profile Diagram Along an Ecocline Extending from the Southern Appalachians through New Mexico to Arizona (After Whittaker, R.H. *Communities and Ecosystems,* MacMillan, N.Y., 1975)

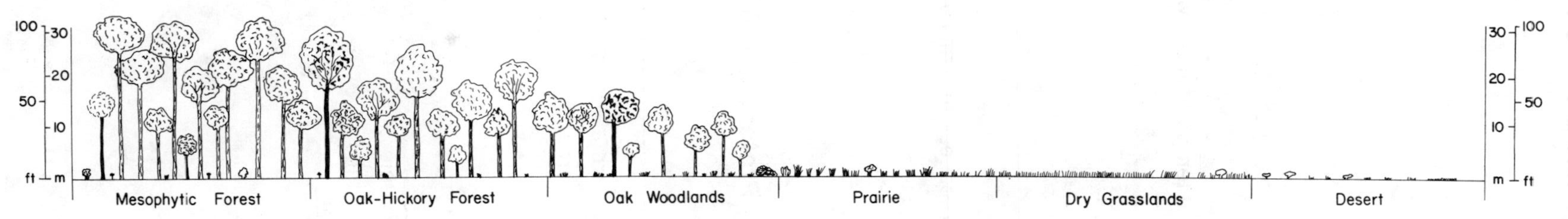

Figure 30.7 Profile Diagram Along an Ecocline from Venezuela through the Caribbean Islands, Florida, the Southern Appalachians, the Great Lakes Region, to the Arctic of Northern Canada (After Whittaker, R.H. *Communities and Ecosystems,* MacMillan, N.Y., 1975)

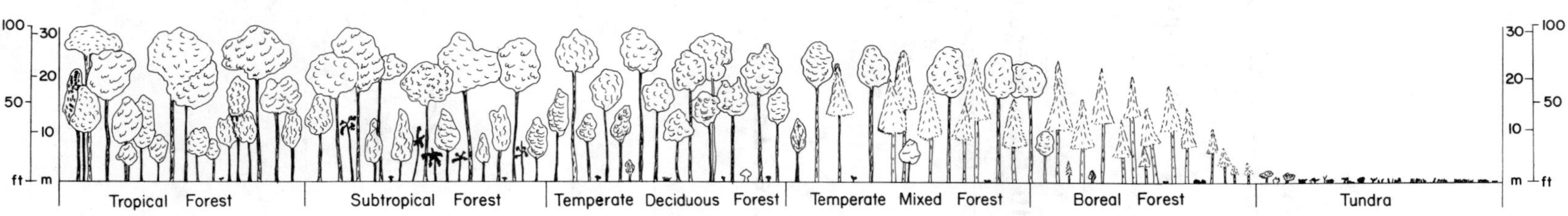

(c) What environmental factors have brought about the change from oak woodlands to prairie visible in Figure 30.6 and the change from boreal forest to tundra in Fig. 30.7?

(d) Based on Figs. 30.6 and 30.7, would you say that temperature or water availability has the greatest effect on biomass?

30.10 Examine Figure 30.8.

(a) Describe the distribution of the plant family Proteaceae in terms of the continents where it is present and those where it is not.

Figure 30.8 World Distribution of the plant family Proteaceae (After Johnson and Briggs, 1975)

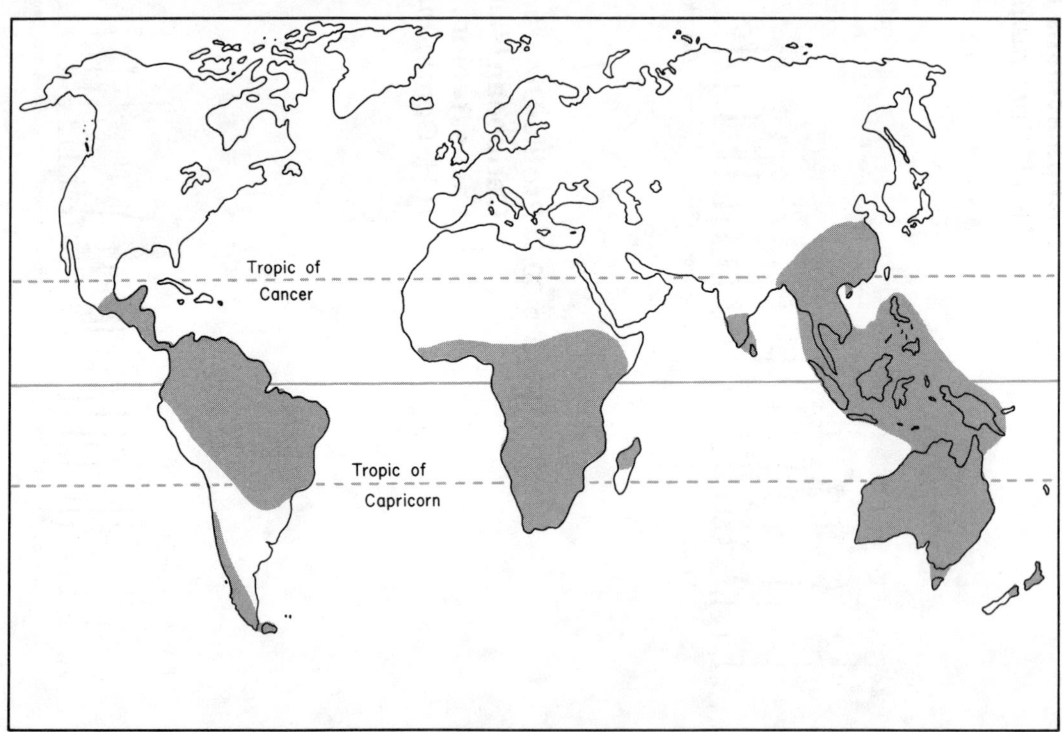

Name______________________________ Date______________

Instructor______________________________ Section______________

(b) Is the family concentrated in the northern or southern hemisphere continents?

(c) Refer to Exercise 16 and suggest one possible reason why the family Proteaceae may be restricted to the southern hemisphere continents.

30.11 What climatic factors appear to control the northward, westward and southern extent of sugar maple in the southeastern U.S.A. (Fig. 30.9). Explain why these factors might limit the geographical range of a plant.

Figure 30.9 Relationship of the geographical range of sugar maple (Acer saccharum) to various climatic indices. (After Dansereau, P., 1957)

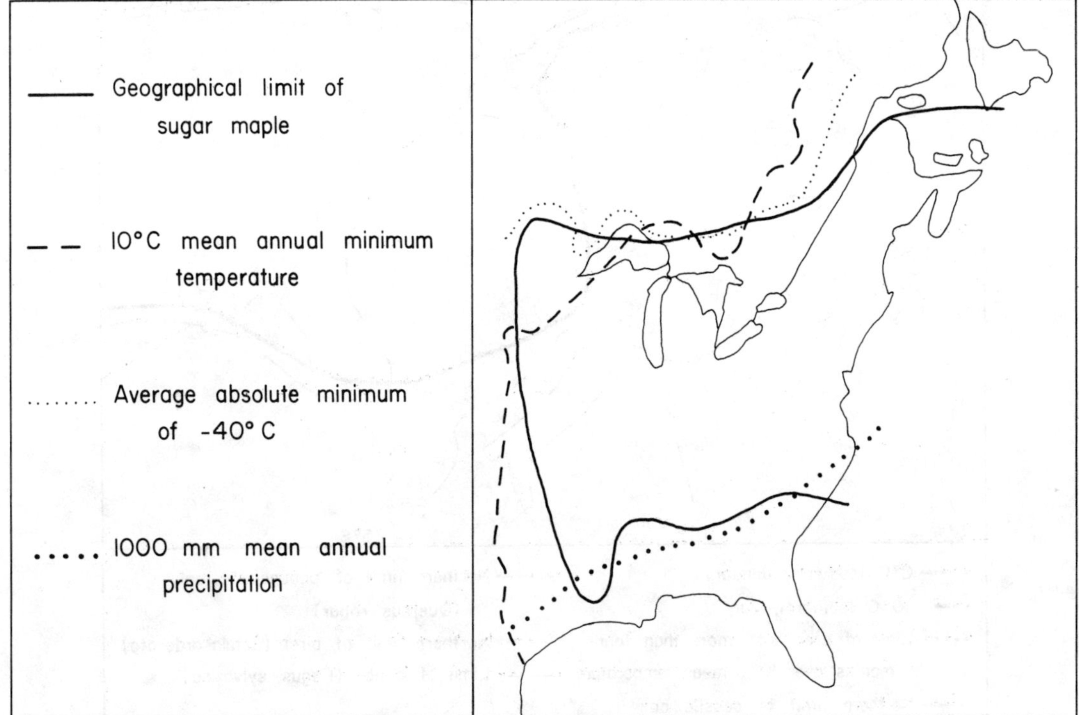

30.12 Examine Figure 30.10.

(a) What climatic temperature variables appear to correspond with the northern limits of beech, birch, and oak?

(b) What does your answer to question 30.12 (a) suggest about the temperature tolerances of these three tree species?

Figure 30.10 Relationship between average temperatures and the poleward limits of beech, birch, and oak. (After Collinson, A.S., 1977)

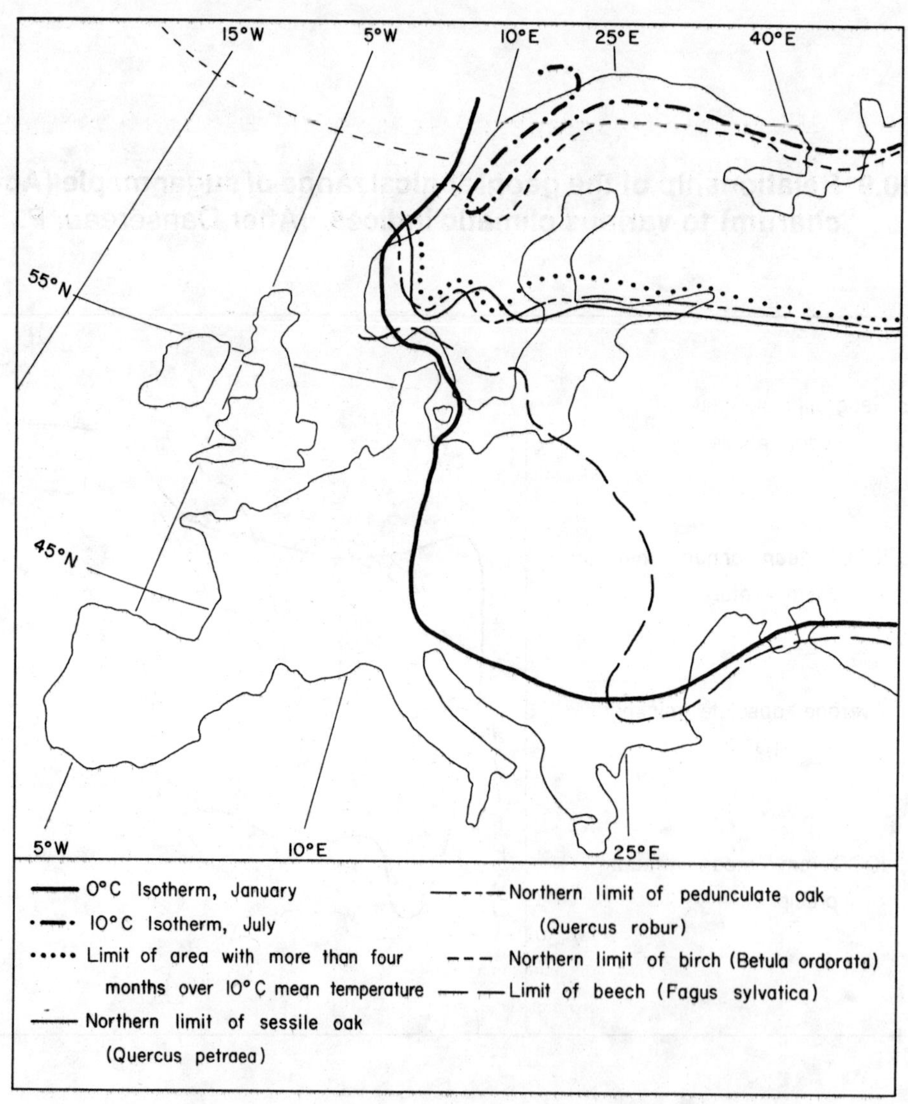

Name________________________________ Date________________

Instructor ___________________________ Section______________

30.13 The Saguaro cactus may reach 15 m in height and 200 years of age. It is conspicuous in much of the Sonoran Desert of southern Arizona and adjacent Sonora, Mexico. The Saguaro cactus lives where winter nighttime frosts are not infrequent.

(a) Although the Saguaro cactus can clearly withstand frost, what does the information in Figure 30.11 suggest as to the tolerance of the cactus to frost conditions?

(b) Why do you think that prolonged frost can restrict the geographic range of the Saguaro cactus?

Figure 30.11 Map of the Geographical Range of the Saguaro Cactus (*Carnegiea gigantea*) (Source: Brown and Gibson. *Biogeography,* The C.V. Mosby Company, St. Louis, 1983)

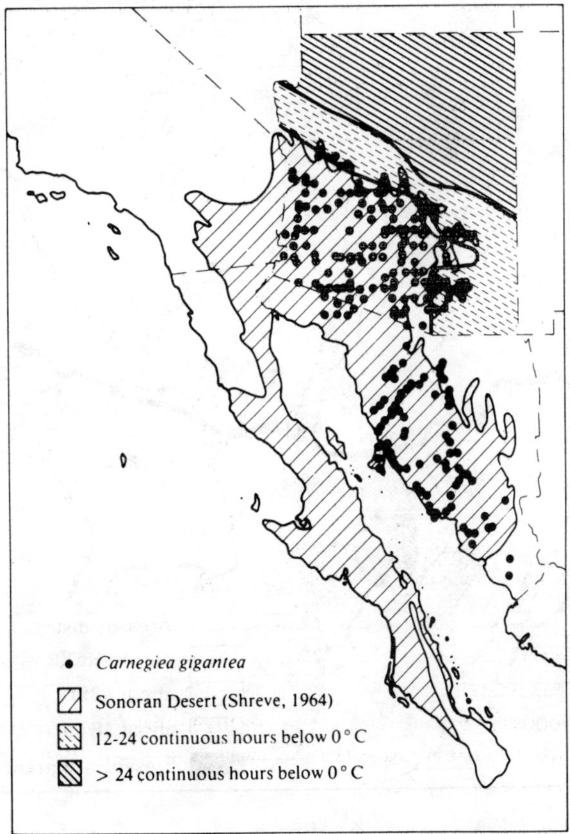

30.14 Prior to European colonization, the North American bison had a population of 60 million in spite of the presence of a small Indian population. By 1875 the distribution of the animal had been severely restricted because of hunting.

(a) Using Figure 30.12 explain how the distribution of the bison changed, and discuss the rate of change, between 1800 and 1850.

(b) How did the Union Pacific Railroad affect the distribution of bison on the North American Plains?

Figure 30.12 The Former and Present Distribution of the North American Bison (After Ziswiler in Illies, 1974)

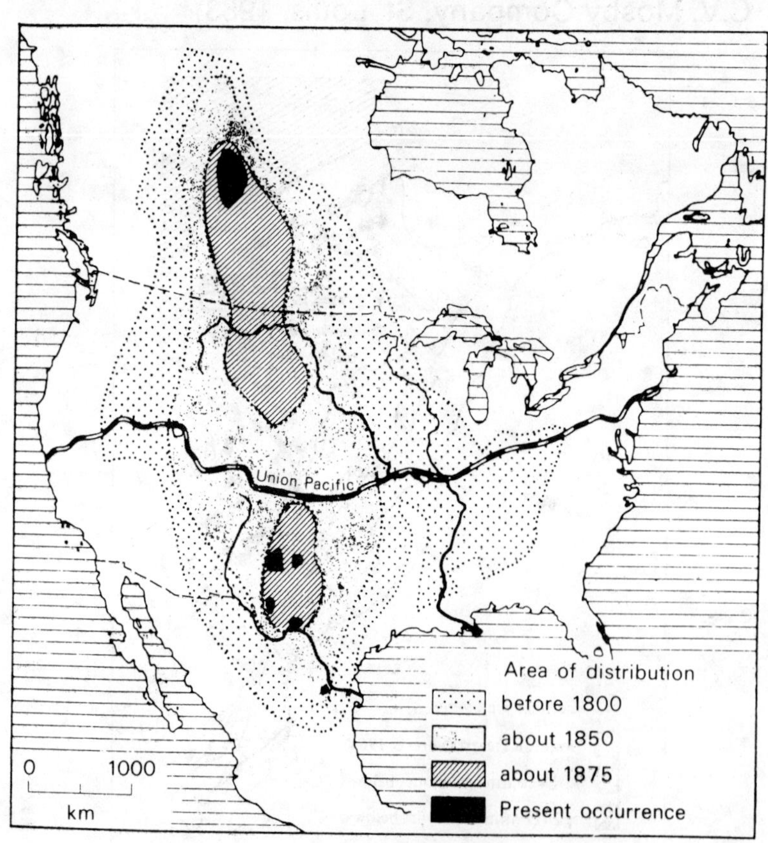

Name ______________________ Date ______________

Instructor ______________________ Section ______________

30.15 The muskrat, a valuable fur bearing animal, escaped from captivity in Czechoslovakia in 1905. This large, semiaquatic, herbivorous rodent inhabits marshes and streams.

(a) Using Fig. 30.13 estimate the eastward rate of expansion of the geographic range of the muskrat since 1905.

(b) If the muskrat had to compete with a species of similar ecological attributes, do you think it could have spread so rapidly?

Figure 30.13 Geographic range of the muskrat in North America where it is native and in Eurasia where it was introduced in 1905. (After Brown and Gibson, 1983)

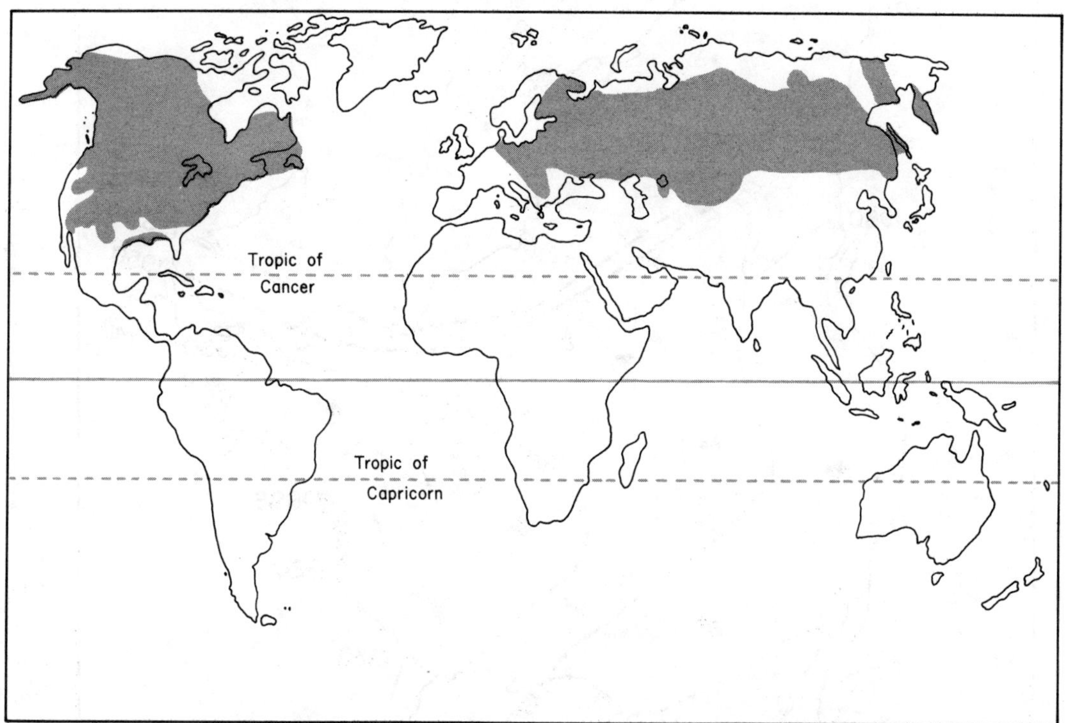

30.16 The European starling was introduced into Central Park, New York, in 1881. Since then, it has spread widely and is now found in all of the states (Fig. 30.14). It is mostly found in urban areas. What have been the rates of westward, southwestward and northwestward expansion? Why do you think the expansion of geographic range has been slower to the north?

Figure 30.14 Expansion of the geographic range of the European Starling following its introduction into North America (After Brown and Gibson, 1983)

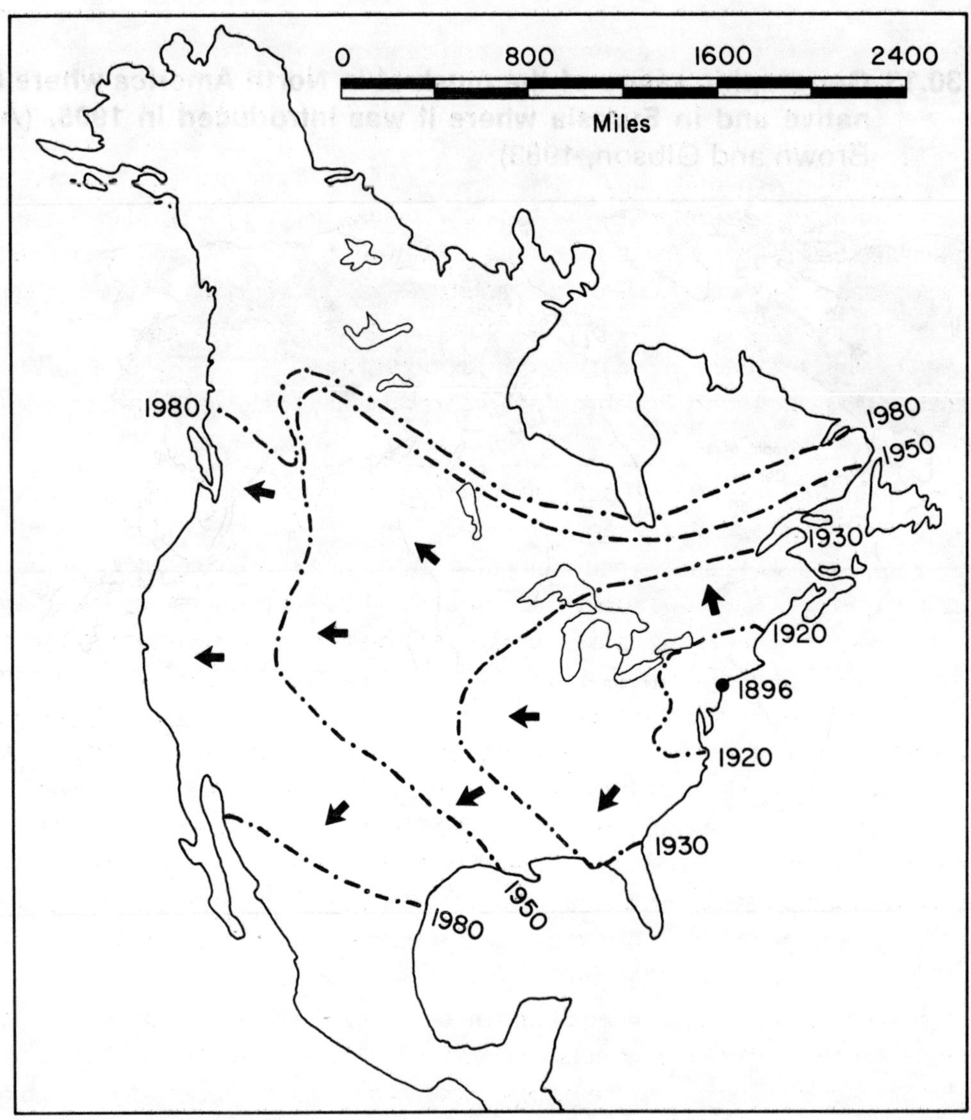

EXERCISE 31

QUATERNARY PALEOENVIRONMENTS

Before man's interference, the distribution of plants and animals on the earth was intimately related to the present climate. However, the climate of the earth has not remained constant over time. The last 2 million years of the earth's history—the Quaternary Period (including the Pleistocene and Holocene phases)—has seen major changes in climate most likely brought about by cyclic variations in the earth's geometry with respect to the sun. These changes in climate have caused variations in the extent of glaciers, have raised and lowered sea level by more than 300 feet, have shifted vegetation belts through 10° of latitude, have caused the expansion and contraction of inland lakes such as the Great Salt Lake in Utah, have changed soils, and have caused significant changes in the animal population (including many extinctions). Only 18,000 yr B.P. the area of the earth covered by glaciers was three times what it is today.

The reconstruction of past Quaternary environments is the field of Quaternary geomorphology and Quaternary paleoecology. The principal approach in reconstructing past vegetation characteristics has been microfossil analysis of sediments, and the principal technique pollen analysis. Pollen analysis relies on the fact that all flowering plants produce pollen, and ferns and mosses produce spores, as part of the process of reproduction. Pollen and spores from different species can be identified by their different shapes and surface characteristics. A resistant outer casing, the exine, allows pollen and spores to be preserved in sediments that are not subject to intense oxidation. Pollen and spores are normally preserved in the sediments of lakes, peat bogs, and soils.

After pollen grains have been extracted from sediments, identification and counting of the number of grains of each species is carried out using a microscope with x330 to x600 lenses. When counting pollen, it is usual to count all pollen grains and spores until a predetermined total has been reached (often 500 grains). Traditionally the counts are expressed in simple percentage terms. Each pollen and spore type is expressed as a percentage of the total pollen sum. When a number of samples from different levels in a sediment profile have been examined, a pollen diagram is constructed. The most common diagram is one in which the values for individual pollen and spore types, over time, are represented in separate "saw edge" graphs (Figure 31.1). Carbon-14 ages of organic material in the sediments provide an age scale for the pollen counts. Vegetational changes, which may be caused by climatic, soil, or biotic factors, may be recorded by pollen preserved in a sediment sequence. Pollen analysis can give a picture of the vegetation at a given site at a given point in time, and can also provide data on changes in vegetation over time. Pollen data have been used to construct regional vegetation maps for different time periods of the Quaternary.

The fossil record has also provided abundant evidence of animal extinctions during the Quaternary. One of the most recent episodes was the elimination of the Pleistocene megafauna of North America, including the ground sloth, mammoth, horse, and camel, between 11,500 and 8,000 yr B.P. Some scientists have argued that the extinction of the larger mammals was caused primarily by human hunters. Other scientists have pointed out that many North American birds, which were not hunted intensively, also became extinct at the end of the Pleistocene, suggesting that changes in environmental conditions (climate and vegetation) may have been more important.

Figure 31.1 Simplified Percentage Pollen Diagram for Rogers Lake, Connecticut, U.S.A. (After Davis, 1967)

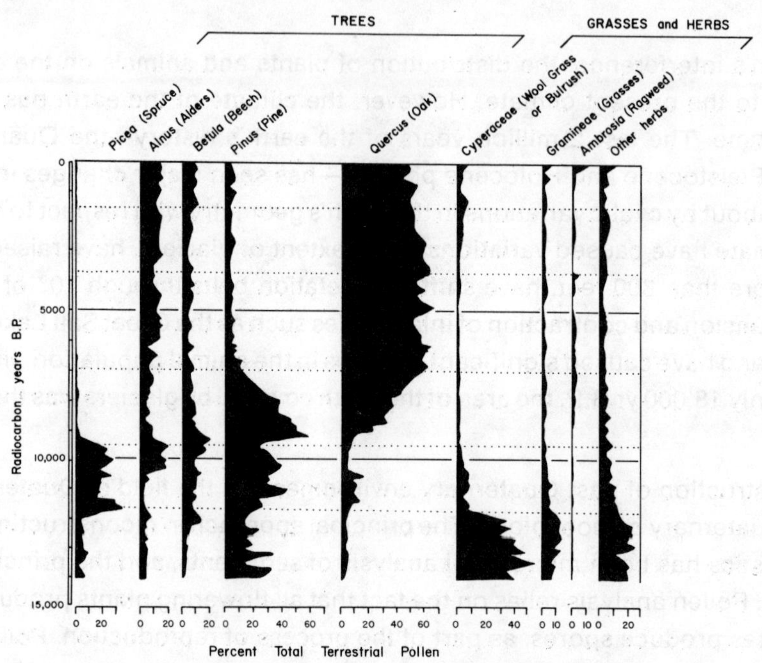

Figure 31.2 Estimated trend of the summer and winter temperatures during the last 20,000 years in the English Midlands (After Manley, 1964)

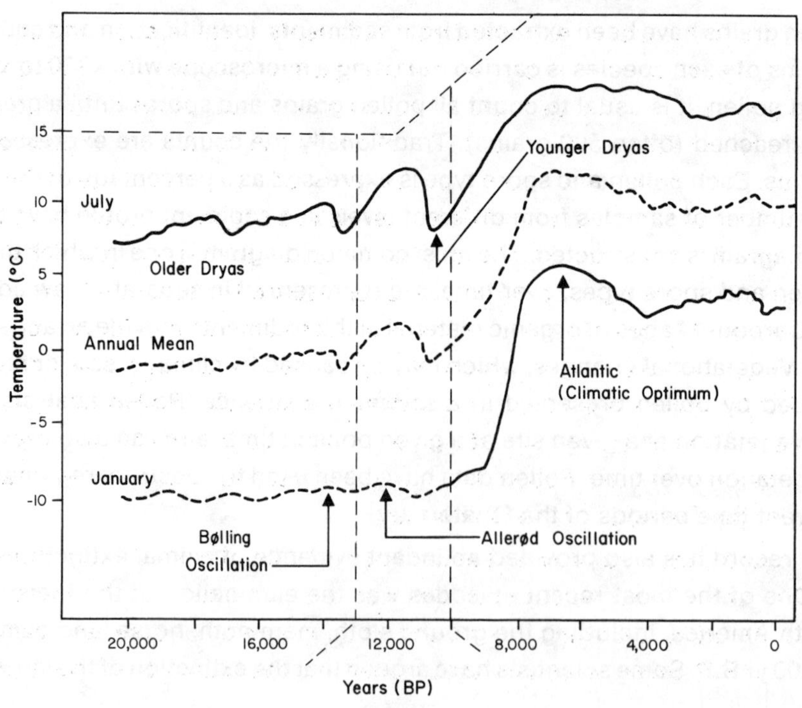

Name ____________________ Date ____________

Instructor ____________________ Section ____________

QUESTIONS

31.1 Figure 31.1 shows variations in abundance of particular terrestrial pollen grains that accumulated in sediments on the floor of Rogers Lake from 15,000 yr B.P. to the present. Figure 31.2 shows estimated temperature fluctuations in the English Midlands since 20,000 yr. B.P. The relatively warm Late Glacial Bϕlling and Allerϕd phases, and the relatively cold Younger and Older Dryas periods are marked on the diagram. The Post-Glacial Atlantic period (the Climatic Optimum), when temperatures were higher than they are today, is also identified.

Examine Figures 31.1 and 31.2 and answer the following questions.

(a) What evidence of the European Bϕlling, Older Dryas, Allerϕd and Younger Dryas periods is there in the Rogers Lake pollen diagram?

(b) Outline the changes in vegetation and climate that occurred at Rogers Lake between 10,000 and 7,000 yr B.P.

(c) What evidence in Figure 31.1 suggests that conditions at Rogers Lake were somewhat warmer than the present in early to mid Holocene times?

(d) Before 12,000 yr. B.P. the percentage pollen diagram shows a dominance of herb pollen over tree pollen. What does this suggest about the vegetation near Rogers Lake at this time, and what had happened to this vegetation by 5,000 yr B.P.?

(e) After 9,000 yr B.P. pine pollen decreased markedly and oak pollen increased markedly. What factors probably brought about these changes in the local vegetation?

31.2 In a recent paper entitled "Vegetation Maps for Eastern North America: 40,000 yr B.P. to the Present" (in *Geobotany II* (ed. R. C. Romans), Plenum Press, N.Y.), Paul and Hazel Delcourt have summarized the available pollen data on late Wisconsin vegetation characteristics. Figure 31.3 shows vegetation characteristics in eastern North America at the peak of Late Wisconsin glaciation (18,000 yr B.P.), at the beginning of the Holocene or Post-Glacial period (10,000 yr B.P.), and prior to European settlement (200 yr B.P.). Examine Figure 31.3 and answer the following questions.

(a) Why is the coastline at 18,000 and 10,000 yr B.P. very different from that at 200 yr B.P.?

(b) Bearing in mind your answer to question (a) above, what do you think happened to ground water table elevations in Florida and in other coastal plain areas between 18,000 and 200 yr B.P.?

(c) In studying past vegetation characteristics, why use a map of vegetation for 200 yr B.P. for comparative purposes rather than a map of the present vegetation?

(d) What was the vegetation like on the higher parts of the Appalachian Mountains 18,000 yr B.P., and what does that tell us about the climate of these areas at that time?

Name ______________________ Date ______________

Instructor ______________________ Section ______________

Figure 31.3 Paleovegetation maps for 18,000, 10,000, and 200 years B.P. (After Delcourt, P.A. and Delcourt, H.R. 1981)

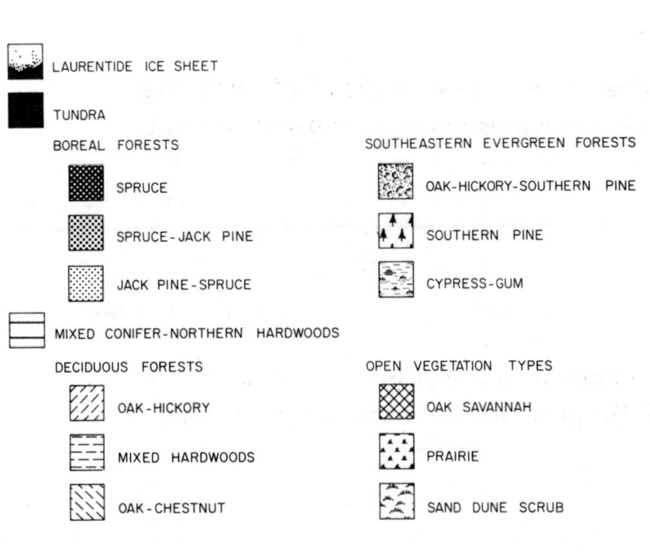

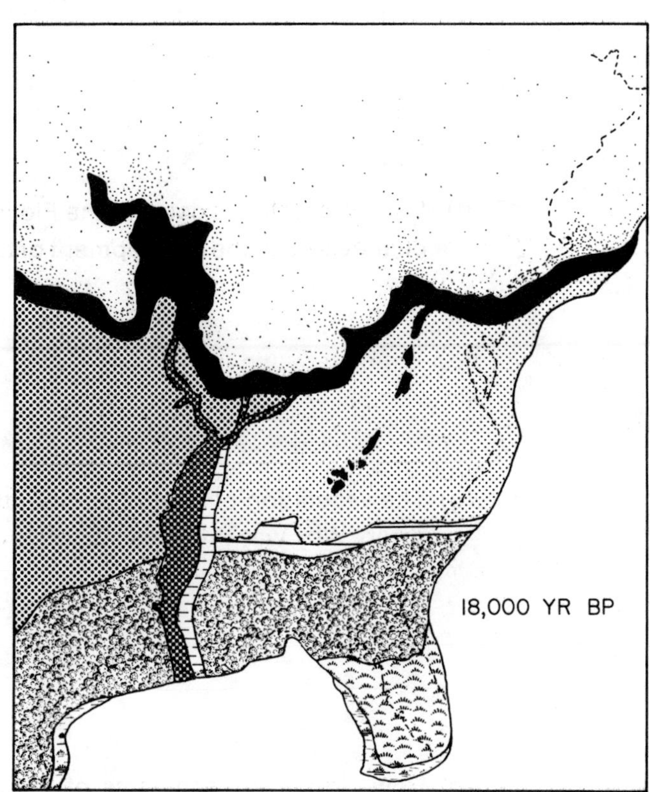

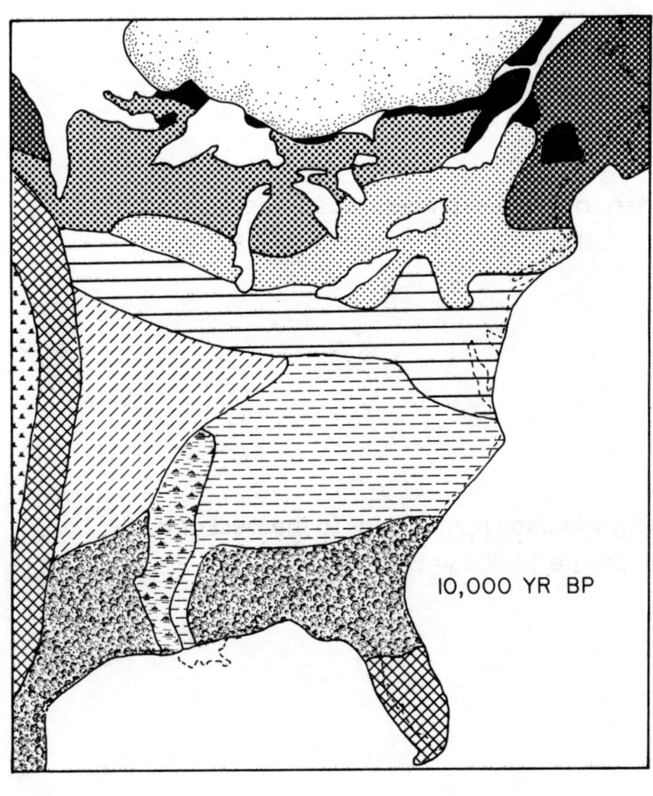

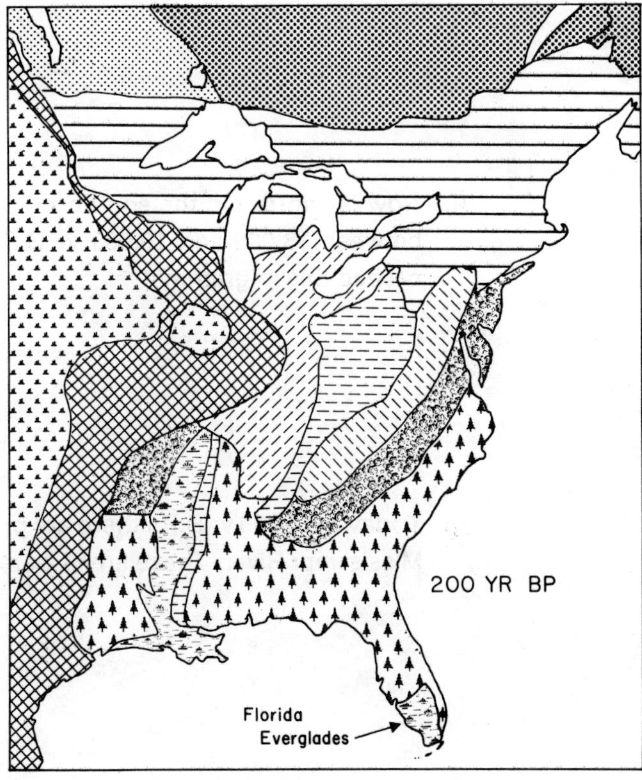

(e) How did the vegetation of the Lake Superior, Michigan, and Huron areas 10,000 yr B.P. differ from that of today?

(f) What vegetation existed in the Florida Everglades region 18,000 yr B.P., and what changes led to the development of the present cypress-gum swamp environment?

(g) The vegetation of central and southern Florida 18,000 yr B.P. was a sand dune scrub. Why was this area so arid during the peak of Late Wisconsin glaciation?

(h) How did the environments of the Adirondack Mountains in New York State at 18,000 and 10,000 yr B.P. differ from those of today?

(i) How far north did the southern boundary of the jackpine-spruce forest migrate between 18,000 and 10,000 yr B.P. and between 10,000 and 200 yr B.P.? What caused this migration?

(j) Today (and 200 yr B.P.) there are prairie grasslands in the plains to the west of the Mississippi River. These areas were not prairie 18,000 yr B.P. Why?

Name ______________________ Date ____________

Instructor ______________________ Section ____________

31.3 As the Scandinavian Ice Sheet expanded and contracted, the pattern of vegetation in Europe was considerably disturbed. Figures 31.4-31.6 illustrate paleogeographic conditions in Europe during full glacial and interglacial times, and during the late glacial Allerφd Interstadial ca. 12,000 B.P. Examine these fiqures and also the map of the world vegetation zones (Figure 30.1) and answer the following questions.

(a) Compare the present-day coastline of northern Europe with those of full glacial and Allerφd Interstadial times.

(b) During full glacial times a series of proglacial lakes extended along the southern boundary of the Scandinavian Ice Sheet. What is a proglacial lake, and why did these proglacial lakes develop?

(c) What happened to the environment of the European Alps during full glacial times?

(d) Compare and contrast the environments of the British Isles during full glacial and Allerφd times.

Figure 31.4 Profiles of the Vegetation Belts of Europe from the Mediterranean Sea to the Arctic Ocean, during Glacial and Interglacial Times (Source: Cox and Moore. *Biogeography*, John Wiley & Sons, N.Y., 1980: 191)

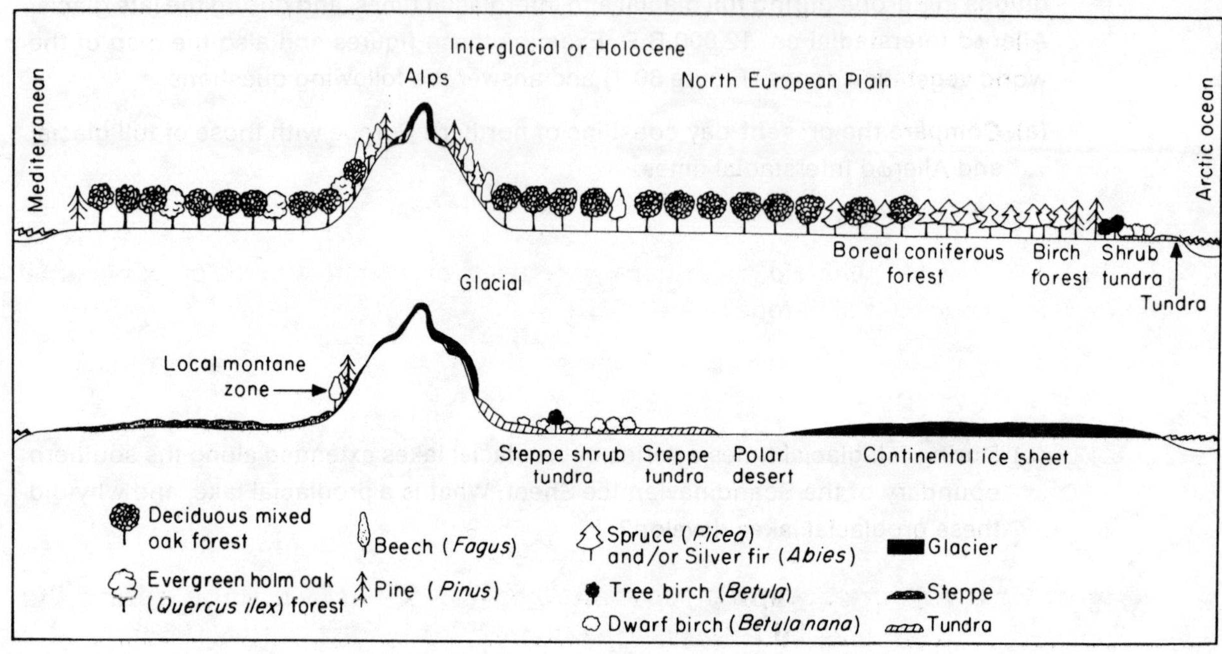

Figure 31.5 Paleogeographic reconstruction of northern Europe during the Maximum of the last glaciation (Source: Gerasimov, 1969)

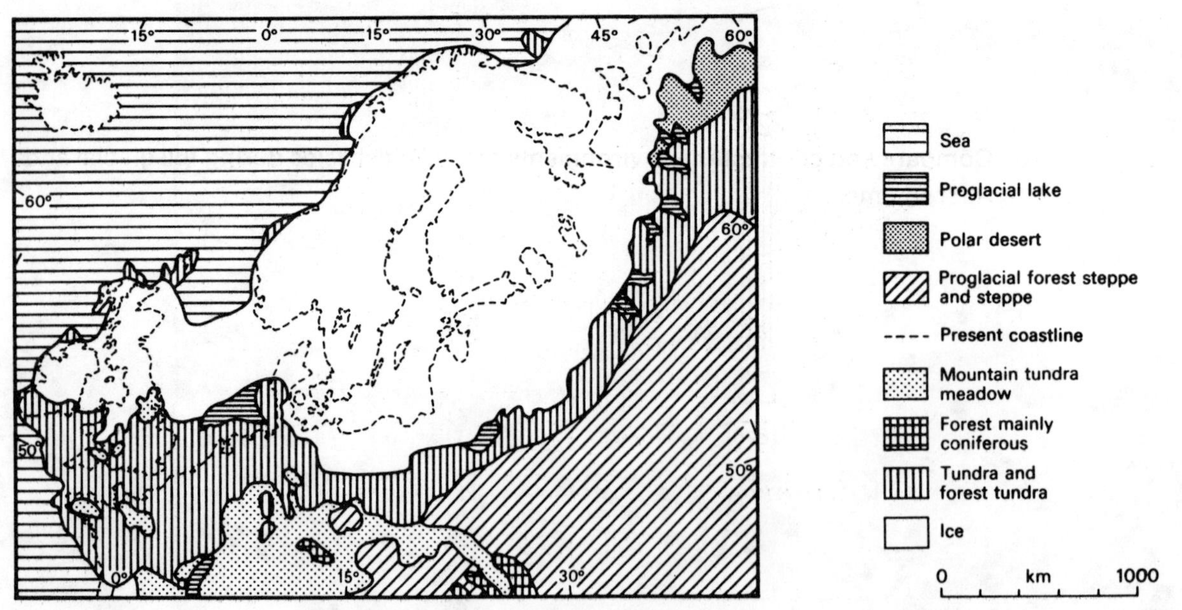

Name ______________________ Date ______________

Instructor ______________________ Section ______________

(e) Estimate the widths of the tundra belts south of the Laurentide and Scandinavian Ice Sheets in full glacial times (Figures 31.3 and 31.5). Explain any differences in terms of paleoclimatic conditions.

(f) How far south did the belt of tundra and birch migrate with the onset of glacial conditions in Europe?

Figure 31.6 Paleogeographic reconstruction of northern Europe during the Lake Glacial Allerφd Interstadial (After Gerasimov, 1969)

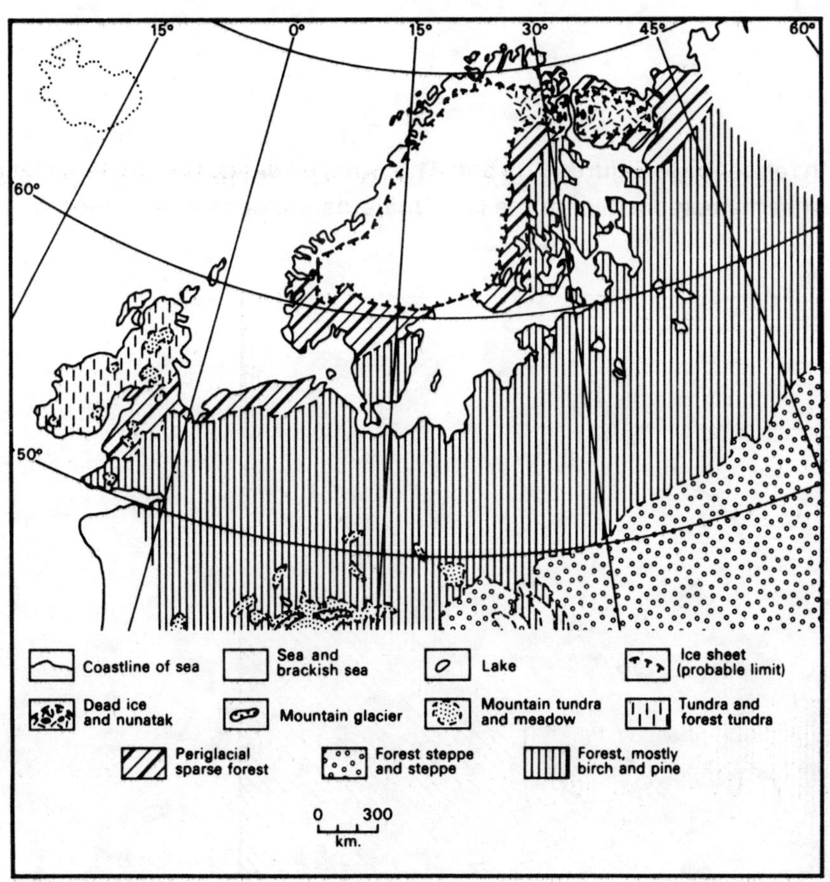

(g) Outline the major differences between the present natural vegetation of northern Europe and the vegetation that existed during full glacial times.

31.4 Examine Figures 31.2, 31.7 and 31.8 and answer the following questions.

(a) What evidence in Figures 31.7 and 31.8 is there for the argument that man was responsible for the massive late Quaternary mammal extinctions in North America?

(b) With reference to Figures 31.2 and 31.8, what evidence is there that climate was at least partly responsible for the late Quaternary mammal extinctions?

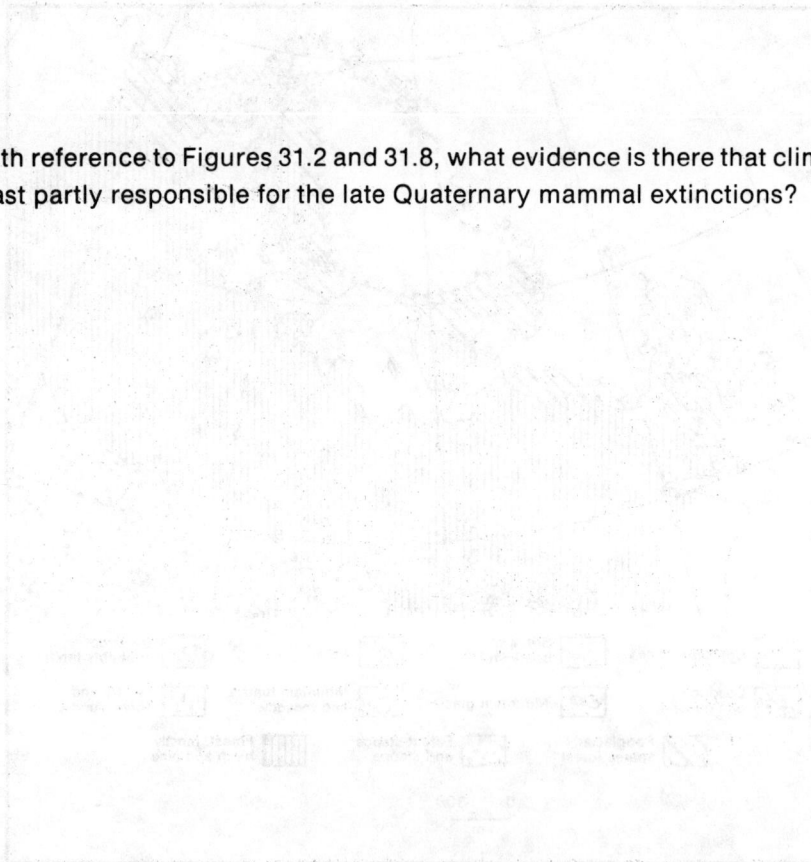

Name ______________________ Date ______________

Instructor ______________________ Section ______________

Figure 31.7 Temporal Sequence of Advancing Populations of Big Game Hunters (After Martin, 1973)

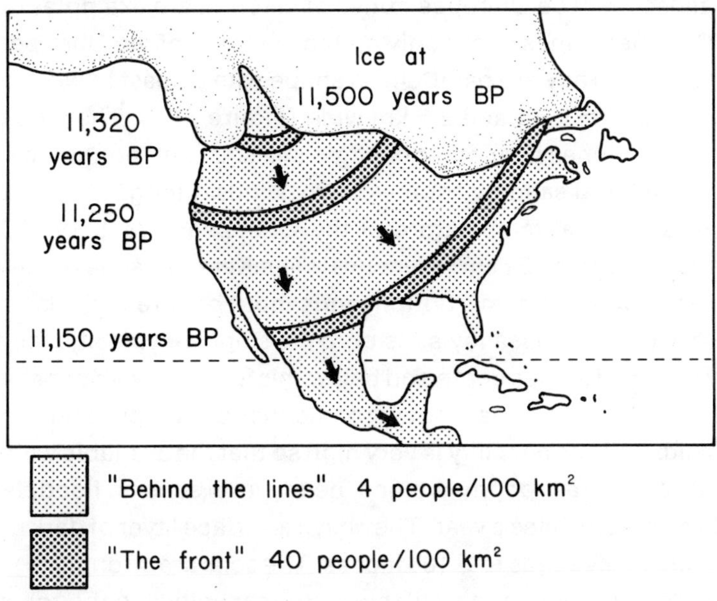

Figure 31.8 Extinctions of North American Mammal Genera During the Last Part of the Quaternary (After Webb, 1969)

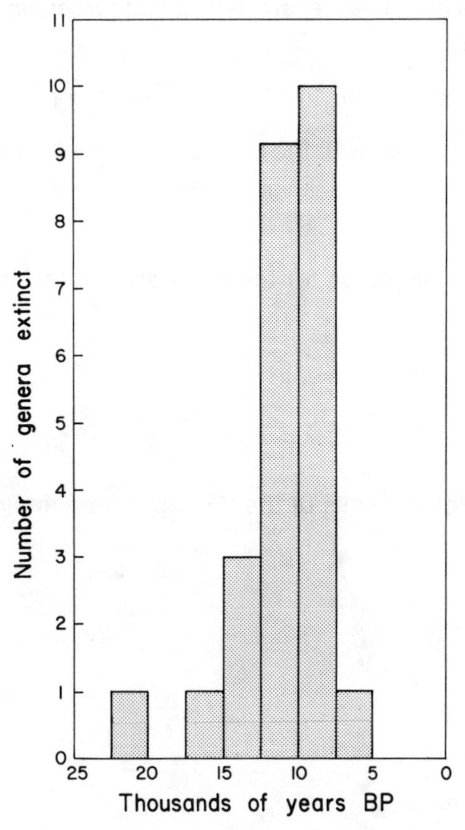

31.5 Many of the earth's landforms were produced under climatic conditions very different from those that exist near them today. These landforms frequently influence the soils and vegetation of their local areas and therefore leave an imprint of the past on the present landscape. The Ohoopee dunes of Georgia are examples of such landforms.

The Ohoopee dunes occur only on the east side of the Ohoopee River in Tattnall County, Georgia. They are parabolic in shape with an east to west long axis, and may be as much as 7 km long and 2.5 km wide (Figure 31.9). They are of aeolian origin, and, like similar dunes near other Georgia streams, are thought to have been formed from riverine alluvial sand by southwesterly winds during full glacial times.

The mean annual temperature of the Ohoopee area is 19.4°C (67°F), and the annual precipitation is 122 cm (48 inches). The Kershaw soils of the dunes differ from the Osier soils of the stream's bottomlands, but both are entisols. The Kershaw soils are excessively drained, sandy soils on 2-8% slopes. Typically, the surface layer is a dark gray sand to 3 inches depth and below this is a yellowish or brownish yellow sand to 83 inches. These soils have low organic matter content and are strongly to very strongly acid. Soil permeability is very high so that the available water capacity is very low. The Osier soils are poorly drained, nearly level soils on floodplains. The soils are flooded two or more times a year. They have a surface layer of dark gray loamy sand to 6 inches above gray sands to $>$ 100 inches. The soils are strongly to very strongly acid. Soil permeability is high and available water capacity low though a seasonally high water table is generally within 12 inches of the surface during winter and spring.

The vegetation in the stream valleys is a bottomland or floodplain hardwood forest. Typical dominants include overcup oak, water hickory, green ash, and American elm. Common understory species include water elm and ironwood. The floodplain hardwoods are very mesic, water-loving species. The dunes are a more xeric environment and support a dwarf oak-evergreen shrub forest, and an evergreen scrub-lichen forest (Figure 31.10).

Examine Figures 31.9 and 31.10 and answer the following questions.

(a) On Figure 31.9 identify with the letters (X, Y, and Z) areas of bottomland hardwood vegetation (X), dwarf oak-evergreen shrubs forest (Y), and evergreen scrub-lichen forest (Z).

(b) What is a typical pH range for the Kershaw and Osier soils?

(c) Why do you think the soils of the Ohoopee area are in the order of entisols?

Name ______________________ Date ______________

Instructor ______________________ Section ______________

Figure 31.9 The Ohoopee Dunes, Tattnall County, Georgia

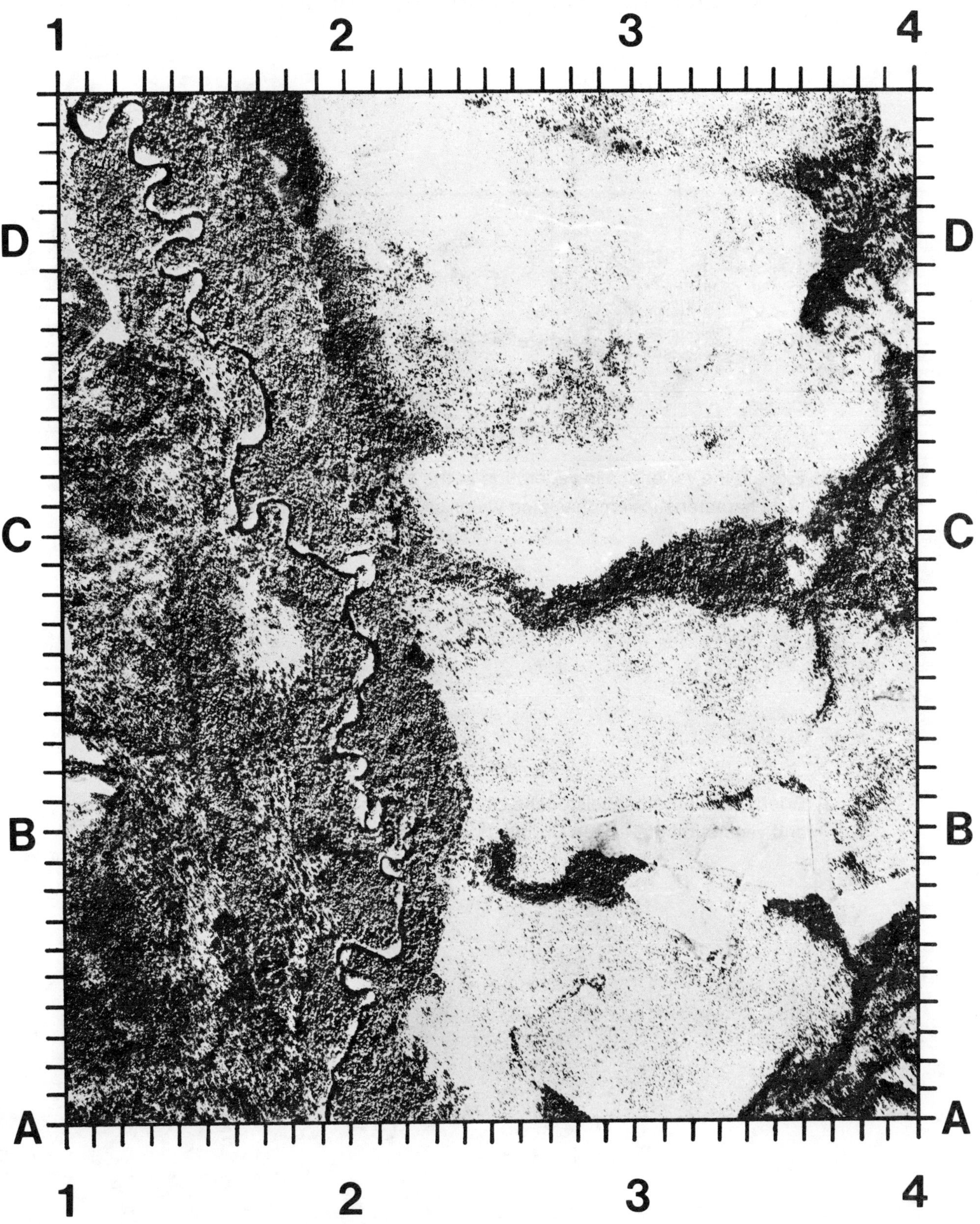

Figure 31.10 Xeric Environments on the Ohoopee Dune Sandhills. (A) dwarf oak-evergreen shrub forest; (B) evergreen shrub-lichen forest; (C) longleaf pine; (D) turkey oak; (E) rosemary (Ceratiola); (F) woody mint; (G) dwarfed live oak; (H) lichen ground cover (After Wharton, The Natural Environments of Georgia)

(d) Explain the variations in vegetation in the area of the Ohoopee dunes in terms of soil characteristics emphasizing water capacity and drainage.

(e) Briefly state how the Pleistocene Ohoopee dunes have influenced the present soils and vegetation of Tattnall County.

APPENDIX A

SELECTED BAR SCALES

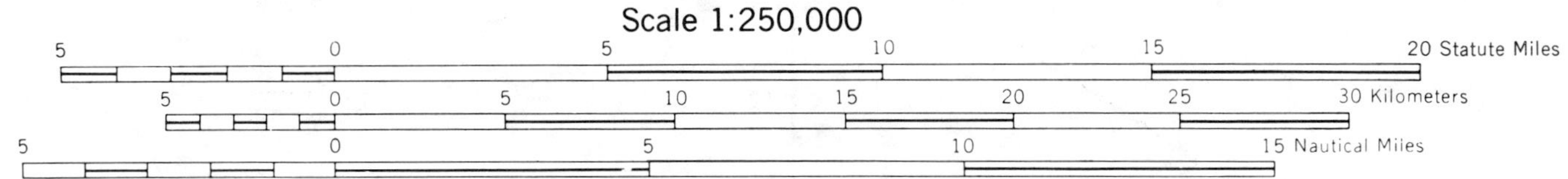

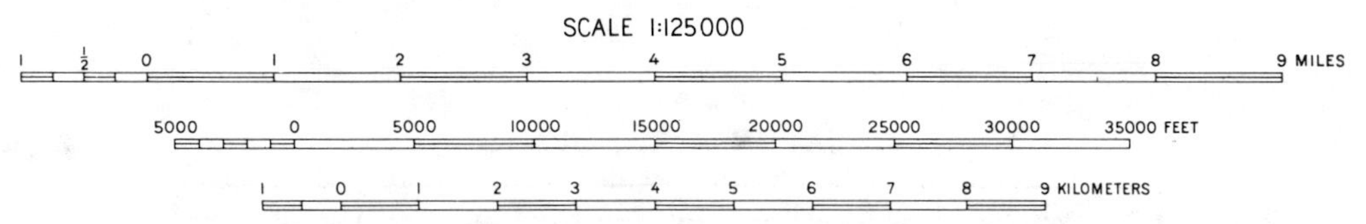

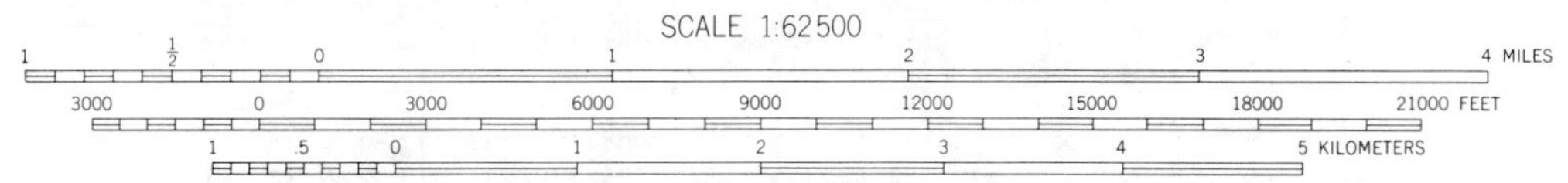

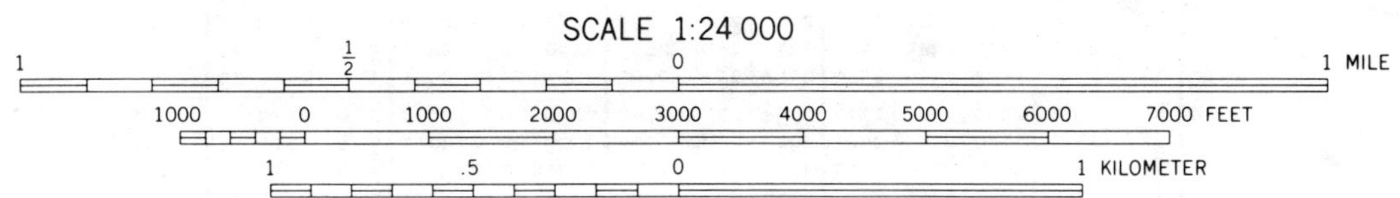

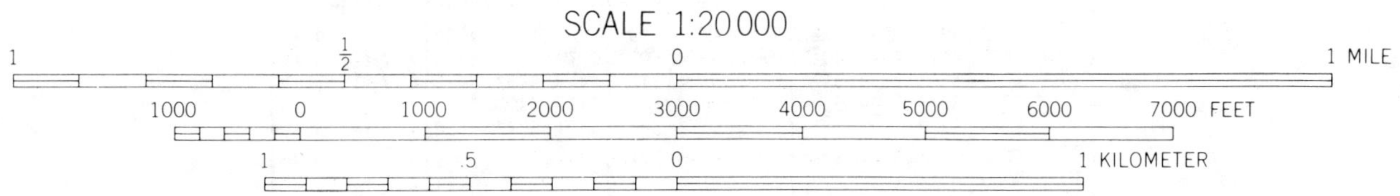

APPENDIX B

TANGENT TABLES

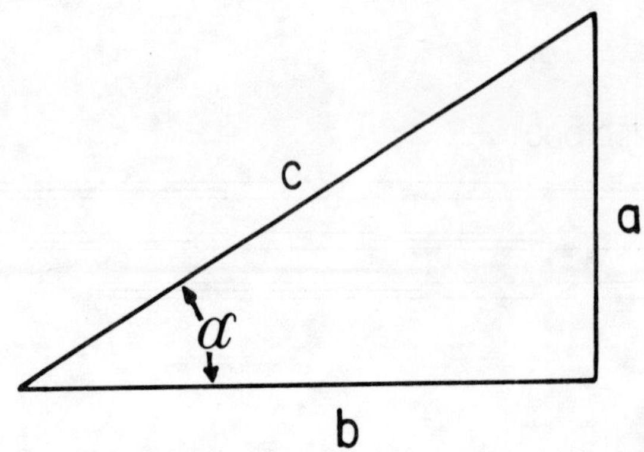

$$\text{Tan } \alpha = \frac{\text{opposite}}{\text{adjacent}} = \frac{a}{b}$$

$$\text{Sin } \alpha = \frac{\text{opposite}}{\text{hypotenuse}} = \frac{a}{c}$$

$$\text{Cos } \alpha = \frac{\text{adjacent}}{\text{hypotenuse}} = \frac{b}{c}$$

NATURAL TANGENTS

Degrees	0′ 0°·0	6′ 0°·1	12′ 0°·2	18′ 0°·3	24′ 0°·4	30′ 0°·5	36′ 0°·6	42′ 0°·7	48′ 0°·8	54′ 0°·9	Mean Differences 1	2	3	4	5
0	·0000	0017	0035	0052	0070	0087	0105	0122	0140	0157	3	6	9	12	15
1	·0175	0192	0209	0227	0244	0262	0279	0297	0314	0332	3	6	9	12	15
2	·0349	0367	0384	0402	0419	0437	0454	0472	0489	0507	3	6	9	12	15
3	·0524	0542	0559	0577	0594	0612	0629	0647	0664	0682	3	6	9	12	15
4	·0699	0717	0734	0752	0769	0787	0805	0822	0840	0857	3	6	9	12	15
5	·0875	0892	0910	0928	0945	0963	0981	0998	1016	1033	3	6	9	12	15
6	·1051	1069	1086	1104	1122	1139	1157	1175	1192	1210	3	6	9	12	15
7	·1228	1246	1263	1281	1299	1317	1334	1352	1370	1388	3	6	9	12	15
8	·1405	1423	1441	1459	1477	1495	1512	1530	1548	1566	3	6	9	12	15
9	·1584	1602	1620	1638	1655	1673	1691	1709	1727	1745	3	6	9	12	15
10	·1763	1781	1799	1817	1835	1853	1871	1890	1908	1926	3	6	9	12	15
11	·1944	1962	1980	1998	2016	2035	2053	2071	2089	2107	3	6	9	12	15
12	·2126	2144	2162	2180	2199	2217	2235	2254	2272	2290	3	6	9	12	15
13	·2309	2327	2345	2364	2382	2401	2419	2438	2456	2475	3	6	9	12	15
14	·2493	2512	2530	2549	2568	2586	2605	2623	2642	2661	3	6	9	12	16
15	·2679	2698	2717	2736	2754	2773	2792	2811	2830	2849	3	6	9	13	16
16	·2867	2886	2905	2924	2943	2962	2981	3000	3019	3038	3	6	9	13	16
17	·3057	3076	3096	3115	3134	3153	3172	3191	3211	3230	3	6	10	13	16
18	·3249	3269	3288	3307	3327	3346	3365	3385	3404	3424	3	6	10	13	16
19	·3443	3463	3482	3502	3522	3541	3561	3581	3600	3620	3	7	10	13	16
20	·3640	3659	3679	3699	3719	3739	3759	3779	3799	3819	3	7	10	13	17
21	·3839	3859	3879	3899	3919	3939	3959	3979	4000	4020	3	7	10	13	17
22	·4040	4061	4081	4101	4122	4142	4163	4183	4204	4224	3	7	10	14	17
23	·4245	4265	4286	4307	4327	4348	4369	4390	4411	4431	3	7	10	14	17
24	·4452	4473	4494	4515	4536	4557	4578	4599	4621	4642	4	7	11	14	18
25	·4663	4684	4706	4727	4748	4770	4791	4813	4834	4856	4	7	11	14	18
26	·4877	4899	4921	4942	4964	4986	5008	5029	5051	·5073	4	7	11	15	18
27	·5095	5117	5139	5161	5184	5206	5228	5250	5272	5295	4	7	11	15	18
28	·5317	5340	5362	5384	5407	5430	5452	5475	5498	5520	4	8	11	15	19
29	·5543	5566	5589	5612	5635	5658	5681	5704	5727	5750	4	8	12	15	19
30	·5774	5797	5820	5844	5867	5890	5914	5938	5961	5985	4	8	12	16	20
31	·6009	6032	6056	6080	6104	6128	6152	6176	6200	6224	4	8	12	16	20
32	·6249	6273	6297	6322	6346	6371	6395	6420	6445	6469	4	8	12	16	20
33	·6494	6519	6544	6569	6594	6619	6644	6669	6694	6720	4	8	13	17	21
34	·6745	6771	6796	6822	6847	6873	6899	6924	6950	6976	4	9	13	17	21
35	·7002	7028	7054	7080	7107	7133	7159	7186	7212	7239	4	9	13	18	22
36	·7265	7292	7319	7346	7373	7400	7427	7454	7481	7508	5	9	14	18	23
37	·7536	7563	7590	7618	7646	7673	7701	7729	7757	7785	5	9	14	18	23
38	·7813	7841	7869	7898	7926	7954	7983	8012	8040	8069	5	9	14	19	24
39	·8098	8127	8156	8185	8214	8243	8273	8302	8332	8361	5	10	15	20	24
40	·8391	8421	8451	8481	8511	8541	8571	8601	8632	8662	5	10	15	20	25
41	·8693	8724	8754	8785	8816	8847	8878	8910	8941	8972	5	10	16	21	26
42	·9004	9036	9067	9099	9131	9163	9195	9228	9260	9293	5	11	16	21	27
43	·9325	9358	9391	9424	9457	9490	9523	9556	9590	9623	6	11	17	22	28
44	·9657	9691	9725	9759	9793	9827	9861	9896	9930	9965	6	11	17	23	29

NATURAL TANGENTS

Degrees	0′ 0°·0	6′ 0°·1	12′ 0°·2	18′ 0°·3	24′ 0°·4	30′ 0°·5	36′ 0°·6	42′ 0°·7	48′ 0°·8	54′ 0°·9	Mean Differences 1	2	3	4	5
45	1·0000	0035	0070	0105	0141	0176	0212	0247	0283	0319	6	12	18	24	30
46	1·0355	0392	0428	0464	0501	0538	0575	0612	0649	0686	6	12	18	25	31
47	1·0724	0761	0799	0837	0875	0913	0951	0990	1028	1067	6	13	19	25	32
48	1·1106	1145	1184	1224	1263	1303	1343	1383	1423	1463	7	13	20	27	33
49	1·1504	1544	1585	1626	1667	1708	1750	1792	1833	1875	7	14	21	28	34
50	1·1918	1960	2002	2045	2088	2131	2174	2218	2261	2305	7	14	22	29	36
51	1·2349	2393	2437	2482	2527	2572	2617	2662	2708	2753	8	15	23	30	38
52	1·2799	2846	2892	2938	2985	3032	3079	3127	3175	3222	8	16	24	31	39
53	1·3270	3319	3367	3416	3465	3514	3564	3613	3663	3713	8	16	25	33	41
54	1·3764	3814	3865	3916	3968	4019	4071	4124	4176	4229	9	17	26	34	43
55	1·4281	4335	4388	4442	4496	4550	4605	4659	4715	4770	9	18	27	36	45
56	1·4826	4882	4938	4994	5051	5108	5166	5224	5282	5340	10	19	29	38	48
57	1·5399	5458	5517	5577	5637	5697	5757	5818	5880	5941	10	20	30	40	50
58	1·6003	6066	6128	6191	6255	6319	6383	6447	6512	6577	11	21	32	43	53
59	1·6643	6709	6775	6842	6909	6977	7045	7113	7182	7251	11	23	34	45	56
60	1·7321	7391	7461	7532	7603	7675	7747	7820	7893	7966	12	24	36	48	60
61	1·8040	8115	8190	8265	8341	8418	8495	8572	8650	8728	13	26	38	51	64
62	1·8807	8887	8967	9047	9128	9210	9292	9375	9458	9542	14	27	41	55	68
63	1·9626	9711	9797	9883	9970	2·0057	2·0145	2·0233	2·0323	2·0413	15	29	44	58	73
64	2·0503	0594	0686	0778	0872	0965	1060	1155	1251	1348	16	31	47	63	78
65	2·1445	1543	1642	1742	1842	1943	2045	2148	2251	2355	17	34	51	68	85
66	2·2460	2566	2673	2781	2889	2998	3109	3220	3332	3445	18	37	55	73	92
67	2·3559	3673	3789	3906	4023	4142	4262	4383	4504	4627	20	40	60	79	99
68	2·4751	4876	5002	5129	5257	5386	5517	5649	5782	5916	22	43	65	87	108
69	2·6051	6187	6325	6464	6605	6746	6889	7034	7179	7326	24	47	71	95	119
70	2·7475	7625	7776	7929	8083	8239	8397	8556	8716	8878	26	52	78	104	131
71	2·9042	9208	9375	9544	9714	9887	3·0061	3·0237	3·0415	3·0595	29	58	87	116	145
72	3·0777	0961	1146	1334	1524	1716	1910	2106	2305	2506	32	64	96	129	161
73	3·2709	2914	3122	3332	3544	3759	3977	4197	4420	4646	36	72	108	144	180
74	3·4874	5105	5339	5576	5816	6059	6305	6554	6806	7062	41	81	122	163	204
75	3·7321	7583	7848	8118	8391	8667	8947	9232	9520	9812	46	93	139	186	232
76	4·0108	0408	0713	1022	1335	1653	1976	2303	2635	2972	53	107	160	213	267
77	4·3315	3662	4015	4374	4737	5107	5483	5864	6252	6646					
78	4·7046	7453	7867	8288	8716	9152	9594	5·0045	5·0504	5·0970					
79	5·1446	1929	2422	2924	3435	3955	4486	5026	5578	6140					
80	5·6713	7297	7894	8502	9124	9758	6·0405	6·1066	6·1742	6·2432					
81	6·3138	3859	4596	5350	6122	6912	7720	8548	9395	7·0264					
82	7·1154	2066	3002	3962	4947	5958	6996	8062	9158	8·0285					
83	8·1443	2636	3863	5126	6427	7769	9152	9·0579	9·2052	9·3572					
84	9·5144	9·677	9·845	10·02	10·20	10·39	10·58	10·78	10·99	11·20					
85	11·43	11·66	11·91	12·16	12·43	12·71	13·00	13·30	13·62	13·95					
86	14·30	14·67	15·06	15·46	15·89	16·35	16·83	17·34	17·89	18·46					
87	19·08	19·74	20·45	21·20	22·02	22·90	23·86	24·90	26·03	27·27					
88	28·64	30·14	31·82	33·69	35·80	38·19	40·92	44·07	47·74	52·08					
89	57·29	63·66	71·62	81·85	95·49	114·6	143·2	191·0	286·5	573·0					
90	∞														

Mean differences cease to be sufficiently accurate.

GEOLOGIC TIME SCALE

APPENDIX C

ERA	Periods and Epochs		Duration (Millions of Years)	Date B.P. (Millions of Years)
CENOZOIC ERA	Quanternary Period	Pleistocene Epoch	1-3	3
	Tertiary Period	Pilocene Epoch	8	11
		Miocene Epoch	14	25
		Oligocene Epoch	15	40
		Eocene Epoch	20	60
		Paleocene Epoch	10	70±2
MESOZOIC ERA	Cretaceous Period		65	135±5
	Jurassic Period		45	180±5
	Triassic Period		45	225±5
PALEOZOIC ERA	Permian Period		45	270±10
	Pennsylvanian Period		80	
	Mississippi Period			350±10
	Devonian Period		50	400±10
	Silurian Period		40	440±10
	Ordovician Period		60	500±15
	Cambrian Period		100	600±20
PROTEROZOIC ERA	Keweenawan "Period"		1650	
	Huronian "Period"			
ARCMAE—OZOIC ERA	Timiskaming "Period"		3000	
	Keewatin "Period" Onverwacht			
AZOIC ERA	Godthaab "Period"			

APPENDIX D

CONVERSION FACTORS FOR UNITS OF MEASURE

Distance

1	inch	=	2.54	centimeters (cm)
1	foot	=	0.3048	meter (m)
1	statute mile	=	1.6093	kilometers (km)
1	nautical mile	=	1.8532	kilometers (km)
1	meter	=	39.37	inches
1	meter	=	3.2808	feet
1	kilometer	=	0.6214	miles (statute)

Area

1	square inch	=	6.4516	cm^2
1	square foot	=	0.0929	m^2
1	square statute mile	=	2.590	km^2
1	square nautical mile	=	3.4348	km^2
1	cm^2	=	0.155	square inches
1	m^2	=	10.764	square feet
1	km^2	=	0.3861	square statute miles
1	km^2	=	0.2912	square nautical miles
1	acre	=	0.405	hectares (ha)
1	hectare	=	10,000	m^2
1	hectare	=	2.469	acres

Volume

1	cubic inch	=	16.3871	cm^3
1	cubic foot	=	0.0283	m^3
1	cm^3	=	0.0610	cubic inch
1	m^3	=	35.315	cubic feet
1	m^3	=	1,000	liters
1	US gallon	=	3.785	liters
1	Imperial gallon	=	4.546	liters
1	US gallon	=	0.833	Imperial gallons
1	Imperial gallon	=	1.20	US gallons
1	liter	=	0.2642	US gallons
1	liter	=	0.220	Imperial gallons
1	US ounce	=	29.57	milliliters
1	Imperial ounce	=	28.41	milliliters
1	US ounce	=	1.041	Imperial ounces

Weight

1	grain	=	0.0648	grams (g)
1	ounce	=	28.35	grams (g)
1	pound	=	0.4536	kilograms (kg)
1	g	=	0.0035	ounces
1	kg	=	2.2046	pounds
1	short ton	=	2,000	pounds
1	short ton	=	907.185	kg
1	short ton	=	0.9072	metric tons
1	metric ton	=	1,000	kg
1	metric ton	=	2204.6	pounds
1	metric ton	=	1.1023	short tons

Speed

1	mile per hour (mph)	=	1.6093	km h^{-1}
1	mph	=	0.447	m s^{-1}
1	mph	=	0.8684	knots
1	knot	=	1.1516	mph
1	knot	=	0.5148	m s^{-1}
1	knot	=	1.8532	km h^{-1}
1	km h^{-1}	=	0.28	m s^{-1}
1	km h^{-1}	=	0.6214	mph
1	m s^{-1}	=	3.6	km h^{-1}
1	m s^{-1}	=	2.237	mph

Air Pressure

1	inch mercury	=	33.864	millibars (mb)
1	inch mercury	=	0.0345	kg cm^{-2}
1	mm mercury	=	1.3332	mb
1	mb	=	0.0295	inch mercury
1	mb	=	0.7501	mm mercury
1	mb	=	0.10	kilopascals
1	mb	=	100	pascals
1	kilopascal	=	10	mb
1	mb	=	0.0010	kg cm^{-2}
1	mb	=	100	newtons m^{-2}

Radiation and Energy

1	Watt	=	1	joule per second
1	Watt m^{-2} (W m^{-2})	=	0.001434	cal cm^{-2} min^{-1}
1	cal cm^{-2} min^{-1}	=	697.3	W m^{-2}
1	BTU ft^{-2} h^{-1}	=	3.155	W m^{-2}
1	BTU ft^{-2} h^{-1}	=	0.00452	cal cm^{-2} min^{-1}
1	W m^{-2}	=	0.317	BTU ft^{-2} h^{-1}
1	cal cm^{-2} min^{-1}	=	221	BTU ft^{-2} h^{-1}
1	cal cm^{-2}	=	41.84	kilojoules m^{-2} (kJ m^{-2})
1	megajoule m^{-2} (MJ m^{-2})	=	23.9	cal cm^{-2}
1	cal cm^{-2}	=	1	langley (ly)

APPENDIX E

SURFACE WEATHER MAP

Name: ______________________ Date: ______________

Instructor: ______________________ Section: ______________

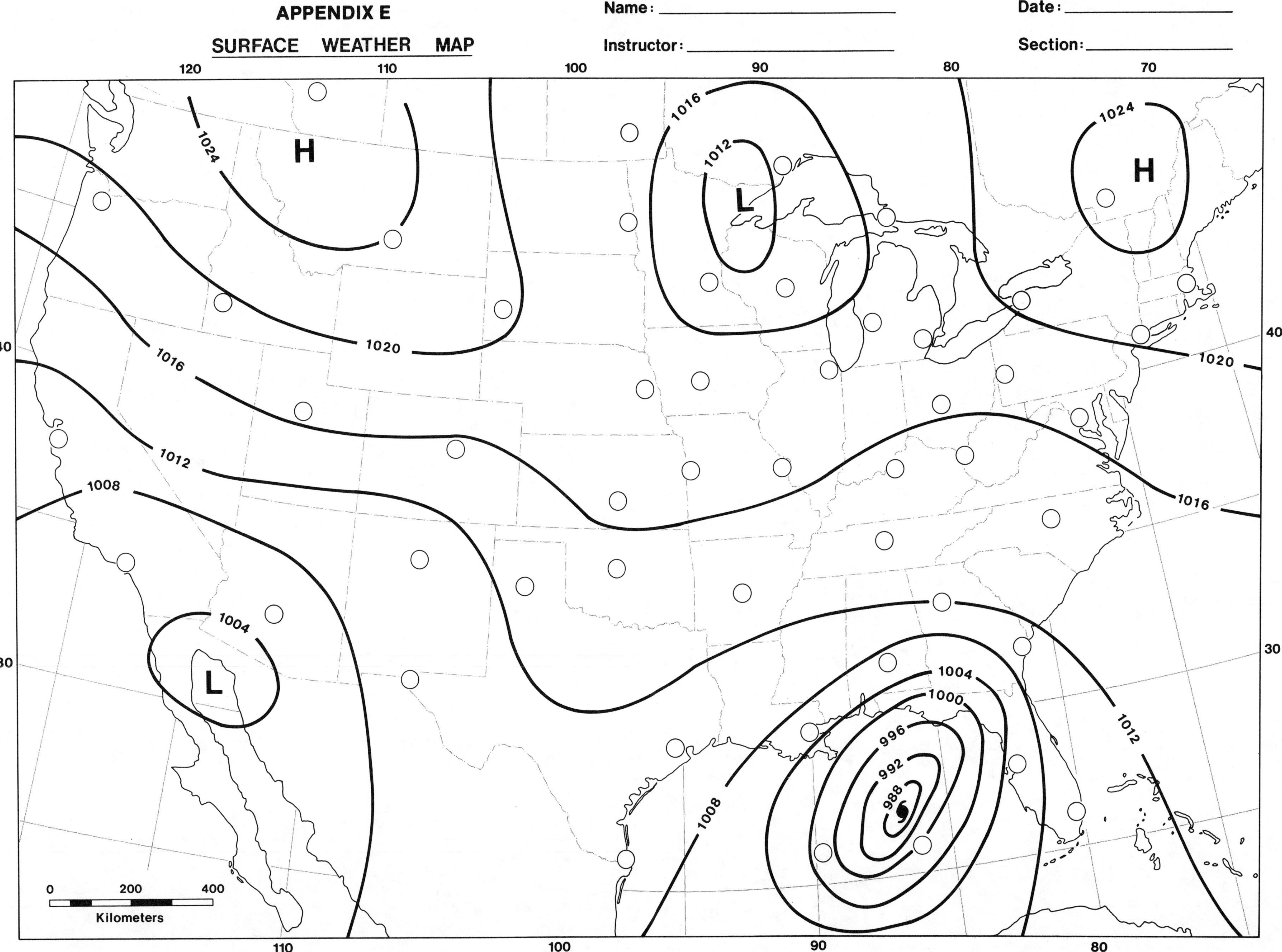

APPENDIX F

GLOBAL CLIMATES ACCORDING TO THE KÖPPEN CLASSIFICATION SYSTEM

(Modified from Strahler; Geiger and Pohl; Trewartha)

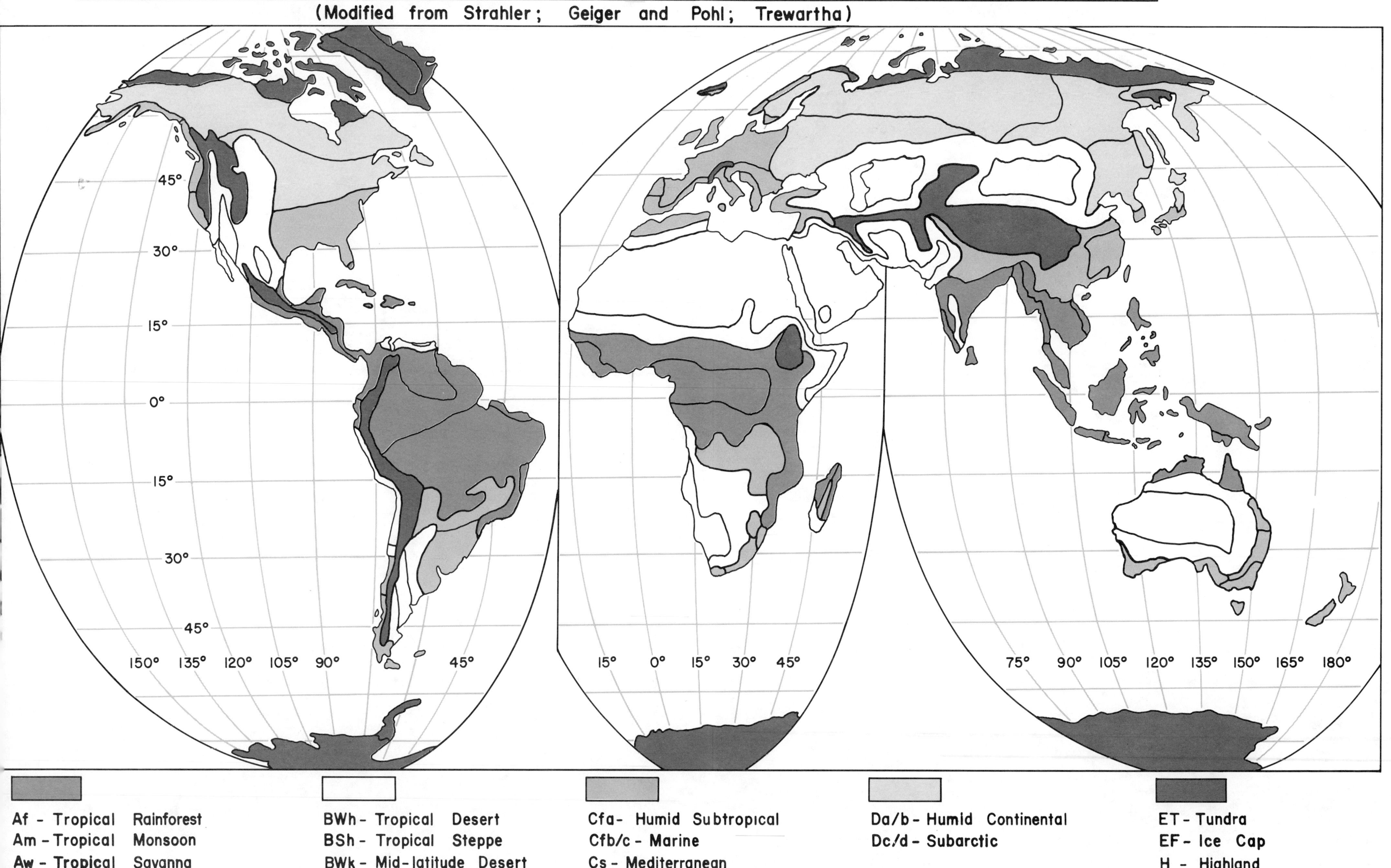

National Mapping Program

Topographic Map Symbols

National Large Scale Series

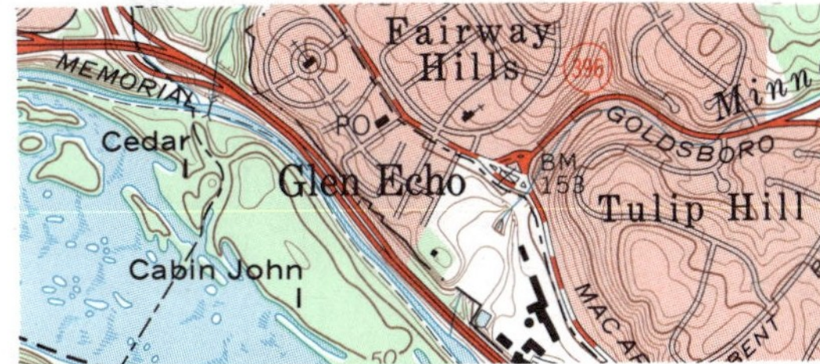

1:24,000 scale – conventional units

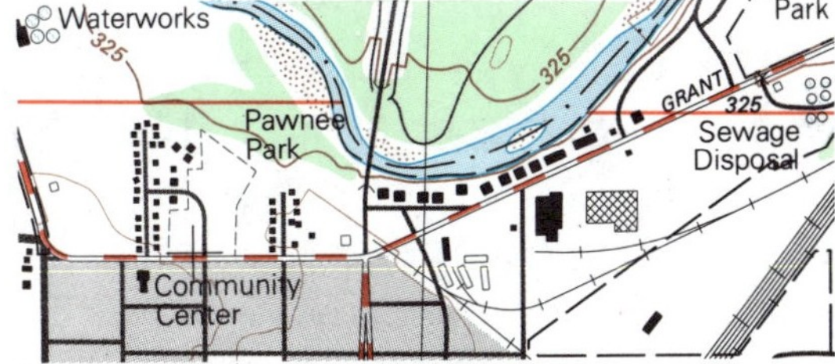

1:25,000 scale – metric units

Provisional edition

U. S. Department of the Interior
Geological Survey
National Mapping Division

Map series and quadrangles

Each map in a U. S. Geological Survey series conforms to established specifications for size, scale, content, and symbolization. Except for maps which are formatted on a County or State basis, USGS quadrangle series maps cover areas bounded by parallels of latitude and meridians of longitude.

Map scale

Map scale is the relationship between distance on a map and the corresponding distance on the ground. Scale is expressed as a ratio, such as 1:25,000, and shown graphically by bar scales marked in feet and miles or in meters and kilometers.

Standard edition maps

Standard edition topographic maps are produced at 1:20,000 scale (Puerto Rico) and 1:24,000 or 1:25,000 scale (conterminous United States and Hawaii) in either 7.5 x 7.5- or 7.5 x 15-minute format. In Alaska, standard edition maps are available at 1:63,360 scale in 7.5 x 20 to 36-minute quadrangles. Generally, distances and elevations on 1:24,000-scale maps are given in conventional units: miles and feet, and on 1:25,000-scale maps in metric units: kilometers and meters.

The shape of the Earth's surface, portrayed by contours, is the distinctive characteristic of topographic maps. Contours are imaginary lines which follow the land surface or the ocean bottom at a constant elevation above or below sea level. The contour interval is the elevation difference between adjacent contour lines. The contour interval is chosen on the basis of the map scale and on the local relief. A small contour interval is used for flat areas; larger intervals are used for mountainous terrain. In very flat areas, the contour interval may not show sufficient surface detail and supplementary contours at less than the regular interval are used.

The use of color helps to distinguish kinds of features:

- Black – cultural features such as roads and buildings.
- Blue – hydrographic features such as lakes and rivers.
- Brown – hypsographic features shown by contour lines.
- Green – woodland cover, scrub, orchards, and vineyards.
- Red – important roads and public land survey system.
- Purple – features added from aerial photographs during map revision. The changes are not field checked.

Some quadrangles are mapped by a combination of orthophotographic images and map symbols. Orthophotographs are derived from aerial photographs by removing image displacements due to camera tilt and terrain relief variations. An orthophotoquad is a standard quadrangle format map on which an orthophotograph is combined with a grid, a few place names, and highway route numbers. An orthophotomap is a standard quadrangle format map on which a color enhanced orthophotograph is combined with the normal cartographic detail of a standard edition topographic map.

Provisional edition maps

Provisional edition maps are produced at 1:24,000 or 1:25,000 scale (1:63,360 for Alaskan 15-minute maps) in conventional or metric units and in either a 7.5 x 7.5- or 7.5 x 15-minute format. Map content generally is the same as for standard edition 1:24,000- or 1:25,000-scale quadrangle maps. However, modified symbolism and production procedures are used to speed up the completion of U.S. large-scale topographic map coverage.

The maps reflect a provisional rather than a finished appearance. For most map features and type, the original manuscripts which are prepared when the map is compiled from aerial photographs, including hand lettering, serve as the final copy for printing. Typeset lettering is applied only for features which are designated by an approved name. The number of names and descriptive labels shown on provisional maps is less than that shown on standard edition maps. For example, church, school, road, and railroad names are omitted.

Provisional edition maps are sold and distributed under the same procedures that apply to standard edition maps. At some future time, provisional maps will be updated and reissued as standard edition topographic maps.

National Mapping Program indexes

Indexes for each State, Puerto Rico, the U. S. Virgin Islands, Guam, American Samoa, and Antarctica are available. Separate indexes are available for 1:100,000-scale quadrangle and county maps; USGS/Defense Mapping Agency 15-minute (1:50,000-scale) maps; U. S. small scale maps (1:250,000, 1:1,000,000, 1:2,000,000 scale; State base maps; and U. S. maps); land use/land cover products; and digital cartographic products.

Series	Scale	1 inch represents approximately	1 centimeter represents	Size (latitude x longitude)	Area (square miles)
Puerto Rico 7.5-minute	1:20,000	1,667 feet	200 meters	7.5 x 7.5 min.	71
7.5-minute	1:24,000	2,000 feet (exact)	240 meters	7.5 x 7.5 min.	49 to 70
7.5-minute	1:25,000	2,083 feet	250 meters	7.5 x 7.5 min.	49 to 70
7.5 x 15-minute	1:25,000	2,083 feet	250 meters	7.5 x 15 min.	98 to 140
USGS/DMA 15-minute	1:50,000	4,166 feet	500 meters	15 x 15 min.	197 to 282
15-minute	1:62,500	1 mile	625 meters	15 x 15 min.	197 to 282
Alaska 1:63,360	1:63,360	1 mile (exact)	633.6 meters	15 x 20 to 36 min.	207 to 281
County 1:50,000	1:50,000	4,166 feet	500 meters	County area	Varies
County 1:100,000	1:100,000	1.6 miles	1 kilometer	County area	Varies
30 x 60-minute	1:100,000	1.6 miles	1 kilometer	30 x 60 min.	1,568 to 2,240
U. S. 1:250,000	1:250,000	4 miles	2.5 kilometers	1° x 2° or 3°	4,580 to 8,669
State maps	1:500,000	8 miles	5 kilometers	State area	Varies
U. S. 1:1,000,000	1:1,000,000	16 miles	10 kilometers	4° x 6°	73,734 to 102,759
U. S. Sectional	1:2,000,000	32 miles	20 kilometers	State groups	Varies
Antarctica 1:250,000	1:250,000	4 miles	2.5 kilometers	1° x 3° to 15°	4,089 to 8,336
Antarctica 1:500,000	1:500,000	8 miles	5 kilometers	2° x 7.5°	28,174 to 30,462

How to order maps

Mail orders. Order by map name, State, and series/scale. Payment by money order or check payable to the U. S. Geological Survey must accompany your order. Your complete address, including ZIP code, is required.

Maps of areas *east* of the Mississippi River, including Minnesota, Puerto Rico, the Virgin Islands of the United States, and Antarctica.

Eastern Distribution Branch
U. S. Geological Survey
1200 South Eads Street
Arlington, VA 22202

Maps of areas *west* of the Mississippi River, including Alaska, Hawaii, Louisiana, American Samoa, and Guam.

Western Distribution Branch
U. S. Geological Survey
Box 25286, Federal Center
Denver, CO 80225

A single order combining both eastern and western maps may be placed with either office.

Residents of Alaska may order Alaska maps or an index for Alaska from the Alaska Distribution Section, U. S. Geological Survey, New Federal Building – Box 12, 101 Twelfth Avenue, Fairbanks, AK 99701.

Sales counters. Maps of the area may be purchased over the counter at the following U. S. Geological Survey offices.

Alaska	Anchorage	Room 108, Skyline Building, 508 Second Avenue
	Fairbanks	Room 126, New Federal Building, 101 Twelfth Avenue
California	Los Angeles	Room 7638, Federal Building, 300 North Los Angeles Street
	Menlo Park	Room 122, Building 3, 345 Middlefield Road
	San Francisco	Room 504, Custom House, 555 Battery Street
Colorado	Denver	Building 41, Federal Center
	Denver	Room 169, Federal Building, 1961 Stout Street
District of Columbia	Washington	Room 1028, General Services Administration Bldg., 19th and F Sts. NW
Missouri	Rolla	1400 Independence Road
Texas	Dallas	Room 1C45, Federal Building, 1100 Commerce Street
Utah	Salt Lake City	Room 8105, Federal Building, 125 South State Street
Virginia	Arlington	1200 South Eads Street
	Reston	Room 1C402, National Center, 12201 Sunrise Valley Drive
Washington	Spokane	Room 678, U. S. Court House, West 920 Riverside Avenue

Commercial dealers. Names and addresses of dealers are listed in each State index. Commercial dealers sell U. S. Geological Survey maps at their own prices.

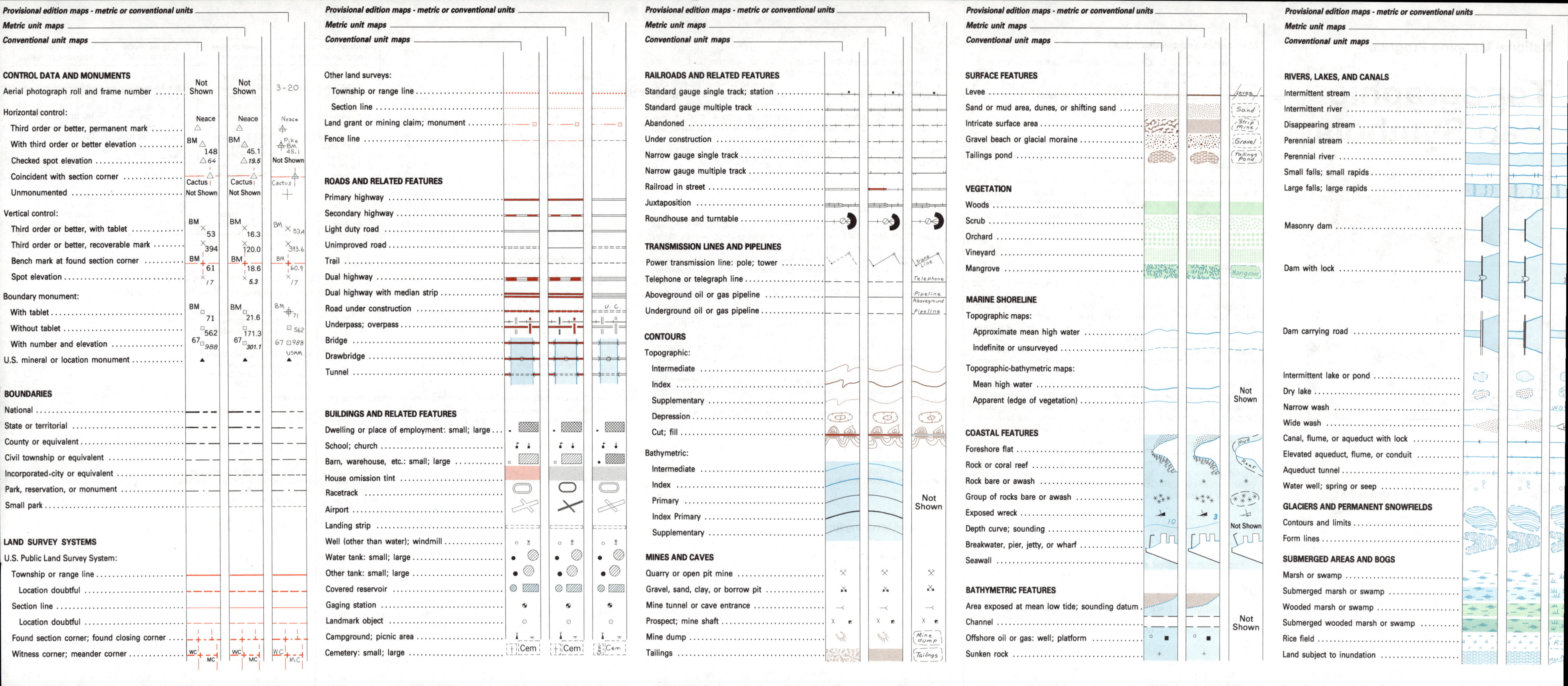
Provisional edition maps - metric or conventional units
Metric unit maps
Conventional unit maps
CONTROL DATA AND MONUMENTS
Aerial photograph roll and frame number
Not Shown
Not Shown
3-20
Horizontal control:
Third order or better, permanent mark
Neace
With third order or better elevation
BM
148
45.1
Pike BM 45.1
Checked spot elevation
64
19.5
Not Shown
Coincident with section corner
Cactus
Unmonumented
Not Shown
Vertical control:
Third order or better, with tablet
53
16.3
53.4
Third order or better, recoverable mark
394
120.0
393.6
Bench mark at found section corner
61
18.6
60.9
Spot elevation
17
5.3
Boundary monument:
With tablet
71
21.6
Without tablet
562
171.3
With number and elevation
67
988
301.1
U.S. mineral or location monument
USMM
BOUNDARIES
National
State or territorial
County or equivalent
Civil township or equivalent
Incorporated city or equivalent
Park, reservation, or monument
Small park
LAND SURVEY SYSTEMS
U.S. Public Land Survey System:
Township or range line
Location doubtful
Section line
Location doubtful
Found section corner; found closing corner
Witness corner; meander corner
WC
MC
Other land surveys:
Township or range line
Section line
Land grant or mining claim; monument
Fence line
ROADS AND RELATED FEATURES
Primary highway
Secondary highway
Light duty road
Unimproved road
Trail
Dual highway
Dual highway with median strip
Road under construction
U. C.
Underpass; overpass
Bridge
Drawbridge
Tunnel
BUILDINGS AND RELATED FEATURES
Dwelling or place of employment: small; large
School; church
Barn, warehouse, etc.: small; large
House omission tint
Racetrack
Airport
Landing strip
Well (other than water); windmill
Water tank: small; large
Other tank: small; large
Covered reservoir
Gaging station
Landmark object
Campground; picnic area
Cemetery: small; large
Cem
RAILROADS AND RELATED FEATURES
Standard gauge single track; station
Standard gauge multiple track
Abandoned
Under construction
Narrow gauge single track
Narrow gauge multiple track
Railroad in street
Juxtaposition
Roundhouse and turntable
TRANSMISSION LINES AND PIPELINES
Power transmission line: pole; tower
Trans Line
Telephone or telegraph line
Telephone
Aboveground oil or gas pipeline
Pipeline Aboveground
Underground oil or gas pipeline
Pipeline
CONTOURS
Topographic:
Intermediate
Index
Supplementary
Depression
Cut; fill
Bathymetric:
Intermediate
Index
Primary
Index Primary
Supplementary
Not Shown
MINES AND CAVES
Quarry or open pit mine
Gravel, sand, clay, or borrow pit
Mine tunnel or cave entrance
Prospect; mine shaft
Mine dump
Mine dump
Tailings
Tailings
SURFACE FEATURES
Levee
Levee
Sand or mud area, dunes, or shifting sand
Sand
Intricate surface area
Strip mine
Gravel beach or glacial moraine
Gravel
Tailings pond
Tailings Pond
VEGETATION
Woods
Scrub
Orchard
Vineyard
Mangrove
Mangrove
MARINE SHORELINE
Topographic maps:
Approximate mean high water
Indefinite or unsurveyed
Topographic-bathymetric maps:
Mean high water
Apparent (edge of vegetation)
Not Shown
COASTAL FEATURES
Foreshore flat
Mud
Rock or coral reef
Reef
Rock bare or awash
Group of rocks bare or awash
Exposed wreck
Depth curve; sounding
10
3
Not Shown
Breakwater, pier, jetty, or wharf
Seawall
BATHYMETRIC FEATURES
Area exposed at mean low tide; sounding datum
Channel
Offshore oil or gas: well; platform
Sunken rock
Not Shown
RIVERS, LAKES, AND CANALS
Intermittent stream
Intermittent river
Disappearing stream
Perennial stream
Perennial river
Small falls; small rapids
Large falls; large rapids
Masonry dam
Dam with lock
Dam carrying road
Intermittent lake or pond
Dry lake
Narrow wash
Wide wash
Canal, flume, or aqueduct with lock
Elevated aqueduct, flume, or conduit
Aqueduct tunnel
Water well; spring or seep
GLACIERS AND PERMANENT SNOWFIELDS
Contours and limits
Form lines
SUBMERGED AREAS AND BOGS
Marsh or swamp
Submerged marsh or swamp
Wooded marsh or swamp
Submerged wooded marsh or swamp
Rice field
Land subject to inundation